Vol. 29. **The Analytical Chemistry of Sulfur and Its Compounds** (*in three parts*). By J. H. Karchmer

Vol. 30. **Ultramicro Elemental Analysis.** By Günther Tölg

Vol. 31. **Photometric Organic Analysis** (*in two parts*). By Eugene Sawicki

Vol. 32. **Determination of Organic Compounds: Methods and Procedures.** By Frederick T. Weiss

Vol. 33. **Masking and Demasking of Chemical Reactions.** By D. D. Perrin

Vol. 34. **Neutron Activation Analysis.** By D. De Soete, R. Gijbels, and J. Hoste

Vol. 35. **Laser Raman Spectroscopy.** By Marvin C. Tobin

Vol. 36. **Emission Spectrochemical Analysis.** By Morris Slavin

Vol. 37. **Analytical Chemistry of Phosphorus Compounds.** Edited by M. Halmann

Vol. 38. **Luminescence Spectrometry in Analytical Chemistry.** By J. D. Winefordner, S. G. Schulman and T. C. O'Haver

Vol. 39. **Activation Analysis with Neutron Generators.** By Sam S. Nargolwalla and Edwin P. Przybylowicz

Vol. 40. **Determination of Gaseous Elements in Metals.** Edited by Lynn L. Lewis, Laben M. Melnick, and Ben D. Holt

Vol. 41. **Analysis of Silicones.** Edited by A. Lee Smith

Vol. 42. **Foundations of Ultracentrifugal Analysis.** By H. Fujita

Vol. 43. **Chemical Infrared Fourier Transform Spectroscopy.** By Peter R. Griffiths

Vol. 44. **Microscale Manipulations in Chemistry.** By T. S. Ma and V. Horak

Vol. 45. **Thermometric Titrations.** By J. Barthel

Vol. 46. **Trace Analysis: Spectroscopic Methods for Elements.** Edited by J. D. Winefordner

Vol. 47. **Contamination Control in Trace Element Analysis.** By Morris Zief and James W. Mitchell

Vol. 48. **Analytical Applications of NMR.** By D. E. Leyden and R. H. Cox

Vol. 49. **Measurement of Dissolved Oxygen.** By Michael L. Hitchman

Vol. 50. **Analytical Laser Spectroscopy.** Edited by Nicolo Omenetto

Vol. 51. **Trace Element Analysis of Geological Materials.** By Roger D. Reeves and Robert R. Brooks

Vol. 52. **Chemical Analysis by Microwave Rotational Spectroscopy.** By Ravi Varma and Lawrence W. Hrubesh

Vol. 53. **Information Theory As Applied to Chemical Analysis.** By Karel Eckschlager and Vladimir Štěpánek

Vol. 54. **Applied Infrared Spectroscopy: Fundamentals, Techniques, and Analytical Problem-solving.** By A. Lee Smith

Vol. 55. **Archaeological Chemistry.** By Zvi Goffer

Vol. 56. **Immobilized Enzymes in Analytical and Clinical Chemistry.** By P. W. Carr and L. D. Bowers

Vol. 57. **Photoacoustics and Photoacoustic Spectroscopy.** By Allan Rosencwaig

Vol. 58. **Analysis of Pesticide Residues.** Edited by H. Anson Moye

Vol. 59. **Affinity Chromatography.** By William H. Scouten

Vol. 60. **Quality Control in Analytical Chemistry.** By G. Kateman and F. W. Pijpers

Vol. 61. **Direct Characterization of Fineparticles.** By Brian H. Kaye

Vol. 62. **Flow Injection Analysis.** By J. Ruzicka and E. H. Hansen

(*continued on back*)

Statistical Methods in Analytical Chemistry

CHEMICAL ANALYSIS

A SERIES OF MONOGRAPHS ON
ANALYTICAL CHEMISTRY AND ITS APPLICATIONS

VOLUME 123

A WILEY-INTERSCIENCE PUBLICATION

JOHN WILEY & SONS, INC.

New York / Chichester / Brisbane / Toronto / Singapore

Statistical Methods in Analytical Chemistry

PETER C. MEIER

CILAG A.G.
Schaffhausen, Switzerland

and

RICHARD E. ZÜND

LONZA A.G.
Visp, Switzerland

A WILEY-INTERSCIENCE PUBLICATION

JOHN WILEY & SONS, INC.

New York / Chichester / Brisbane / Toronto / Singapore

This text is printed on acid-free paper.

Library of Congress Cataloging in Publication Data:

Meier, Peter C., 1945–
Statistical methods in analytical chemistry / Peter C. Meier and Richard E. Zünd.
p. cm.—(Chemical analysis; v. 123)
"A Wiley-interscience publication."
Includes bibliographical references and index.
ISBN 0-471-58454-1
1. Chemistry, Analytic—Statistical methods. I. Zünd, Richard E. II. Title. III. Series.
QD75.4.S8M45 1993
543′.072—dc20 92-27288

Printed in the United States of America

10 9 8 7 6 5 4 3 2 1

To our wives, respectively, Therese and Edith, who granted us the privilege of "book" time, and spurred us on when our motivation flagged.

PREFACE

Both authors are analytical chemists. Our cooperation dates back to those happy days we spent getting educated and later instructing undergraduates and Ph.D. candidates in the late Professor W. Simon's laboratory at the Swiss Federal Institute of Technology in Zürich (ETH-Z). Interests ranged far beyond the mere mechanics of running and maintaining instruments. Designing experiments and interpreting the results in a wider context were primary motives, and the advent of computerized instrumentation added further dimensions. Masses of data awaiting efficient and thorough analysis on the one hand, and introductory courses in statistics slanted toward pure mathematics on the other, drove us to the autodidactic aquisition of the necessary tools. Mastery was slow in coming because texts geared to chemistry were rare, such important techniques as linear regression were relegated to the "advanced topics" page, and idiosyncratic nomenclatures confused the issues. Having been through dispiriting experiences, we happily accepted, as, on the suggestion of Dr. Simon, the opportunity arose to submit a manuscript. We were guided in this present enterprise by the wish to combine the cookbook approach with the consequent use of PCs and programmable calculators. Furthermore, then when-and-how of tests was to be explained in both simple and complex examples of the type a chemist understands. Because many analysts are involved in quality control work, we felt the consequences statistics have for the accept/reject decision would have to be spelled out. The formalization that the analyst's habitual quest for high-quality results has undergone—the keywords are GMP and ISO 9000—is increasingly forcing the use of statistics.

PETER C. MEIER

Schaffhausen, Switzerland

RICHARD E. ZÜND

Visp, Switzerland
December 1992

ACKNOWLEDGMENTS

Our employers were very generous in allowing us to divert some time and means toward writing this book; most numerical examples not specifically marked as simulated were drawn from actual cases. Many unnamed individuals helped us by unintentionally drawing our attention to a problem; some examples and, in scattered places, a comment or an advice to the reader reflects these experiences.

TECHNICAL DETAILS

The text was prepared using WordPerfect 5.1 on a PS/2. All calculations were originally performed on a Hewlett-Packard HP-71B BASIC-programmable pocket calculator fitted with 22K RAM, a Curve and a Math ROM, and a HP-9114B microfloppy disc drive. A HP-2225B ThinkJet matrix printer was used to graph the majority of the figures at a resolution of 320 × 320 pixels, respectively, 0.25 mm/0.01 inch per line pair. A zooming photocopier allowed us to retouch and assemble the enlarged figures before the final reduction. A number of figures (**Chapter**.Figure number) were plotted on a Kyocera F-800 laser printer using GW-BASIC (**1**.20; **4**.4, 22-25), Symphony™ (**2**.22), Lotus 1-2-3™ (**4**.27–29), respectively Freelance™ (**1**.1, 4–6, 8, 27–29; **2**.1, 3, 5, 7, 9, 14, 16, 17, 19, 20, 23, 24; **3**.3, **4**.11). As soon as PCs became available at home, a program package was written on the GW-BASIC platform contained in MS-DOS 3.0, see Section 5.3 and the disc that forms part of this package.

AFFILIATIONS

PCM: CILAG A.G., CH-8201 Schaffhausen, Switzerland. Cilag is a member of the Johnson&Johnson Group. Functions: QA work in analytics and R&D; GMP/GLP auditing; project management. Member of the Advisory Board of *Analytica Chimica Acta*.

REZ: LONZA, A.G., CH-3930 Visp, Switzerland. Lonza is a member of the Alusuisse-Lonza Group. Function: formerly R&D in the Central Analytical Laboratory, now head of the Process Analytics Section.

CONTENTS

INTRODUCTION 1

CHAPTER 1 UNIVARIATE DATA 7

1.1. Mean and Standard Deviation 7
1.1.1. The Most Probable Value 7
1.1.2. The Dispersion 9
1.1.3. The Independency of Measurements 14
1.1.4. Reproducibility and Repeatibility 16
1.1.5. Reporting the Results 18
1.1.6. Interpreting the Results 19
1.2. Distributions and the Problem of Small Numbers 21
1.2.1. The Normal Distribution 22
The Normal Distribution: Equations 23
1.2.2. Student's *t* Distribution 26
1.3. Confidence Limits 27
1.3.1. Confidence Limits of the Distribution 30
1.3.2. Confidence Limits of the Mean 31
1.4. Simulation of a Series of Measurements 32
1.5. Testing for Deviations 36
1.5.1. Examining Two Series of Measurements 38
Student's *t* Test: Equations 39
1.5.2. The *t* Test 40
1.5.3. Extension of the *t* Test to More than Two Series of Measurements 45
1.5.4. Multiple Range Test 47
1.5.5. Outlier Tests 49
1.5.6. Analysis of Variance (ANOVA) 51
1.6. Number of Determinations 55

1.7. Width of a Distribution 58
1.7.1. The F test 58
1.7.2. Confidence Limits for a Standard Deviation 61
1.7.3. Bartlett Test 63
1.8. Charting a Distribution 64
1.8.1. Histograms 64
1.8.2. χ^2 Test 66
1.8.3. Probability Charts 70
1.8.4. Conventional Control Charts (Shewhart Charts) 72
1.8.5. CUMSUM Charts 75
1.9. Errors of the First and Second Kind 76

CHAPTER 2 BI- AND MULTIVARIATE DATA 81

2.1. Correlation 81
2.2. Linear Regression 84
2.2.1. Standard Approach 86
2.2.2. Slope and Intercept 88
2.2.3. Residual Variance 90
2.2.4. Testing Linearity and Slope 92
2.2.5. Interpolating $Y(x)$ 94
2.2.6. Interpolating $X(y)$ 98
2.2.7. Limit of Detection 103
2.2.8. Minimizing the Cost of Calibration 107
2.2.9. Standard Addition 109
2.2.10. Weighed Regression 111
2.2.11. Validating an Analytical Method 116
2.2.12. Intersection of Two Linear Regression Lines 120
2.3. Nonlinear Regression 120
2.3.1. Linearization 121
2.3.2. Nonlinear Regression and Modeling 122
2.4. Multidimensional Data 124
2.4.1. Visualizing Data 125
2.4.2. Full Factorial Experiments 129

CHAPTER 3 ANCILLARY TECHNIQUES 135

3.0. Introduction 135
3.1. Optimization Techniques 135
3.2. Exploratory Data Analysis 138
3.3. Error Propagation and Numerical Artifacts 139
3.4. Ruggedness and Suitability of a Method 141
3.5. Smoothing and Filtering Data 143
3.6. Monte Carlo Technique 145
3.7. Computer Simulation 148
3.8. Programs 150

CHAPTER 4 COMPLEX EXAMPLES 155

4.0. Introduction 155
4.1. To Weigh or Not To Weigh 155
4.2. Nonlinear Fitting 159
4.3. UV-Assay Cost Structure 164
4.4. Process Validation 169
4.5. Regulations and Realities 173
4.6. Diffusing Vapors 176
4.7. Stability à la Carte 178
4.8. Secret Shampoo Switch 180
4.9. Tablet Press Woes 181
4.10. Sounding Out Solubility 184
4.11. Exploring a Data Jungle 186
4.12. Sifting through Sieved Samples 193
4.13. Controlling Cyanide 199
4.14. Ambiguous Automation 202
4.15. Mistrusted Method 206
4.16. Quirks of Quantitation 207
4.17. Pursuing Propagating Errors 211
4.18. Content Uniformity 213
4.19. How Full Is Full? 218
4.20. Warranty or Waste 220
4.21. Arrhenius-Abiding Aging 222
4.22. Facts or Artifacts 225
4.23. Proving Proficiency 228

CHAPTER 5 APPENDICES **239**

5.0. Introduction 239

5.1. Numerical Approximations to Some Frequently Used Distributions 239

5.1.1. Normal Distribution 239

Calculation of CP from z 240

Calculation of z from CP 241

5.1.2. Student's t Distributions 243

Calculation of Student's t from df and p 243

Calculation of p from Student's t and f 245

5.1.3. F Distributions 246

5.1.4. χ^2 Distributions 248

5.2. Software Instructions 250

5.3. Software Reference 263

5.4. Technical Notes 300

5.5. List of Symbols and Abbreviations 303

REFERENCES **305**

INDEX **315**

Statistical Methods in Analytical Chemistry

INTRODUCTION

Analytical chemistry and statistics are relative newcomers to their respective fields—chemistry and mathematics: Modern instrumental analysis is an outgrowth of the technological advances made in physics and electronics since the middle of this century. Statistics have been with us somewhat longer, but were impractical until the advent of powerful EDP equipment in the late sixties and early seventies. Both have been perfected for much the same reason: Technical and economical decisions have to be made, and one wants to increase the probability of making the right one. Running chemical operations the way a creative chef invents new sauces is hardly a recipe for success; methods had to be devised to quantify and objectively characterize the broth that was to yield, say, a dyestuff. So, in this way results in the form of numbers soon became everyday fare for the chemist. It became apparent that one (uncertain) number is no result, that the "true" value would have to be extracted from several such measurements. Other industries were confronted by much the same troubles, with one difference, perhaps. While crop yields and the strength of steel constructions are amenable to such "natural" measures as weights and distances, and the influencing factors are intuitively felt, chemical reactions afford no such easy access to their secrets: Numbers pertain to abstract or invisible quantities and are generally expensive to obtain. It therefore does not come as a surprise that statistical techniques were first introduced in areas where data was more readily obtainable. In the end, though, the marriage between high-tech analytical chemistry and statistics was unavoidable.

This book, written by two passionate analysts, treats the application of statistics to analytical chemistry[1] in a very practical manner. A minimum of tools are explained and then applied to everyday, that is, complex situations. The computer and especially computer graphics are viewed as valuable aids.

In an undertaking such as this there is no room for proving theory, however trivial. The examples should be illuminating to both beginners and specialists from other fields. The reader should be shown how to make sensible use of statistics to support imminent decisions and actions. What are the circumstances of this decision process? The following scheme might serve as an illustration:

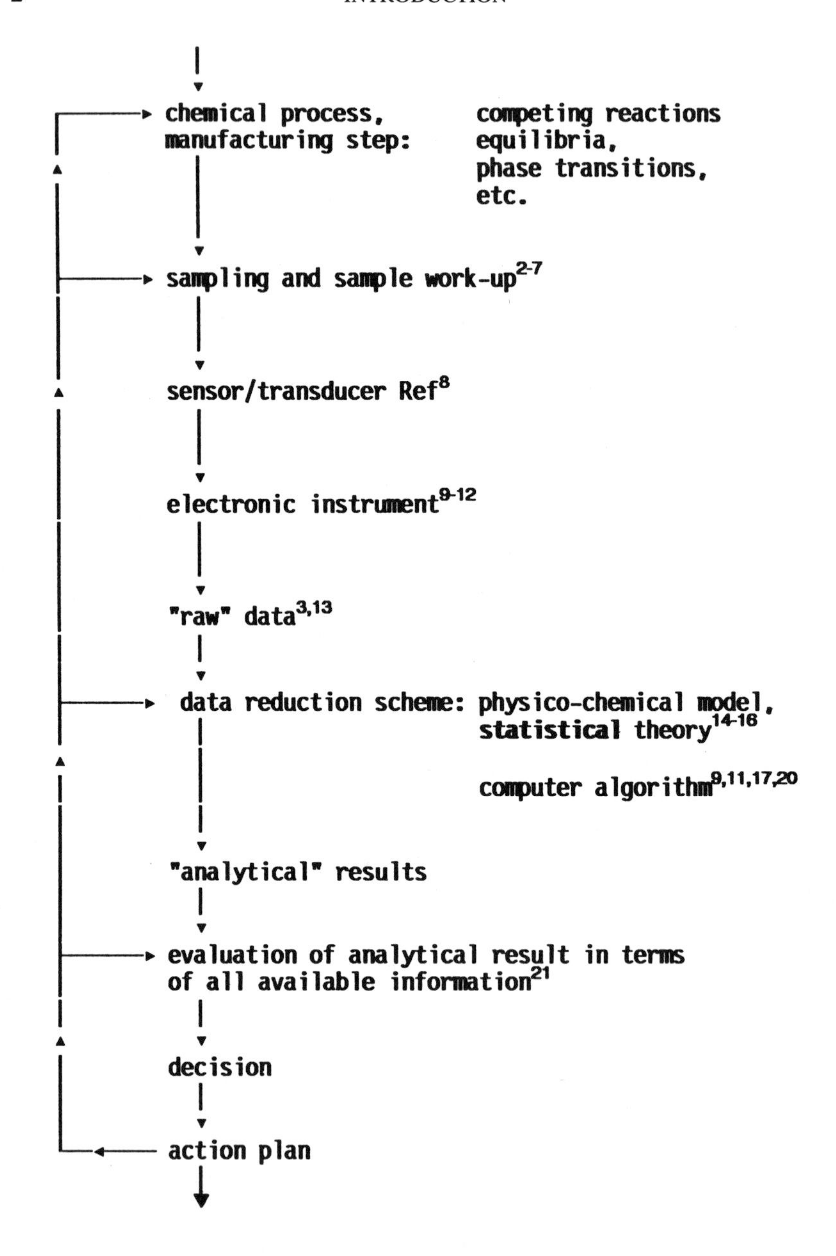
chemical process,
manufacturing step:
competing reactions
equilibria,
phase transitions,
etc.
sampling and sample work-up[2-7]
sensor/transducer Ref[8]
electronic instrument[9-12]
"raw" data[3,13]
data reduction scheme:
physico-chemical model,
statistical theory[14-16]
computer algorithm[9,11,17,20]
"analytical" results
evaluation of analytical result in terms
of all available information[21]
decision
action plan

Every step is fraught with uncertainties and is subject to artifacts.[22] Sampling,[2,5] a weak point for various reasons, is the hinge on which everything else depends: Carelessness in this area makes the best intentions meaningless. Lack of maintenance and instrument calibration endangers the relevance of the decisions taken. Note that statistics are only part of the picture; the decision process has to be viewed as a coherent whole; a decision can only be passed by taking into account the complex interrelations among the chemical species being consumed and formed, and the legal, economic, scientific, and environmental characteristics of the analytical process.

The sensor only incompletely maps physico-chemical reality into an electrical signal that is then subject to signal conditioning hardware, and the instrument again transforms the individual measurements into time averages, signal areas,[11] or other user-oriented information. Just filtering out noise, strictly speaking, already constitutes an adulteration. The instrument configuration (hard- and software) is to be regarded as just one element that the analyst has to put to use without exceeding its inherent limitations (not necessarily the design specifications quoted in the prospectus!).[9]

Thus one can be far from the ideal world often assumed by statisticians: tidy models, theoretical distribution functions, and independent, essentially uncorrupted measured values with just a bit of noise superimposed.

Furthermore, because of the costs associated with obtaining and analyzing samples, small sample numbers are the rule. On the other hand, linear ranges upwards of 1 : 100 and relative standard deviations of usually 2% and less compensate for the lack of data points. In this respect, analytical chemistry, as a relatively high-precision science, is in a class different from the subject matters that provided the examples for statistical theory and gave impetus towards its development. Trace and biochemical analysis are, in some ways, exceptions.

For these reasons only a small number of statistical techniques are really of relevance in day-to-day operations. These shall be the scope of this book. Those that are applied only in very specialized areas are disregarded, as are those for which a much deeper understanding is necessary than can be communicated between the covers of this book.

There are many excellent texts available on statistics, some only introductory, others very advanced. The same is true for texts on analytical chemistry, or computer programming, for that matter. The addressed audience is, more often than not, specialized. Modern homo analyticus, an all-arounder in the best sense of the word, is, to our best knowledge, rarely among these addresses. The adjective "modern" is necessary and needs to be explained: It is used to circumscribe the chemist at home in a computerized laboratory and in contact with a world around him that places many organizational, financial, and other nonchemical demands on him that he should all consider in his work. This is obviously a person with sufficient knowledge in diverse disciplines, and able to integrate the various aspects of a problem, but who is unwilling to acquire the statistical armament by reading through thousands of pages of sometimes esoteric literature, replete with, in his eyes, extensive proofs, superfluous indices, and worked examples that bear no resemblance to situations experienced in the lab. Also, since there are nearly as many simplifying assumptions to a given question as there are statisticians studying it, the earnest analyst who wants to prepare a defensible support for his decision, and is not out to practice art for its own sake, is shaken in his resolve. We hope this book will provide him or her with an easily digestible introduction.

The practicing statistician knows that hundreds of tests have been described in the literature, and more are being developed, but only a few dozen have gained acceptance for practical—as opposed to artificial—situations. The authors have found the tests and procedures described in Chapters 1 to 3 to be the most useful for the constellation of a few precise measurements of law-abiding parameters prevalent in analytical chemistry, but this does not disqualify other perspectives and procedures. For many situations routinely encountered several solutions of varying theoretical rigor are available. A case in point is linear regression, where the assumption of error-free abscissa values is often violated. Is one to propagate formally more correct approaches, such as the maximum likelihood one, or is a weighted, or even an unweighted least-squares regression sufficient? The exact numerical solutions found by these three models will differ: Any practical consequences thereof must be reviewed on a case-by-case basis. The position taken by the authors, in the context of day-to-day decision making, is that chemists should first become comfortable with the confidence-limits concept and be able to fully utilize the mechanics of plain linear regression before they savor theoretically sounder, but more complicated, fare. This book, as all such endeavors, is an attempt to create the impossible, namely, the text that satisfies all readers, bores none, and leaves everyone's self-esteem intact. The choice of subjects, the detail which they are presented, and the opinions implicitly rendered, of course, reflect the particular experiences and outlook of the authors.

The approach toward writing a book can vary from a formal start in axioms and a theoretical example on the last page to an unstructured collage of recipes that, due to lack of cross-references, introductions, and hints leave the reader wondering what to use when. Here a structure was chosen that is believed to be useful to novices and to experts in the making: The first three chapters set out in detail how a test or technique is carried out, with numerical examples, but without theory. The fourth chapter brings in the criminalistic aspect, all surprise and combination. The last chapter is a conglomerate of useful algorithms and programs.

Section 1.1. serves as a first introduction that by necessity is not in all accounts underpinned by previously presented theory; cross-references to later sections are given, however. The intention is that the reader recognizes some informal tools he or she has applied to simple data sets. Only then are the Normal and the *t* Distributions touched upon to supply the know how for the calculation of confidence limits and their interpretation. All told, this section is a hodgepodge of unavoidable subjects that must appear somewhere near the beginning.

Chapters 1 and 2 present the classical statistical techniques using simple situations for illustration. Chapter 3 expands into areas accessible only since computing became an everyday experience.

Chapter 4 contains a number of examples that might confront the analyst in this or a similar guise nearly every day: A problem's setting is presented so as to convey a sense of the often conflicting demands imposed on the solution, and of the role statistical craftwork is to play in this context. Core sections of BASIC programs are presented in Chapter 5 to complement the equations or to place a ready-made tool at the disposition of the inclined reader. A 3.5″ microfloppy disc is included with the book to provide the reader with ready-made programs and data files: Preaching without practice is unsound. Details are found in Section 5.3.

Many figures illustrate abstract concepts; heavy use is made of numerical simulation to evade the textbook style "constructed" examples that, due to reduction to the bare essentials, answer one simple question, but do not tie in to the reader's perceived reality of messy numbers. Furthermore, many texts still assume the practitioner has little more than a pencil and an adding machine available, and thus propagate involved schemes for designing experiments to ease the number-crunching load, instead of having the computer take care of not-so-round numbers and curved calibration functions, and giving the experimenter free rein. Major developments of the early 80s, namely, programmability, large amounts of random-access memory, and the scientific functions of hand-held calculators and, more recently, cheap PCs, now accessible even to penny-pinching students, are fully integrated into the concept of this book: There are only two worked examples that make use of integer numbers to ease calculations (see Section 1.1.1); no algebraic or numerical shortcuts are taken or calculational schemes presented to evade divisions or roots, as was so common in texts only a decade old.

Chapter 4 and the paragraph on Technical Details prove that a programmable calculator is the revolutionary element. Add-ons, such as printers, plotters, and magnetic mass storage devices are a boon to productivity and take the drudgery out of tabulating numbers and drawing diagrams. Full-fledged PCs are more comfortable, of course, but are also more expensive. In this connection a chuckle is in order: All numerical data and examples provided by one author were recalculated independently by the other, using a different machine. The experience was sobering in that mistakes due to hasty programming, the use of computers of unequal numerical precision, and truncation errors when continuing with intermediate results, were uncovered. The ensuing heckling and attempts to find the root cause made for fun and frustration. If this can happen to two scientists well versed both in chemistry and computing, some doubt is cast on efforts by (teams of) narrowly focused individuals.

Terminology was chosen to reflect recent guides[23] or, in the case of statistical symbols, common usage.[24]

There are innumerable references that cover theory, and still many more that provide practical applications of statistics to chemistry in general and analytical chemistry in particular. Articles from *Analytical Chemistry* were chosen as far as possible to provide worldwide availability. Where necessary, articles in English that appeared in *Analytica Chimica Acta*, *Analyst*, or *Fresenius Zeitschrift für Analytische Chemie* were cited.

There are a number of authorative articles the reader is urged to study that amplify on issues central to analytical understanding.[25,27]

CHAPTER

1

UNIVARIATE DATA

The title implies that in this first chapter techniques are dealt with that are useful when the observer concentrates on a single aspect of a chemical system and repeatedly measures the chosen characteristic. This is a natural approach, first because the treatment of one-dimensional data is definitely easier than that of multidimensional data, and second, because a useful solution to a problem can very often be arrived at in this manner.

Section 1.1 treats the calculation of the mean, the standard deviation, and the standard deviation of the mean without recourse to the underlying theory. It is intended as a quick introduction under the tacit assumption of normally distributed values.

1.1. MEAN AND STANDARD DEVIATION

The simplest and most frequent question is "what is the typical value that best represents these measurements, and how reliable is it?"

1.1.1. The Most Probable Value

Given that the assumption of normally distributed data (see Section 1.2.1) is valid, several useful and uncomplicated methods are available for finding the most probable value and its confidence interval, and for comparing such results.

When only a few measurements of a given property are available, and especially if an asymmetry is involved, the median is often more appropriate than the mean. The *median*, x_m, is defined as that value which bisects the set of n ordered observations; i.e.,

- if n is odd, $(n - 1)/2$ observations are smaller than the median and the next higher value is reported as the median:

$$(\textbf{example}: n = 9: x_i = 4, 5, 5, 6, 7, 8, 8, 9, 9; x_m = 7).$$

- if n is even, the average of the middle two observations is reported:

$$(\textbf{example}: n = 8: x_i = 4,5,5,6,7,8,8,9; x_m = 6.5).$$

The most useful characteristic of the median is the small influence exerted on it by extreme values, that is, its robust nature. The median can thus serve as a check on the calculated mean (see below).

The *mean*, $\bar{x}$, can be shown to be the best estimate of the true value μ; it is calculated as the arithmetic mean of n observations:

$$\bar{x} = \Sigma(x_i)/n, \tag{1.1}$$

where Σ means "obtain the arithmetic sum of all values x_i, with $i = 1 \cdots n$."

A numerical **example**: A set of ordered observations reads:

$$x_i: 4,4,4,5,5,6,6,6,6,7,7,8,9,9,17$$

$$\text{first 14 values: } x_m = 6,\ \bar{x} = 6.143,$$

$$\text{all 15 values: } x_m = 6,\ \bar{x} = 6.867.$$

Notice that by the inclusion of x_{15} the mean is strongly influenced whereas the median is not. The value of such comparisons lies in the automatic processing of large numbers of small data sets, in order to pick out the suspicious ones for manual inspection (Fig. 1.1; see also the next section).

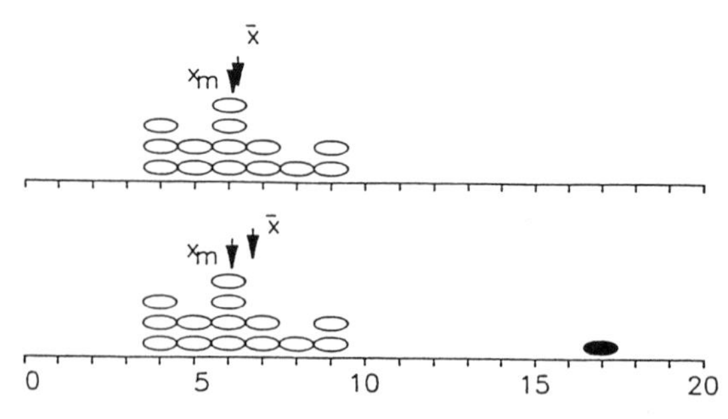

Figure 1.1. The median x_m and the average $\bar{x}$ are given for a set of observations. This figure is a simple form of a histogram; see Section 1.8.1. An additional measurement at $x = 17$ would shift $\bar{x}$ but not x_m.

1.1.2. The Dispersion

The reliability of a mean is judged by the distribution of the individual measurements about the mean. There are two generally used measures of the spread (the scatter) of a set of observations, namely, the range R and the standard deviation s_x.

The *range* R is the difference between the largest and the smallest observation:

$$R = x_{\max} - x_{\min}. \tag{1.2}$$

Precisely because of this definition, the range is very strongly influenced by extreme values. Typically, for a given number of repetitions n, the range $R(n)$ will come to a certain expected (and tabulated) multiple of the true standard deviation. In Fig. 1.2 the ranges R obtained for 390 simulations are depicted. It is apparent that the larger the sample size n, the more likely the occurrence of extreme values: For $n = 4$ the extremes are expected to be around ± 1 standard deviation from the mean, and the

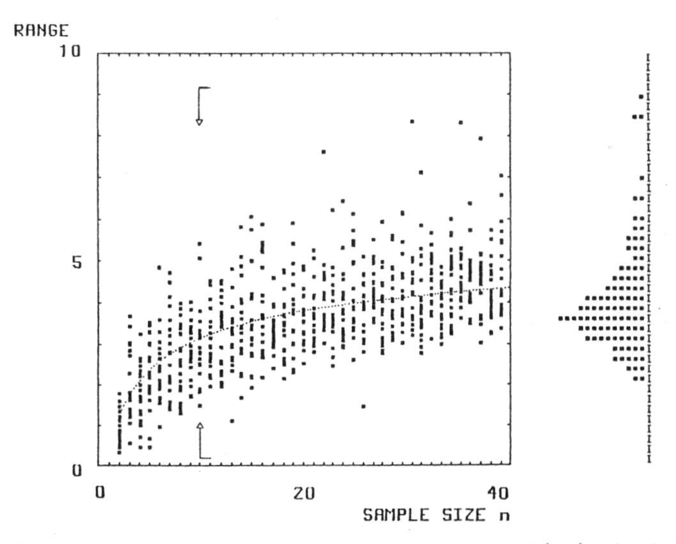

Figure 1.2. The range $R(n)$ for size of sample n, with $n = 2 \cdots 40$ (left). The dotted line gives the tabulated values.[24] The range R is given as $y = R/s_x$ in units of the experimental standard deviation. A total of 8190 normally distributed values with mean 0 and standard deviation 1 were simulated (see Section 3.6). The right-hand figure gives the distribution of ranges found after simulating 100 sets of $n = 10$ normally distributed values.

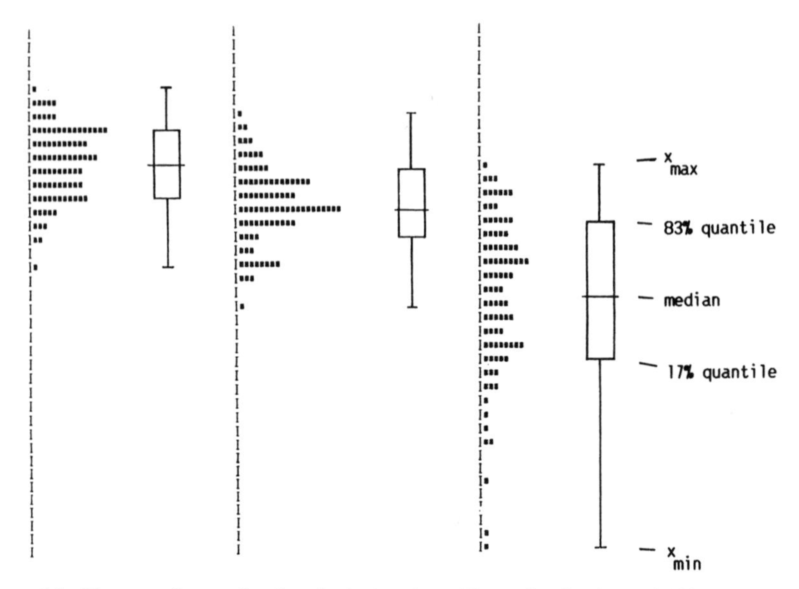

Figure 1.3. The use of quantiles for displaying data. Three distributions of 100 events each are shown in histogram (left) and in quantile (right) form. The reason for choosing the 17th and the 83rd quantiles is explained in the text.

other two values in between. At $n = 40$ the range is expected to be twice as large.

There is no alternative to using the full range if there are only few observations available. But with n larger than, say, 9, the concept of *quantiles* (or percentiles) can be used to buffer the calculated range against outliers: For $n = 10$, casting out the highest and the lowest observations leaves $n' = 8$; the corresponding range is termed the 10–90% range (difference between the 10th and the 90th percentiles). This process of eliminating the extremes can be repeated to yield, e.g., the 20–80% range, etc. A very useful application is in the graphical presentation of data, for instance, by drawing a box around the central $\frac{2}{3}$ of all observations and extensions to mark the overall range: See Fig. 1.3.

The advantage of this technique is that no assumptions need to be made about the type of distribution underlying the data and that many sets of observations can be visually compared without distraction by the individual numbers. For further details, see Section 1.8. Note that the fraction of all observations bounded by the 17th and the 83rd quantiles encompasses two thirds of all values, which is very close to the 68.3% expected in the interval $\pm 1 \cdot s_x$ around the mean $\bar{x}$.

The *standard deviation* s_x is the most commonly used measure of dispersion. Theoretically, the parent population from which the n observations are drawn must meet the criteria set down for the Normal Distribution (cf. Section 1.2.1); in practice, the requirements are not as stringent, because the standard deviation is a relatively robust statistic. The almost universal implementation of the standard deviation algorithm in calculators and program packages certainly increases the danger of its misapplication, but this is counterbalanced by the observation that the consistent use of a somewhat inappropriate statistic can also lead to the right conclusions.

The standard deviation s_x is by definition the square root of the variance V_x,

$$S_{xx} = \sum (x_i - \bar{x})^2 = \sum (r_i)^2, \tag{1.3a}$$

$$S_{xx} = \sum (x_i^2) - \left(\sum (x_i)\right)^2 / n, \tag{1.3b}$$

$$V_x = S_{xx}/(n - 1), \tag{1.3c}$$

$$s_x = \sqrt{V_x}, \tag{1.3d}$$

where $f = (n - 1)$ is the number of degrees of freedom, by virtue of the fact that, given $\bar{x}$ and $n - 1$ freely assigned x_i values, the nth value is fixed. S_{xx} is the sum of squares of the residuals r_i that are obtained when the average value $\bar{x}$ is subtracted from each observation x_i.

Example: For a data set $x(\ ) = 99.85$, 100.36, 99.75, 99.42, and 100.07 one finds $\Sigma(x_i) = 499.45$, $\Sigma(x_i^2) = 49890.5559$; S_{xx} according to Eq. (1.3b) is thus $49890.5559 - 49890.0605 = 0.4954$. Here the five significant digits "49890" are unnecessarily carried along with the effect that the precision, which is limited by the computer's word length, is compromised. The average $\bar{x}$ is found as 99.89 and the standard deviation s_x as ± 0.3519 [via Eqs. (1.3c) and (1.3d); see also Table 1.1 in the next example).

The calculation via Eq. (1.3b) is the one implemented in most calculators[19,28] because no x_i values need to be stored. The disadvantage inherent in this approach lies in the danger of digit truncation (cf. error propagation, Section 3.3), as is demonstrated in Table 1.1 in the following example. This can be avoided by subtracting a constant from each observed x_i, $x_i' = x_i - c$, so that fewer significant digits result before doing the calculations according to Eq. (1.3b); this constant could be chosen to be $c = 99.00$ in the example above.

Table 1.1. Reliability of Calculated Standard Deviations.

[In all cases the correct mean 99.85608 was found. The digits given in *italics* deviate from the correct value given in **bold** numbers. The square of the result in the second line differs from the correct value 0.000022727 only by +0.0000000000000688, or less than 1 in $3.3 \cdot 10^8$. The result obtained GW-BASIC in double-precision mode is even closer and is assumed to be the most accurate of all answers. As is pointed out in Section 1.7.2, only the first one or two nonzero digits (rounded) are to be reported (e.g., "0.005" or "0.0048"); all available digits were printed here to demonstrate the limitations inherent in the employed algorithms. The number of significant digits carried along internally (where available) and the those displayed are given in columns 2 and 3. The notes show how, by way of example, the first data point was typed in. A prime indicates a multiplication by 10,000 after the subtraction, so as to eliminate the decimal point. The HP-71 displays the last three digits "226" either if cases a′, b′, etc. apply, or if the calculated SDEV is multiplied by 1000. The TI-95 has a feature that allows 13 significant places to be displayed. The TI-30D fails in one case and displays "negative difference." The difference between specifying [c:\ bp \ msd/D] instead of just [c:\ bp \ msd] when calling program MSD under GW-BASIC is quite evident: one gains at least another seven significant digits in the double-precision mode; in this case, because the intermediate results are accessible, the fault can be unequivocally assigned to the SQR function.]

Calculator Model	Internal Digits	Digits Displayed	Standard Deviation As Displayed	Note
TI-95	?	13	**0.004767***074574621*	a
			0.00476728434226*5*	b
HP-71B	15	12	**0.00476728434226**	a
HP-32S	15	11	**0.00476***6235412*	a
			0.004767284342	b, c, d
			0.00476728434 2*3*	d′
HP-11 and 41C	?	10	**0.004***878012*	a
			0.00476*6235*	b
			0.004767284	c, d, e
			0.004767284342	c′, d′, e′
HP-55	12	10	**0.00***500000000*	a
			0.004767*075*	b
			0.004767284	c
TI-30D	10	8	**0.0***104403*	a
			0.004*878*	b
			0.00476*62*	c
			no result	d
			0.0047672843	d′
GW-BASIC		16	**0.004767284342264767**	f
		16	**0.004767284***262925386*	g

Table 1.1. *(Continued).*

Note	Amount c Subtracted	Digits Typed In	Number of Significant Digits
a	0.0000	99.8536	6
b	90.0000	9.8536	5
c	99.0000	0.8536	4
d	99.8000	0.0536	3
e	99.8500	0.0036	2 (3)
f	99.85608	−0.00248	Double-precision mode, Eq. (1.3a); see program MSD
g	99.85608	−0.00248	Single-precision mode, Eq. (1.3a); see program MSD

Example: The exact volume of a 100-ml graduated flask is to be determined by five times filling it to the mark, weighing the contents to the nearest 0.1 mg, correcting for the density to transform grams to milliliters, and averaging; the density-corrected values are 99.8536, 99.8632, 99.8587, 99.8518, and 99.8531 ml. For the purpose of demonstration, this task is solved using GW-BASIC and seven different models of calculators with the results in Table 1.1 (it of course would be more appropriate to round all volumes to four or at most five significant digits because a difference in filling height of 0.2 mm, which might just be discernible, amounts to a volume difference of $0.02 \cdot (0.8)^2 \cdot \pi/4 \approx 0.01$ ml; diameter of flask's neck, 8 mm).

In effect, this approximates the situation of Eq. (1.3a). All told, the user of calculators and software should be aware that the tools at his disposal might not be up to the envisaged task if improperly employed.

The *relative standard deviation*, RSD (also known as the *coefficient of variation* c.o.v.), which is frequently used to compare reproducibilities, etc., is calculated as

$$\mathrm{RSD} = 100 \cdot s_x/\bar{x}, \tag{1.4}$$

and is given in percent.

For reasons that will not be detailed here the *standard deviation of the mean* is found as

$$s_{\bar{x}} = \sqrt{V_x}/\sqrt{n}, \tag{1.5a}$$

$$s_{\bar{x}} = \sqrt{V_x/n}, \tag{1.5b}$$

$$= s_x/\sqrt{n}. \tag{1.5c}$$

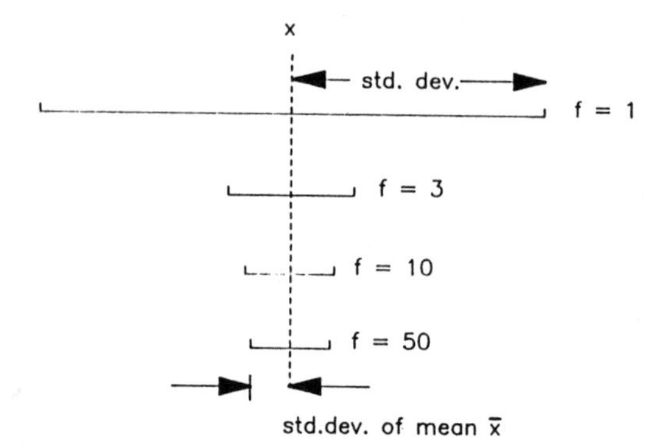

Figure 1.4. The standard deviation of the mean $s_{\bar{x}}$ converges toward zero for a large number of measurements n (schematic; cf. Fig. 1.16).

The difference between s_x and $s_{\bar{x}}$ is crucial: While the first describes the population as such and tends with increasing n toward a positive constant, the latter describes the quality of the determination of the population mean, and tends toward zero. See Fig. 1.4.

1.1.3. Independency of Measurements

A basic requirement, in order that the above results "mean" and "standard deviation" are truly representative of the sampled population, is that the individual measurements should be independent of each other. Two general cases must be distinguished:

(a) Samples are taken for classical off-line processing; e.g., a 10-ml aliquot is withdrawn from a reaction vessel every hour and measurements are conducted thereupon.

(b) The sensor is immersed in the reaction medium and continuously transmits values.

In case (a) the different samples must be individually prepared. In the strictest interpretation of this rule, every factor that could conceivably contribute to the result needs to be checked for bias, i.e., solvents, reagents, calibrations, and instruments. That this is impractical is immediately apparent, especially because many potential influences are eliminated by careful experimental design, and because the experienced analytical chemist can often identify the major influences beforehand. Three examples will illustrate the point.

In UV spectroscopy the weighing and dilution steps usually introduce more error than does the measurement itself and thus the wish to obtain a replicate measurement involves a second weighing and dilution sequence.

In contrast, in HPLC assays the chromatographic separation and the integration of the resulting analyte peak normally are more error prone than is the preparation of the solutions; here it would be acceptable to simply reinject the same sample solution in order to obtain a quasi-independent measurement. Two independent weighings and duplicate injection for each solution is a commonly applied rule.

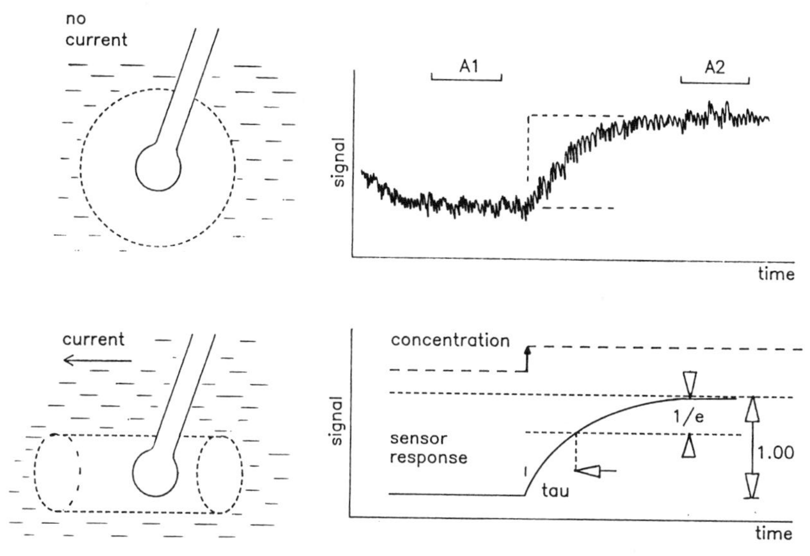

Figure 1.5. Under stagnant conditions a sensor will sample a volume; i.e., the average response for that volume is obtained. A sensor in a current yields an average reading over time and cross-section. The observed signal S over time t is the convolute of the local concentration with the sensor's sampling volume and time constant. Two measurements are only then independent when they are separated at least by five time constants and/or a multiple of the sampling volume's diameter. At the left, the sampled volumes are depicted. At the right, a typical signal-versus-time record (e.g. strip-chart recorder trace) and the system response to a step change in concentration are shown. Tau (τ) is the time constant defined by an approximately 63.2% change $(1 - 1/e) = 0.63212$, with $e = 2.71828 \cdots$. "$A1$" and "$A2$" indicate valid averages taken at least $5 \cdot \tau$ after the last disturbance.

In flame photometry, signal drift and lamp flicker require that one or a few unknowns be bracketed by calibrations. Here, independent measurements on the same solutions means repeating the whole calibration and measurement cycle.

In other words, those factors and operations that contribute the most toward the total variance (see additivity of variances, next section) need to be individually repeated for two measurements to be independent.

In case (b) the independent variables are time or distance of movement of the sensor. Repeat measurements should only be taken after the sensor has had enough time to adjust to new conditions (delay larger than about five time constants). Thus if a continuous record of measurements is available (strip-chart recorder or digitized readings), an "independent" measurement constitutes the average over a given time span at least five time constants τ after the last such average. The time spans from which measurements are drawn for averaging may not overlap. The time constant is determined by provoking a step response. See Figure 1.5.

1.1.4. Reproducibility and Repeatability

Both measures refer to the random error introduced every time a given property of a sample is measured. The distinction between the two must be defined for the specific problem at hand. Examples for continuous (Fig. 1.6) and discrete (Fig. 1.7) records are given below.

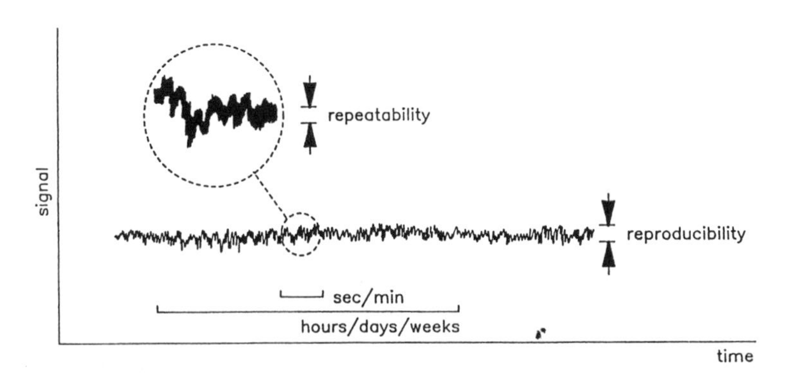

Figure 1.6. Repeatability and reproducibility are defined using historical data. The length of the time interval over which the parameter is reviewed is critical: The shorter it is, the better the experimental boundary conditions tend to be defined. The repeatability sets the limit on what could potentially be attained; the reproducibility defines what is attained in practice, using a given set of instrumentation and SOPs.

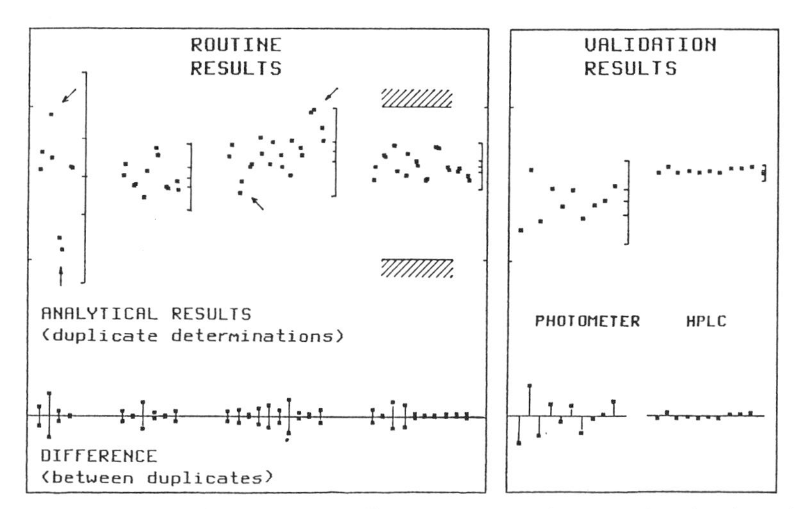

Figure 1.7. Reproducibility and repeatability. For a cream the assay data for the active principle is shown for retrospective surveys (left) and validation runs (right). This particular product is produced in about 20 batches a year. At the end of every year, product release analysis data for a number of randomly picked batches is reviewed for an overall picture of the performance of the laboratory. In four successive years, 30 batches were investigated; the repeat determinations (HPLC method) are given by squares and the respective mean, CL(x), and CL($\bar{x}$) are indicated by bars (top); for definitions of CL, see Section 1.3. The residuals for the double determinations are shown below. The following conclusions can be drawn. (a) All data are within the $\pm 9.1\%$ specifications (hatched bars), because otherwise the releases would not have been granted. (b) The mean of the third group is higher than the others ($p < 0.025$, 95% CL being shown). (c) Four pairs of data points are marked with arrows: Because the individual points within a pair give typical residuals, either one of three artifact-causing mechanisms must be investigated: (1) over- or underdosing during production, (2) inhomogeneity, and (3) errors of calibration. Points 1 and 2 can be cleared up by taking more samples and checking the production records; point 3 is a typical problem found in routine testing laboratories (dead lines, motivation). This is a reason why GLP (Good Laboratory Practices) Regulations mandate that reagent or calibration solutions be marked with the date of production, the shelf life, and the signature of the technician, in order that such questions can be retrospectively cleared. In the right panel, validation data for an outdated photometrical method (left) and the HPLC method (right) are compared. HPLC is obviously much more reliable. The HPLC residuals in the right-hand panel (repeatability: same technician, day, and batch) should be compared with those in the left-hand panel (reproducibility: several technicians, different days and batches) to gain a feeling for the difference between a research and a routine lab.

Repeatability is most commonly defined as the standard deviation obtained using a given SOP (standard operating procedure) in connection with a particular sample, and repeatedly measuring a parameter in the same laboratory, on the same hardware, and by the same technician in a short period of time. Thus boundary conditions are as controlled as possible. The standard deviation so obtained could only be improved upon by changing the agreed-upon analytical method (column type, eluent, instrument, integration parameters, etc.).

Reproducibility is understood to be the standard deviation obtained for the same SOP over a longer period of time. This time frame, along with other particulars, has to be defined. For example, similar but not identical HPLC configurations might be involved, as well as a group of laboratory technicians; the working standard and key reagents might have been replaced, and seasonal/diurnal temperature and/or humidity excursions could have taken their toll. The only thing that one has to be careful to really exclude is batch-to-batch variation in the sample. This problem can be circumvented by stashing away enough of a typical (and hopefully stable) batch, so as to be able to run a sample during every analysis campaign. Incidentally, this doubles as a form of a system suitability test; cf. Sections 1.8.4 and 2.2.11.

In mathematical terms, using the additivity of variances rule,

$$V_{\text{reprod}} = V_{\text{repeat}} + V_{\text{temp}} + V_{\text{operator}} + V_{\text{chemicals}} + V_{\text{workup}} + V_{\text{population}} + \cdots \tag{1.6}$$

Each of these variances is the square of the corresponding standard deviation and describes the effect of one factor on the uncertainty of the result.

1.1.5. Reporting the Results

As will be shown in Section 1.7.2, the standard deviations determined for the small sets of observations typical for analytical chemistry are trustworthy only to one or two significant digits. Thus, for $x(\ )$: 1.93, 1.92, 2.02, 1.97, 1.98, 1.96, and 1.90, an ordinary pocket calculator will yield

$$\bar{x}: 1.954285714, \qquad s_x = 0.04\underline{0}766469.$$

The second significant digit in s_x (underlined) corresponds to the third decimal place of $\bar{x}$. In reporting this result, one should round as follows:

$$\bar{x}: 1.954 \pm 0.041 \qquad (n = 7).$$

Depending on the circumstances, it might even be advisable to round to one digit less; i.e.,

$$\bar{x}: 1.95 \pm 0.04 \qquad (n = 7)$$

or

$$\bar{x}: 1.95 \pm 2.1\% \text{ RSD} \qquad (n = 7) \qquad (\text{RSD: relative standard deviation}).$$

Notice that a result of this type, in order to be interpretable, must comprise three numbers: the mean, the (relative) standard deviation, and the number of measurements that went into the calculation.

1.1.6. Interpreting the Results

The inevitability of systematic and random errors in the measurement process, somewhat loosely circumscribed by "drift" and "noise," means that $\bar{x}$ and s_x can only be approximations to the true values. Thus the results found in the above section can be viewed under three different perspectives:

(a) Does the found mean $\bar{x}$ correspond to expectations? The expected value $E(\bar{x})$, written as μ (Greek mu), is either a theoretical value or an experimental average underpinned by so many measurements that one is very certain of its numerical value. The question can be answered by the t test explained in Section 1.5.2. A rough assessment is obtained by checking to see whether μ and $\bar{x}$ are separated by more than $2 \cdot s_{\bar{x}}$ or not: If the difference Δx is larger, $\bar{x}$ is probably not a good estimate for μ.

(b) Does the found standard deviation s_x correspond to expectations? The expected value $E(s_x)$ is σ (Greek sigma), again either a theoretical value or an experimental average. This question is answered by the F test explained in Section 1.7.1. Proving s_x to be different from σ is not easily accomplished, especially if n is small.

(c) Is the mean $\bar{x}$ significant? The answer is the same as for question (a), but with $\mu = 0$. If the values $(\bar{x} - 2 \cdot s_x)$ and $(\bar{x} + 2 \cdot s_x)$ bracket zero, it is improbable that μ differs from zero.

The standard deviation as defined above relates to the repeatability of measurements on the same sample. When many samples are taken from a large population, "sampling variability" and "population variability" terms have to be added to Eq. (1.6) and the interpretation will reflect this.

For analytical applications it is important to realize that three distributions are involved, namely, one that describes the measurement process, one that brings in the sampling error, and another that characterizes the sampled population. In a thought experiment the difference between the population variability (which not necessarily follows a symmetrical distribution function), and the errors associated with the measurement process (repeatability, reproducibility, both usually normally distributed) is explored.

In chemical operations (see Fig. 1.8) a synthesis step is governed by a large number of variables, such as the concentration ratios of the reactants, temperature profiles in time and space, presence of trace impurities, etc. The outcome of a single synthesis operation (one member of the

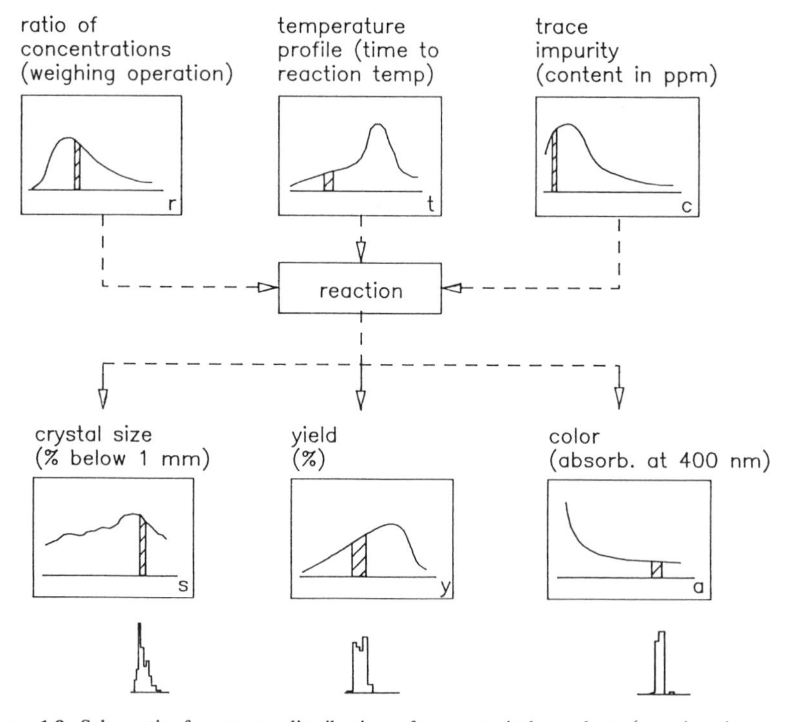

Figure 1.8. Schematic frequency distributions for some independent (reaction input or control) resp. dependent (reaction output) variables to show how non-Gaussian distributions can obtain for a large population of reactions (i.e., all batches of one product in five years), while approximate Normal Distributions are obtained for repeat measurements on one single batch. For example, the hatched areas correspond to the process parameters for a given run, while the histograms give the distribution of analytical results obtained from repeat determinations on one (several) sample(s) from this run.

sampled population) will yield a set of characteristic results, such as yield, size distribution of crystals, or purity. If the synthesis is redone may times, it is improbable that the governing variables will assume exactly the same values every time. The small variations encountered in temperature profiles, for example, will lead to a variation in, say, the yield that might well follow a skewed distribution. Repetition of the analyses on one single sample will follow a Normal Distribution, however.

1.2. DISTRIBUTIONS AND THE PROBLEM OF SMALL NUMBERS

If a large number of repeat observations on one and the same sample are plotted, most fall within a narrow interval around the mean, and a decreasing number is found further out. The familiar term "bell curve" is appropriate; see Fig. 1.9.

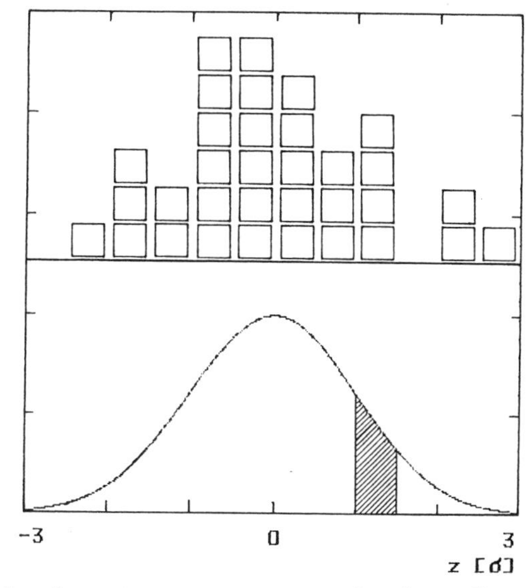

Figure 1.9. A large number of repeat measurements x_i are plotted according to the number of observations per x interval. A bell-shaped distribution can be discerned. The corresponding probability densities PD are plotted as a curve versus the z value. The probability that an observation is made in the shaded zone is equal to the zone's area relative to the area under the whole curve.

1.2.1. The Normal Distribution

It would be of obvious interest to have a theoretically underpinned function that describes the observed frequency distribution shown in Fig. 1.9. A number of such distributions (symmetrical or skewed) are described in the statistical literature in full mathematical detail; apart from the normal and the *t* distributions, none is used in analytical chemistry except under very special circumstances, e.g., the Poisson and the Binomial Distributions. For a long time it was widely believed that experimental measurements accurately conformed to the Normal Distribution. On the whole this is a pretty fair approximation, perhaps arrived at by uncritical extrapolation from a few well-documented cases. It is known that real distributions are wider than the Normal one; *t* Distributions for 4 to 9 degrees of freedom (see Section 1.2.2) are said to closely fit actual data.[14]

Does this mean that one should abandon the Normal Distribution? As will be shown in Sections 1.8.1–1.8.3, the practicing analyst rarely gets together enough data points to convincingly demonstrate adherence to one

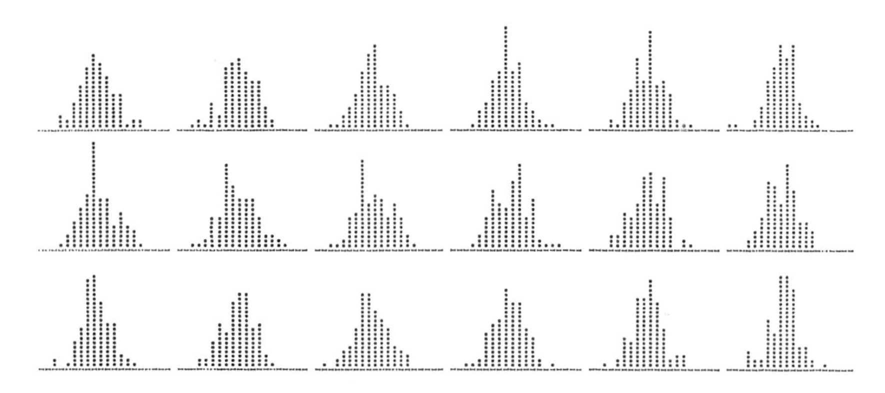

Figure 1.10. The figure demonstrates what is obtained when the Monte Carlo Method (cf. Section 3.6) is used to simulate normally distributed values: Each histogram (cf. Section 1.8.1) summarizes 100 "measurements." Obviously many do not even come close to what one expects under the label bell curve. For practical purposes, distributions over less than 50–100 measurements must be regarded as belonging to the Normal Distribution class, even if small deviations are observed, because the contrary cannot be proven. The only real exceptions consist in (1) manifest asymmetry, and (2) the a priori knowledge that another model applies. For example, if the outcome of an observation can only be of the type 0 or 1, a Binomial Distribution must be used. Since nearly all types of measurements in analytical chemistry belong to the class yielding continuous values, however, it is a defensible approach to assume a Normal Distribution. If results are obtained in digitized form, the Gaussian approximation is valid only if the true standard deviation is at least 3–5 times greater than the digitizer resolution.

or the other distribution model. So, for all practical purposes, the Normal Distribution remains a viable alternative to unwieldy but "better" models. See Fig. 1.10.

For general use, the Normal Distribution has a number of distinct advantages over other distributions, some of the more important being

- its efficiency
- its lack of bias
- its wide acceptance
- the many programs and tests that incorporate it

Its characteristics are described in detail in the following box.

NORMAL DISTRIBUTION (ND)

The Normal or Gaussian Distribution describes a bell-shaped frequency profile defined by the function

$$\text{PD} = \frac{1}{\sigma \cdot \sqrt{2\pi}} \cdot \exp\left\{ \frac{-1}{2} \cdot \left[\frac{x - \mu}{\sigma} \right]^2 \right\}, \qquad (1.7)$$

where μ: true average, as deduced from theory or through a very large number of measurements

σ: true standard deviation, as deduced from theory or through a very large number of measurements

x: observed value

PD: the probability density as a function of x; i.e., the expected frequency of observation at x

Since it is impractical to tabulate PD(x) for various combinations of μ and σ, the Normal Distribution is usually presented in a normalized form where $\mu = 0$ and $\sigma = 1$; that is,

$$\text{PD} = 0.39894 \cdot \exp(-z^2/2), \qquad (1.8)$$

where $z = (x - \mu)/\sigma$; this state of affairs is abbreviated ND(0, 1) as opposed to ND(μ, σ). Because of the symmetry inherent in PD $= f(z)$, the ND(0, 1) tables are only given for positive z values usually over the range $z = 0 \ldots 4$; with entries for 0.05 or smaller increments of z see Table 1.2.

Table 1.2. The Probability Density of the Normal Distribution. (Because of the symmetry, the density values are identical for z and $-z$.)

z	PD
$z = 0$	PD = 0.3989
1	0.2420
2	0.0540
3	0.0060

The corresponding statistical table is known as the probability density table; a few entries given are for identification purposes in Table 1.2, see also program PDVAL.

When many observations are made on the same sample, and these are plotted in histogram form the bell-shaped curve becomes apparent for n larger than about 100. Five such distributions calculated according to the Monte Carlo method (see Section 3.6) for $n = 100, 300, 1000, 3000$, and 10,000 are shown in Fig. 1.11; a scaling factor was introduced to yield the same total area (1000 points) per distribution. The z-axis scale is $-4 \cdots 4 \cdot \sigma$, resp., $C = 80$ classes (bins); that is, each bin is $\sigma/10$ wide. A rule of thumb for plotting histograms suggests $C = \sqrt{n}$ classes (bins); that would mean about 8–12, 15–20, 30–35, 50–60, respectively, 100 bins. The number C is often chosen so as to obtain convenient boundaries, such as whole numbers. A constant bin width was chosen here for illustrative purposes, thus the left two figures do not represent the optimum in graphical presentation: One

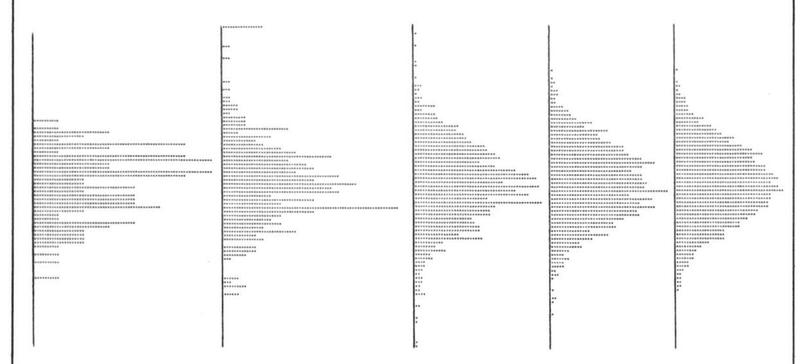

Figure 1.11. Simulated Normal Distributions for $n = 100$ to 10,000 events. For details see text.

could either fuse 5–10 adjacent bins into one, e.g., bins 1–10, 11–20, etc., and plot the average, or then one could plot a moving average (cf. Section 3.5).

Since one is only rarely interested in the density at a precise point on the z axis, the cumulative probability (cumulative frequency) tables are more important: in effect, the integral from $-\infty$ to $+z$ over the probability density function for various $z \geq 0$ is tabulated; again a few entries are given in Table 1.3.

Table 1.3. The Cumulative Probability of the Normal Distribution. (The hatched area corresponds to the difference ΔCP in the CP plot.)

z	CP
$z = 0$	CP = 0.5000
1	0.8413
2	0.9773
3	0.9987

The integral function is symmetrical about the coordinate ($z =$ 0, CP = 0.5000); for this reason only the right half is tabulated, the other values being obtained by subtraction from 1.000, i.e., for $z = -2$, CP = 1 − 0.97725 = 0.02275.

Some authors adopt other formats, for instance, the integral $z = 0$ to $+z$ is given with CP = 0.0000–0.4987 (at $z = 3$), or then the integral $-z$ to $+z$ is given with CP = 0.0000–0.9973 (at $z = 3$; $1 - 2 \cdot 0.00135 = 0.9973$).

In lieu of Normal Distribution tables, fairly accurate approximations to the entires can be made by using Eqs. (1.9a) and (1.9b).

The cumulative probability table can be presented in two forms, namely,

$$1 - \mathrm{CP} = P(z) \tag{1.9a}$$

and

$$z = P'(\mathrm{lgt}(1 - \mathrm{CP})), \tag{1.9b}$$

where (1 − CP) is the area under the curve between $+z$ and $+\infty$.

P and P' are functions that involve polynomials of order 6. The coefficients and measures of accuracy are given in the Section 5.1.1. Both functions are used in sample programs in Chapter 5 and on the diskette (programs CPVAL and ZVAL).

There are instrumental methods of analysis that have *Poisson*-distributed noise, e.g., optical and mass spectroscopy. For an introduction to parameter estimation under conditions of linked mean and variance, see Ref. 29.

1.2.2. Student's *t* Distribution

The Normal Distribution is the limiting case ($n = \infty$) for the Student's t Distribution. Why a new distribution? The reason is simply this: If the number of observations n becomes small, the mean's confidence interval $\mathrm{CI}(\bar{x})$ can no longer be ignored. The same is true for the uncertainty associated with the calculated standard deviation s_x. What is sought, in effect, is a modification of the Normal Distribution that provides for a normally distributed $\bar{x}$ (instead of a fixed μ) and a variance V_x following a χ^2 distribution (instead of a fixed σ^2). This can be visualized as follows:

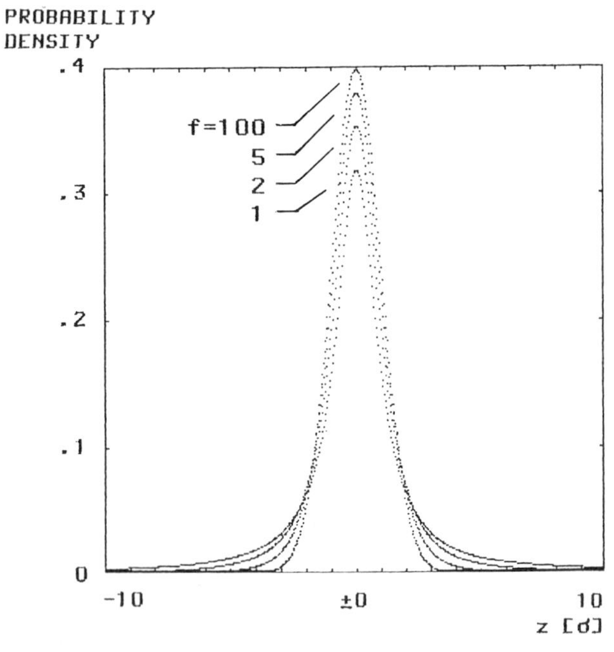

Figure 1.12. The probability density functions for several Student's t Distributions ($f = 1, 2, 5$, resp. 100) are shown. The t Distribution for $f = 100$ already very closely matches a Normal Distribution.

Pick two values $\mu + \Delta\mu$ and $\sigma + \Delta\sigma$, and calculate the Normal Distribution according to Eq. (1.7). Repeat the procedure many times with different deviations $\Delta\mu$ and $\Delta\sigma$ (cf. algorithm, Section 3.6). Add the calculated distributions; this results in a new distribution that is similar in form to the Normal one, but only wider and lower in height for the same area. The Student's t is used exactly as is the variable z in the Normal Distribution; see Table 1.4 and (Section 1.3) Fig. 1.12.

1.3. CONFIDENCE LIMITS

If a result is quoted as having an uncertainty of ± 1 standard deviation, an equivalent statement would be "the 68.3% confidence limits are given by $\bar{x} \pm 1 \cdot \sigma$," the reason being that the area under a Normal Distribution curve between $z = -1.0$ to $z = 1.0$ is 0.683. Now, confidence limits on the 68% level are not very useful for decision making because in one third of all cases, on the average, values outside these limits would be found. What is sought is a confidence level that represents a reasonable compromise between these narrow limits and wide limits:

- **wide limits**

  the statement "the result is within limits" would carry a very low risk of being wrong; the limits would be so far apart as to be meaningless.

Example: $\pm 3.5 \cdot s_{\bar{x}}$: Probability of error 0.047%, confidence level 99.953% for $n = 2$.

- **narrow limits**

  any statement based on a statistical test would be wrong very often, which fact would certainly not augment the analyst's credibility. Alternatively, the statement would rest on such a large number of repeat measurements that the result would be extremely expensive and perhaps out of date.

Example: $\pm 0.5 \cdot s_{\bar{x}}$:

Probability of error	confidence level	for $n =$
61.7%,	38.3%	2
11.4%	88.6%	10
0.6%	99.4	30

Depending on the risks involved, one would like to choose a higher or lower confidence level; as with the many measures of length in use up to the nineteenth century—nearly every principality defined its own "mile"—confusion would ensue. Standardization is reflected in the confidence levels commonly listed in statistical tables: 90, 95, 98, 99, 99.5, ... %. There is no hard-and-fast rule for choosing a certain confidence level, but one has to take into account such things as the accuracy and precision of the analytical methods, the price of each analysis, time and sample constraints, etc. A fair compromise has turned out to be the 95% level; i.e., 1 in 20 tests will suggest a deviation (too high or too low) where none is expected. Rerunning 1 test out of 20 to see whether a real or a statistical outlier had been observed is an acceptable price to pay. In effect, the 95% confidence level comes close to being an agreed-upon standard. Because confidence limits and the number of measurements n are closely linked (see Figs. 1.16 and 1.21) opting for a higher confidence level, such as 99.9%, sharply increases the workload necessary to prove a hypothesis. While this may not be all that difficult if a method with a RSD of $\pm 0.1\%$ were available, in trace analysis, where the RSD is often around $\pm 20\%$ (or more), series of seven or more replicates would be needed just to reduce the confidence limits to $\pm 100\%$ of the estimate $\bar{x}$. The effect is illustrated in Fig. 1.18 and Section 1.6.

Assuming for the moment that a large number of measurements went into a determination of a mean $\bar{x}$ and a standard deviation s_x, what is the width of the 95% confidence interval; what are the 95% confidence limits?

A table of *cumulative probabilities* CP lists an area of 0.975002 for $z = 1.96$; that is, 0.025 (2.5%) of the total area under the curve is found between $+1.96$ standard deviations and $+\infty$. Because of the symmetry of the Normal Distribution function, the same applies for negative z values. Together $p = 2 \cdot 0.025 = 0.05$ of the area, read "probability of observation," is outside the 95% *confidence interval* (outside the 95% *confidence limits* of $\bar{x} - 1.96 \cdot s_x \cdots \bar{x} + 1.96 \cdot s_x$). The answer to the above questions is thus:

$$95\% \text{ confidence limits } \mathrm{CL}(x)\colon \bar{x} \pm z \cdot s_x, \qquad (1.10a)$$

$$95\% \text{ confidence interval } \mathrm{CI}(x)\colon 2 \cdot z \cdot s_x \text{ centered on } \bar{x}. \qquad (1.10b)$$

With $z = 1.96 \approx 2$ for the 95% confidence level, this is the explanation for the often-heard term "$\pm$two sigma" about some mean. [Unless otherwise

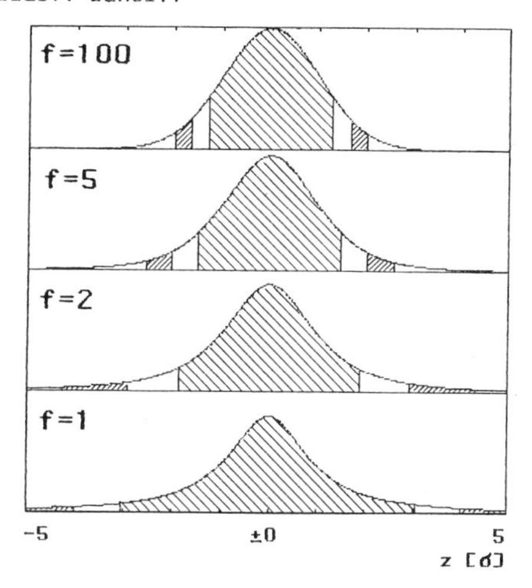

Figure 1.13. Student's t Distributions for 1 (bottom), 2, 5, and 100 (top) degrees of freedom f. The hatched area between the innermost marks is in all cases 80% of the total area under the respective curve. The other marks designate the points at which the area reaches 90, resp. 95%, of the total area. This shows how the t factor varies with f. The t Distribution for $f = 100$ already very closely matches the Normal Distribution. The Normal Distribution, which is equal to $t(f = \inf)$, does not depend on f.

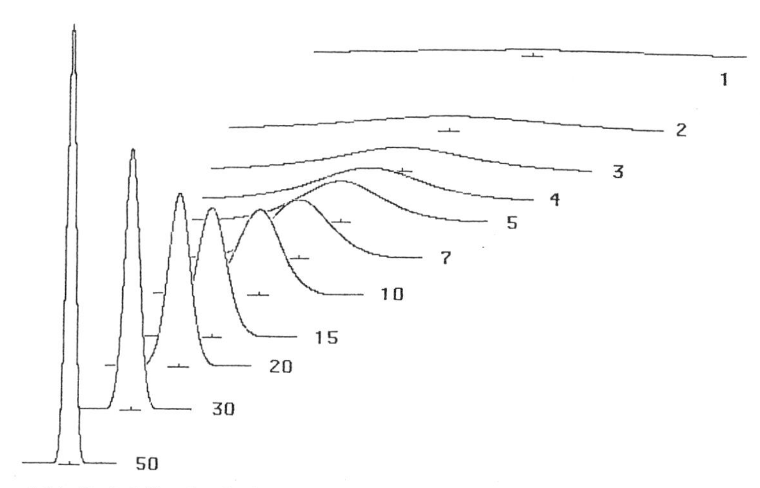

Figure 1.14. Probability density for a mean $\bar{x}$ with $f = n - 1 = 1 \cdots 50$ degrees of freedom. The areas under the curves are equal. If points demarking, say, 95% of the area (cf. Fig. 1.13) were connected, Fig. 1.15 (right) would result.

Table 1.4. Critical Student's t Factors for the One- and Two-Sided Cases for Three Values of the Error Probability p and Six Degrees of Freedom f.

	Two-Sided			One-Sided		
$p =$	0.1	0.05	0.01	0.1	0.05	0.01
$f =$ 1	6.314	12.706	63.66	3.078	6.314	31.821
2	2.920	4.303	9.925	1.886	2.920	6.965
5	2.015	2.571	4.032	1.476	2.015	3.365
10	1.812	2.228	3.169	1.372	1.812	2.764
20	1.725	2.086	2.845	1.325	1.725	2.528
∞	1.645	1.960	2.576	1.282	1.645	2.326

stated, the expressions CL( ) and CI( ) are forthwith assumed to relate to the 95% confidence level.]

In everyday analytical work it is improbable that a large number of repeat measurements is performed; most likely one has to make do with less than 20 replications of any determination. No matter which statistical standards are adhered to, such numbers are considered to be "small," and hence, the law of large numbers, i.e., the Normal Distribution, does not strictly apply. The t Distributions will have to be used; the plural derives from the fact that the probability density functions vary systematically with the number of degrees of freedom f (cf. Figs. 1.12–1.14).

In connection with the above problem one looks for the list "two-tailed (sym.) Student's t factors for $p = 0.05$"; sample values are given for identification. See Table 1.4, and program TVAL.

In Section 5.1.2 an algorithm is presented that permits one to approximate the t tables with sufficient accuracy for everyday use.

1.3.1. Confidence Limits of the Distribution

After having characterized a distribution by using n repeat measurements and calculating $\bar{x}$ and s_x, an additional measurement will be found within the following limits 19 out of 20 times on the average:

$$95\%\ \mathrm{CL}(x) = \bar{x} \pm t \cdot s_x, \tag{1.11a}$$

$$95\%\ \mathrm{CI}(x) = 2 \cdot t \cdot s_x \text{ centered on } \bar{x}. \tag{1.11b}$$

For large n the confidence interval for the distribution converges toward

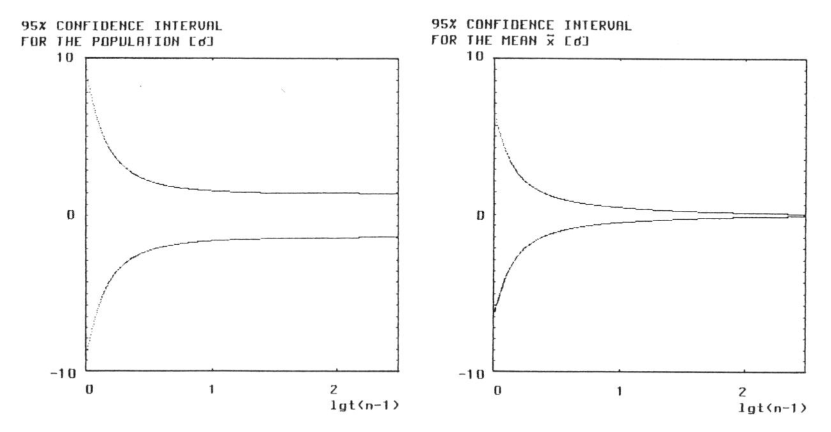

Figure 1.15. The 95% confidence intervals for x and $\bar{x}$ are depicted. The curves were plotted using the approximations given in Section 5.1.2; the f axis was logarithmically transformed for a better overview. Note that solid curves are plotted as if the number of degrees of freedom could assume any positive value. This was done to show the trend; f is always a positive integer. The ordinates are scaled in units of the standard deviation.

the $\pm 1.96 \cdot s_x$ range familiar from the Normal Distribution; cf. Fig. 1.15 (left) and Fig. 1.16 (top).

1.3.2. Confidence Limits of the Mean

If, instead of the distribution as such, the calculated mean $\bar{x}$ is to be qualified:

$$95\%\ \mathrm{CL}(\bar{x}) = \bar{x} \pm t \cdot s_x/\sqrt{n}, \qquad (1.12a)$$

$$95\%\ \mathrm{CI}(\bar{x}) = 2 \cdot t \cdot s_x/\sqrt{n} \quad \text{centered on } \bar{x}. \qquad (1.12b)$$

It is apparent that the confidence interval for the mean rapidly converges toward very small values for increasing n, because both $t(f)$ and $1/\sqrt{n}$ become smaller; see Fig. 1.15 (right) and Fig. 1.16 (bottom).

Example: for $p = 0.05$, $\bar{x} = 10$, $s_x = 1$, and different n:

n	$t(f)$	CL(x)	CI(x)	CL($\bar{x}$)	CI($\bar{x}$)
2	12.71	−2.71 ⋯ 22.71	25.42	1.01 ⋯ 18.99	17.97
3	4.303	5.70 ⋯ 14.30	8.61	7.52 ⋯ 12.48	4.97
5	2.776	7.22 ⋯ 12.78	5.55	8.76 ⋯ 11.24	2.48
10	2.262	7.74 ⋯ 12.26	4.52	9.28 ⋯ 10.72	1.43
100	1.984	8.02 ⋯ 11.98	3.97	9.80 ⋯ 10.20	0.40

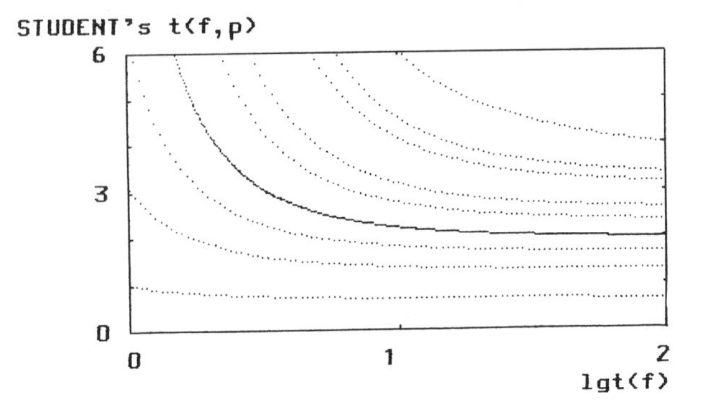

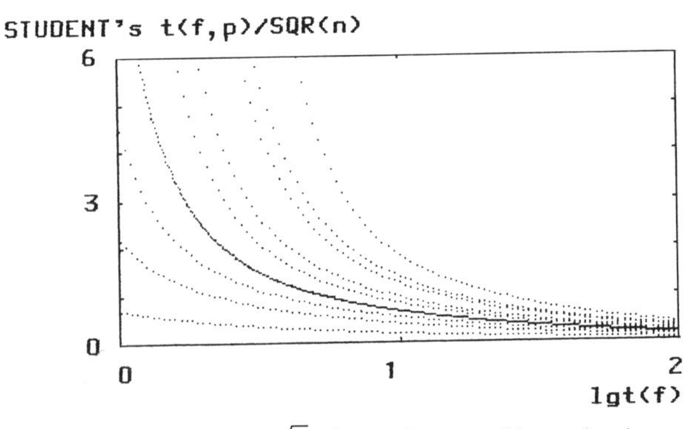

Figure 1.16. The Student's t, resp. $t/\sqrt{n}$, for various confidence levels are plotted; the curves for $p = 0.05$ are enhanced. The other curves are for $p = 0.5$ (bottom), 0.2, 0.1, 0.02, 0.01, 0.002, 0.001, and 0.0001 (top). By plotting a horizontal at, say, $y = 3$, the number of measurements necessary to obtain the same confidence intervals for different confidence levels can be estimated. A vertical at, say, $f = \text{lgt}(8)$ shows how far the confidence limits will be apart for different confidence levels and a given $n = 9$.

1.4. SIMULATION OF A SERIES OF MEASUREMENTS

Simulation by means of the digital computer has become an extremely useful technique (see Section 3.7) that goes far beyond classical interpolation/extrapolation. The reasons for this are

- Very complex systems of equations can be handled; this allows interactions to be studied that elude those who simplify equations to make them manageable on the paper-and-pencil level.[30]

- Fast iterative root-finding algorithms do away with the necessity of algebraically solving for "buried" variables, an undertaking that often enough does not yield closed solutions anyway [a solution is closed when the equation has the form $x = f(a, b, c, \ldots)$ and x does not appear in the function f].
- Nonlinear and discontinuous equations can be easily implemented, e.g., to simulate the effects of a temperature-limiting device or a digital voltmeter.[11]
- Not only deterministic aspects can be modeled, but random ones as well, cf. Refs. 4 and 24 and Section 3.6.

This important technique is introduced at this elementary level to demonstrate characteristics of the confidence-level concept that would otherwise remain unrecognized. Two models are necessary, one for the deterministic, the other for the stochastic aspects.

As an example, the following very general situation is to be modeled: A physico-chemical sensor in contact with an equilibrated chemical system is used to measure the concentration of an analyte. The measurement has noise superimposed on it, so that the analyst decides to repeat the measurement process several times and to evaluate the mean and its confidence limits after every determination. (Note: this modus operandi is forbidden under GLP; the necessary number of measurements and the evaluation scheme must be laid down before the experiments are done.) The simulation is carried out according to the scheme depicted in Fig. 1.17: The computer program that corresponds to the scheme principally contains all of the simulation elements; some simplifications can be

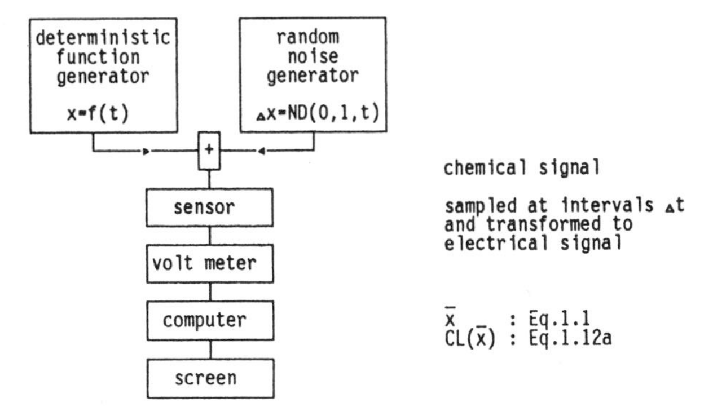

Figure 1.17. Scheme for numerical simulation of measurements.

introduced, though:

- For the present purposes the deterministic function generator yields a constant signal $x = 0$, which means the summation output is identical with that of the noise generator.
- The sensor is taken to be of the linear type; i.e. , it transduces the incoming chemical information into electrical output according to the equation el.signal = constant + slope · (chemical signal); without loss of clarity, the constant can be set to zero and the slope to 1.00.
- One noise generator in parallel to the chemical function generator suffices for the present purposes; if electrical noise in the sensor electronics is to be separately simulated, a second noise generator in parallel to the sensor and a summation point between the sensor and the voltmeter would become necessary. The noise is assumed to be normally distributed with $\mu = 0$ and $\sigma = 1$.
- The computer model does nothing but evaluate the incoming "assay values" in terms of Eqs. (1.1) and (1.12a).

The output of the simulation will be displayed on a "screen" as follows (see Fig. 1.18): The common abscissa is the sample number i. The ordinates are in signal units; the top window shows the individual measurements as points; the bottom window shows how the derived standard deviation converges toward its expected value, $E(s) = 1.00$; in the middle window the mean and the $\mathrm{CL}(\bar{x})$ are shown to rapidly, although erratically, converge toward the expected value 0 ± 0. Equations (1.1), (1.3), (1.5), and (1.12a) (middle widow), and (1.3), (1.5), and (1.42) (bottom window) were used. The simulation covers 50 successive "measurements." Many such simulations were run, and two were picked out: the typical situation (left side) and an interesting one (right side). It is definitely possible that the $\mathrm{CI}(\bar{x})$ does not include the expected value, viz., the section around $i = 4$–12 in the right panel. In a large number of simulations, it turned out, similar but less dramatic situations occur on the average once every 13th trial: The first few points suggest a satisfyingly small scatter. In this particular simulation, this is due to the operation of "pure chance' as defined in the Monte Carlo algorithm. However, inadequate instrument configurations, poor instrument maintenance, improper procedures, or a knowledge of what one is looking for can lead to similar observations. Analysts and managers may (subconciously) fall prey to the last, the psychological, trap because without a rigid plan they are enticed to act selectively, either by stopping an experiment at the "right" time or by replacing apparent "outliers" by more well-behaved repeat results.

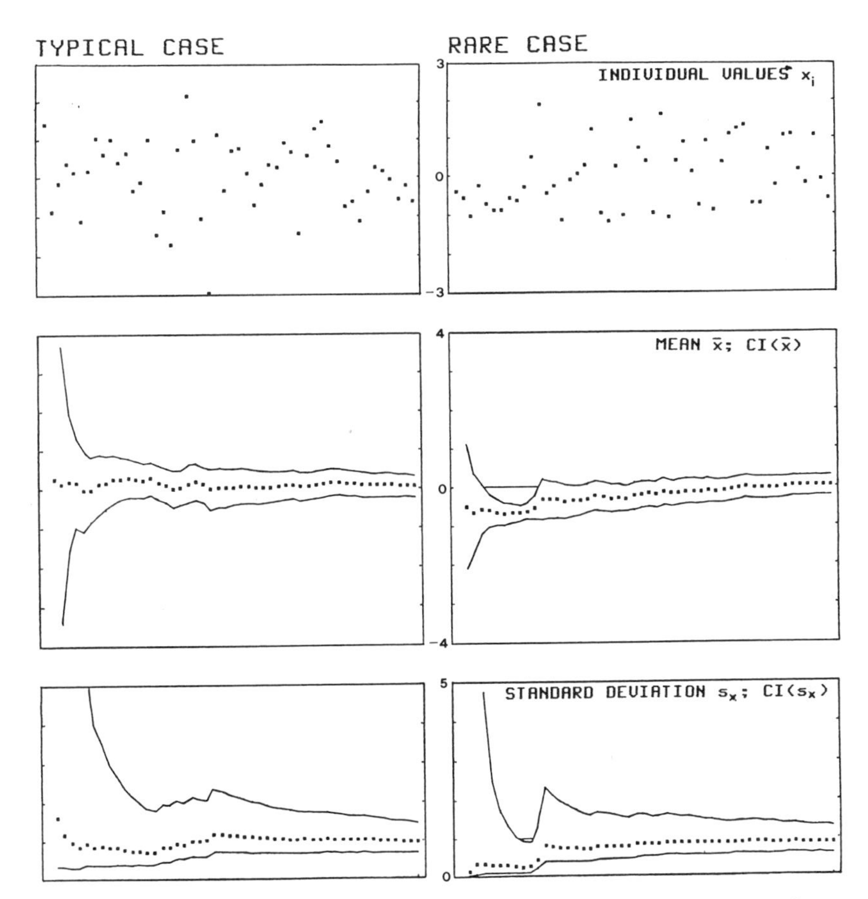

Figure 1.18. Monte Carlo simulation of 50 normally distributed measurements: raw data are depicted as x_i vs. i (top); the mean (dots) and its upper and lower confidence limits (full lines, middle); the confidence limits CL(s_x) of the standard deviation converge toward $\sigma = 1$ [bottom, (Eq. 1.42)]. The vertical divisions are in units of $1 \cdot \sigma$. On the left side a fairly typical case is given, and on the right a selected case is shown. Several hundred simulations were necessary to find this spectacular event: 10 successive points are clustered closely on the low side, which means that the average is low and the standard deviation small at the beginning of the series. The two horizontal lines between points 4 and 12, resp. 7 and 9, indicate the expected $\mu = 0$ and $\sigma = 1$, which are outside the calculated confidence interval. The individual points were generated according to the Monte Carlo method (Section 3.6). The error probability is $p = 0.05$. See also program CONVERGE.

1.5. TESTING FOR DEVIATIONS

The comparison of two results is a problem often encountered by the analyst. Intuitively, two classes of problems can be distinguished:

(a) A systematic difference is found, supported by indirect evidence that from experience precludes any explanation other than "effect observed." This case does not necessarily call for a statistical evaluation, but an example will nonetheless be provided: In the elemental analysis of organic chemicals (CHN analysis) reproducibilities of ± 0.2–0.3% are routine (for a mean of 38.4 wt.% C, for example, this gives a true value within the bounds $38.0 \cdots 38.8$ wt.% for 95% probability). It is not out of the ordinary that traces of the solvent used in the last synthesis step remain in the product. So, a certain pattern of deviations from the theoretical element-by-element percentage profile will be indicative of such a situation, whereas any single element comparison, e.g., $C_{exp} - C_{theor}$, would result in a rejection of the solvent-contamination hypothesis. In practice, varying solvent concentrations would be assumed, and for each one the theoretical elemental composition would be calculated, until the best fit with the experimental observations is found (see χ^2 test).

(b) A measurement technique such as titration is employed that provides a single result which on repetition scatters somewhat about the expected value. If the difference between expected and observed value is so large that a deviation must be suspected, and no other evidence such as gross operator error or instrument malfunction is available to reject this notion, a statistical test is applied. (Note: under GLP, a deviant result may be rejected if and when there is sufficient *documented* evidence of such an error.)

If a statistical test is envisioned, some preparative work is called for: Every statistical test is based on

- a model,
- a confidence level, and
- a set of hypotheses.

It must be realized that these three prerequisites of statistical testing must be established and justified before any testing or interpreting is done; since it is good practice to document all work performed, these apparent "details" are best set down in the experimental plan or the relevant SOP, unless one wants the investigator (even oneself) to obtain undue freedom to bend the conclusions by choosing the three elements to suit his needs (exploratory data analysis is an exception discussed in Section 3.2). A point that cannot be stressed enough is that statistics provides a way of quantizing hidden information and organizing otherwise unmanageable amounts of data in a manner accepted and understood by all parties involved. The outcome of a statistical test is never a hard fact, but always a statement to the effect that a certain interpretation has a probability of $x\%$ or less of not correctly representing the truth. For lack of a more convincing *model*, the t Distribution is usually accepted as a description of measurement variability, and a *confidence level* in the 95–99% range is more or less tacitly assumed to fairly balance the risks involved. If a difference is found, the wording of the result depends on the confidence level, namely, "the difference is 'significant' (95%) or 'highly significant' (99%)." The setting up and testing of *hypotheses* is the subject of Section 1.9. What hypotheses are there?

The "Null" Hypothesis $\mathbf{H_0}$

Given that the measured content for a certain product has been within 2% of the theoretical amount over the past, say, 12 batches, the expectation of a further result conforming with previous ones constitutes the so-called "null" hypothesis, H_0; i.e., no deviation is said to be observed.

The Alternate Hypothesis $\mathbf{H_1}$

Since it is conceivable that some slight change in a process might lead to a different content, a mental note is made of this by stating that if the new result differs from the old one, the alternate hypothesis H_1 obtains. The difference between H_0 and H_1 might be due to $\mu_A \neq \mu_B$ and/or $\sigma_A \neq \sigma_B$. The first possibility is explored in the following section; the second one will be dealt with in Section 1.7.1.

The situation of H_0 and H_1 differing solely in the true averages $\bar{x}_A$ and $\bar{x}_B$ is summed up in Fig. 1.19. Assuming one wants to be certain that the risk of falsely declaring a good batch B to be different from the previous one A is less than 10%, the symmetrical 90% confidence limits are added to A (see Fig. 1.19): Any value B in the shaded area results in the judgment "B different from A, H_1 accepted, H_0 rejected," whereas any result, however suspect, in the unshaded area elicits the comment "no difference between A and B detectable, H_1 rejected, H_0 retained." Note that the expression "A identical to B" is not used; by statistical means only

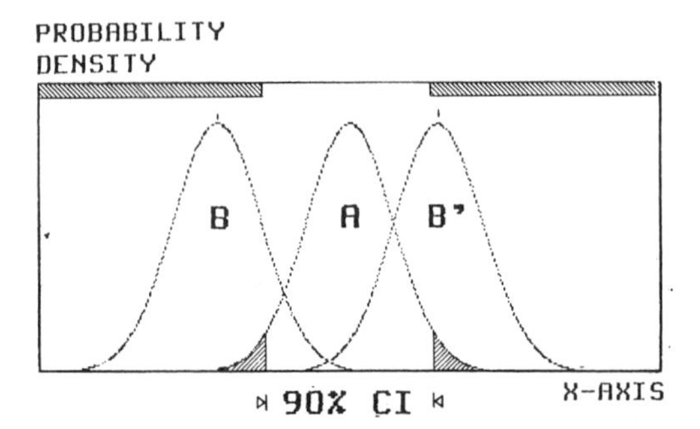

Figure 1.19. The null and the alternate hypotheses H_0 resp. H_1. The Normal Distribution probability curves show the expected spread of results. Since the alternate distribution $ND(\mu_B, \sigma)$ might be shifted towards higher or lower values, two alternative hypotheses H_1 and H_1' are given. The unshaded portion of the x axis represents the 90% confidence interval of A, cf. program HYPOTHES.

deviations can be demonstrated, and similarities must be inferred from their absence.

1.5.1. Examining Two Series of Measurements

Testing requires that a certain protocol be adhered to: This protocol is set forth here in a how-to format of questions and instructions. The actual calculations are shown below.

Data

Two series of measurements are made, such as, for example, n repetitive determinations of an analyte concentration in the same sample by two different methods: $x_{11}, x_{12}, \ldots, x_{1n}, x_{21}, x_{22}, \ldots, x_{2n}$.

Question 1

Are the group variances V_1 and V_2 indistinguishable?

Perform the F test (Section 1.7.1; most authors find the F test to be a prerequisite for the t test). If no significant effect is found (H_0 retained), the two sample variances may be pooled to decrease the uncertainty associated with V through inclusion of a higher number of measurements: The models given in Table 1.5, cases (b1) or (b3), as appropriate, both give the same degree of freedom f, but different variances. However, if a significant difference between V_1 and V_2 is found (H_1 accepted), they may,

Table 1.5. Various Forms of the Student's t Test.
(For details and examples see text.)

Test Statistic	Conditions	Variance, Degrees of Freedom	Equation	References Comments
$t = \dfrac{\bar{x} - \mu}{\sqrt{V_{\bar{x}}}}$	None [case a]	$V_{\bar{x}} = V_x/n \quad f = n - 1$ ($V_{\bar{x}}$: variance of average $\bar{x}$; V_x: variance of x)	(1.13)	$V_x = \dfrac{S_{xx}}{f}$
$t = \dfrac{\Delta\bar{x}}{\sqrt{V_d}}$	$V_1 = V_2$	$V_d = V_p \cdot k$	(1.14)	$\Delta\bar{x} = \bar{x}_1 - \bar{x}_2$
	$n_1 \neq n_2$ [case (b3)]	$V_p = ((n_1 - 1) \cdot V_1 + (n_2 - 1) \cdot V_2)/f$ $k = 1/n_1 + 1/n_2$		Refs. 24, 34, 35 V_p: pooled variance
		$V_d = V_1/n_1 + V_2/n_2$	(1.15)	Refs. 32, 36
		$V_d = V_p' \cdot k'$ $V_p' = (n_1V_1 + n_2V_2)/f$ $k' = 1/n_1 + 1/n_2$	(1.16)	Ref. 37 Ref. 33
		$f = n_1 + n_2 - 2$	(1.17)	Refs. 24, 32–37
	$V_1 = V_2$ $n_1 = n_2 (= n)$ [case (b1)]	$V_d = V_1/n_1 + V_2/n_2$ $V_d = (V_1 + V_2)/n$ $f = 2n - 2$	(1.18) (1.19) (1.20)	Refs. 33, 36 Ref. 34 Ref. 24, 34, 36
	$V_1 \neq V_2$	$V_d = V_1/n_1 + V_2/n_2$	(1.21)	Not permitted according to Ref. 32
	[case (c)]	$f = \dfrac{\left(\dfrac{V_1}{n_1} + \dfrac{V_2}{n_2}\right)^2}{\dfrac{(V_1/n_1)^2}{n_1 \pm 1} + \dfrac{(V_2/n_2)^2}{n_2 \pm 1}} - 2$	(1.22 −) (1.22 +) (1.22 +)	Ref. 34 $(n - 1)$ Ref. 35 $(n + 1)$ Ref. 37 $(n + 1)$
		$f = \dfrac{1}{k^2/f_1 + (1 - k)^2/f_2}$, $k = \dfrac{n_2V_1}{n_2V_1 + n_1V_2}$	(1.23)	Ref. 24
$t = \dfrac{d}{\sqrt{V_d}}$	$n_1 = n_2$ (case b2)	$d_i = x_{1i} - x_{2i} \quad f = n - 1$ d, V_d: average resp. variance of differences d_i	(1.24)	Refs. 32, 35, 36 Refs. 34, 37

according to some authors, be pooled nonetheless, with different models yielding the same variance, but different degrees of freedom f [case (c)].

Question 2
Are the two means $\bar{x}_1$ and $\bar{x}_2$ distinguishable?

If H_0 obtains for question 1, all authors agree that a t test can be performed. If H_1 obtains, opinions diverge as to the propriety of such a t test. In practice a t test will be performed: If the outcome is clear, that is, t is much different from t_c, the differences between the models are negligible; if t is close to t_c, more tests should be performed to tighten the confidence limits, or judgment should be suspended until other evidence becomes available.

Obviously, the t test also involves its own set of null and alternative hypotheses; these are also designated H_0 and H_1, but must not be confused with the hypotheses associated with the F test.

1.5.2. The *t* Test

The most widely used test is that for detecting a deviation of a test object from a standard by comparison of the means, the so-called t test. Note that, before a t test is decided upon, the confidence level must be declared and a decision made to the effect whether a one- or a two-sided test is to be performed. For details, see below. Three levels of complexity are distinguishable (the necessary equations are assembled in Table 1.5 and are all included in program TTEST):

(a) the standard is a precisely known mean, a theoretical average μ, or a preordained value, such as a specification limit.
(b) the test sample and the standard were measured using methods that yield indistinguishable standard deviations s_1 resp. s_2 (cf. F test, Section 1.7.1).
(c) The standard deviations s_1 resp. s_2 are different.

In case (a) only the standard deviation estimated from the experimental data for the test sample is needed, which is then used to normalize the difference $\bar{x} - \mu$. The quotient, the so-called Student's t, is compared with the critical t_c for a chosen confidence level and $f = n - 1$:

$$t = \frac{|\bar{x} - \mu|}{s_{\bar{x}}} = \frac{|d|}{s_{\bar{x}}}; \qquad \text{see Eq. (1.13).}$$

Example: $\bar{x} = 12.79$, $s_x = 1.67$, $n = 7$, $s_{\bar{x}} = 0.63$, $\mu = 14.00$. Chosen confidence level: 95%:

$$t = |12.79 - 14.00|/0.63 = 1.92, \qquad \text{Table 1.4,}$$

$$t_c(6, p=0.05) = 2.45, \qquad \text{two-sided test,}$$

$$t_c(6, p/2=0.05) = 1.94, \qquad \text{one-sided test.}$$

Interpretation: If the alternate hypothesis is stated as "H_1: $\bar{x}$ is different from μ" a two-sided test is applied with 2.5% probability being provided for each possibility "$\bar{x}$ smaller than μ," resp., "$\bar{x}$ larger than μ." Because 1.92 is smaller than 2.45, the test criterion is not exceeded, so H_1 is rejected. On the other hand, if it was known beforehand that $\bar{x}$ can only be smaller than μ, the one-sided test is conducted under the alternate hypothesis "H_1: $\bar{x}$ smaller than μ"; in this case the result is close, with 1.92 almost exceeding 1.94.

In cases (b) and (c) the standard deviation s_d of the average difference $d = \bar{x}_1 - \bar{x}_2$ and the number of degrees of freedom must be calculated.

Case (b) must be divided into three subcases (for equation, see Table 1.5):

(b1) $n_1 = n_2$, one test, repeatedly performed on each of two samples, one test, performed once on every sample from two series of samples, or two tests, each repeatedly performed on the same sample, $1 \cdots n$, $n + 1 \cdots 2n$; see Eqs. (1.18)–(1.20).

(b2) $n_1 = n_2$, two tests being performed pairwise on each of n samples, $1 \cdots n$; see Eq. (1.24).

(b3) $n_1 \neq n_2$, same situation as (b1) but with different n, $1 \cdots n_1$, $n_1 + 1 \cdots n_1 + n_2$; see Eqs. (1.14)–(1.17).

Subcase (b1)

This case is encountered, for example, when batch records from different production campaigns are compared, and the same number of samples was analyzed in each campaign. (Note: under GMP, trend analysis has to be performed regularly to stop a process from slowly, over many batches, drifting into a situation where each parameter on its own is within specifications, but collectively there is the risk of sudden, global loss of control. "Catastrophe theory" has gained a foothold in physical and biological literature to describe such situations; cf. Section 4.14.)

The variance V_d of the average difference is calculated as $(V_1 + V_2)/n$ [see Eqs. (1.18) and (1.19)] with $f = 2n - 2$ degrees of freedom. s_d, the square root of V_d, is used in the calculation of the Student's t statistic $t = d/s_d$.

Example: $\bar{x}_1 = 17.4$, $s_1 = 1.30$, $\bar{x}_2 = 19.5$, $s_2 = 1.15$, $n_1 = n_2 = 8$, $p = 0.05$. The standard deviations are not significantly different (F test, Section 1.7.1); the standard deviation of the mean difference is $s_d = 0.61$ and $t = |17.4 - 19.5|/0.61 = 3.42$; $f = 14$. Since the critical t value is 2.145, the two means $\bar{x}_1$ and $\bar{x}_2$ can be distinguished on the 95% confidence level.

Subcase (b2)

This case, called the paired t test, is often done when two test procedures, such as methods A and B, are applied to the same samples, for instance, when validating a proposed procedure with respect to the accepted one. In particular, an official content uniformity assay might prescribe a photometric measurement (extract the active principle from a tablet and measure the absorbance at a particular wavelength); a new HPLC procedure, which would be advantageous because it would be selective and the extraction step could be replaced by a simple dissolution/filtration step, is proposed as a replacement for the photometric method. In both methods the raw result (absorbance resp. peak area) is converted to concentration units to make the values comparable (Section 2.2.6). Tests are conducted on 10 tablets. The following statistics are calculated:

$$d_i = x_{\mathrm{PM},i} - y_{\mathrm{HPLC},i}, \qquad \text{for } i = 1 \cdots 10.$$

The average d and s_d are calculated according to Eq. (1.1) [resp. (1.5)]. $t = d/s_d$ is compared to the tabulated t value for $f = n - 1$.

Consider the following data: $x(\ ) = 1.73, 1.70, 1.53, 1.78, 1.71$; $y(\ ) = 1.61, 1.58, 1.41, 1.64, 1.58$. One finds $\bar{x} = 1.690$, $s_x = 0.0946$, $\bar{y} = 1.564$, and $s_y = 0.0896$; The F value is 1.11, certainly not higher than the $F_c = 3.2$ critical value for $p = 0.05$; the two standard deviations are recognized as being indistinguishable.

If, despite the fact that the x and the y values are paired, a t test were applied to the difference $(\bar{x} - \bar{y})$ using the standard deviation $s_d = 0.0583$ according to Eq. (1.18), a t value of 2.16 would be found, which is smaller than the critical $t_c(2n - 2) = 2.306$. The two series thus could not be distinguished. The overall mean, found as 1.63, would be advanced.

An important aspect, that of pairing, had been ignored above; if it is now taken into account [proviso: $x(i)$ and $y(i)$ are related as in "ith sample pulled from ith powder mixture and subjected to both PM and HPLC analysis, $i = 1 \cdots n$," etc.], a standard deviation of the differences of $s_d = 0.0089$ is found according to Eq.

(1.3). With the average difference being $d = 0.126$, this amounts to a relative standard deviation of about 7.1% or a $t_d = 14.1$, which is clearly much more significant than what was erroneously found according to case (b1). $t_c(4, 0.05) = 2.77$.

The example demonstrates that all relevant information must be used; ignoring the fact that the PM and HPLC measurements for $i = 1 \cdots 5$ are paired results in a loss of information. The paired data should under all circumstances be plotted (Youden plot, Fig. 2.1) to avoid a pitfall: It must be borne in mind that the paired t test yields insights only for the particular (additive) model examined; see Section 1.5.6, that is, for the assumption that the regression line for the correlation PM vs. HPLC has a slope $b = 1$, and an intercept $a_{\text{PM, HPLC}}$ that is either indistinguishable from zero (H_0), or is significantly different (H_1). If there is any systematic, nonadditive difference between the methods (e.g., interferences, slope $b \neq 1$, nonlinearity, etc.), a regression might be more appropriate (Section 2.2.1). Indeed, in the above example, the data are highly correlated ($s_{\text{res}} = \pm 0.0083$, $r^2 = 0.994$; see Chapter 2). Using case (b1), the overall variance is used without taking into account the correlation between x and y; for this reason the standard deviation of the difference $\bar{x} - \bar{y}$ according to case (b1) is much larger than the residual standard deviation (± 0.0583 vs. ± 0.0083), or the standard deviation of the mean difference ± 0.0089. A practical example is given in Section 4.14. The same is true if another situation is considered: If in a batch process a sample is taken before and after the operation under scrutiny, say, impurity elimination by recrystallization, and both samples are subjected to the same test method, the results from, say, 10 batch processes can be analyzed pairwise. If the investigated operation has a strictly additive effect on the measured parameter, this will be seen in the t test; in all other cases both the difference $\Delta \bar{x}$ and the standard deviation s_x will be affected.

Subcase (b3)

When n_1 and n_2 are not equal the degrees of freedom are calculated as $f = n_1 + n_2 - 2$ for the variance of the difference. Up to this point a random pick of statistics textbooks[24,32–37] shows agreement among the authors. The pooled variance is given as

$$V_p = ((n_1 - 1) \cdot V_1 + (n_2 - 1) \cdot V_2)/f; \qquad \text{see Eq. (1.14)},$$

where the numerator is the sum of the squares of the residuals, taken relative to $\bar{x}_1$ or $\bar{x}_2$, as appropriate. Some authors simplify the equation by dropping the -1 in $n - 1$, under the assumption $n \gg 1$ [Eq. (1.16)].

In order to get the variance V_d of the difference of the means d, a way must be found to multiply the pooled variance V_p by a number akin to

$1/n$, as in Eq. (1.19). A formula is proposed, namely, $V_d = V_p \cdot (1/n_1 + 1/n_2)$. Other authors take the sum of the variances of the means, $V_d = V_1/n_1 + V_2/n_2$ [Eq. (1.15)].

It is evident that there is no simple, universally agreed-upon formula for solving this problem. For practical applications, then, if t is much different from t_c, any one of the above equations will do, and if t is close to t_c, or if high stakes are involved in a decision, more experiments (to achieve equality $n_1 = n_2$) might be the best recourse [see subcase (b1)].

Case (c)

Subcase (b3) is an indication of the difficulties associated with testing hypotheses. There is no theory available if $s_1 \neq s_2$; testing for differences under these premises is deemed improper by some,[32] while others[34,35,37] propose a simple equation for the variance of the difference:

$$V_d = V_1/n_1 + V_2/n_2; \qquad \text{see Eq. (1.21)},$$

and a very complicated one for f:

$$f = V_d^2/(Q1 + Q2) - 2; \qquad \text{see Eq. (1.22)},$$

where V_d as above and $Q_i = (V_i/n_i)^2/(n_i \pm 1)$, $i = 1, 2$.

The ambiguity over the sign in the denominator is due to the fact that both versions, $(n_i - 1)$[34] and $(n_i + 1)$,[35,37] are found in the literature (+ resp. − behind the equation number (1.22) indicate the sign used in the denominator). Equation (1.22 −) appears to be correct; Eq. (1.22 +) might have arisen from transcription errors.

Some numerical examples will illustrate the discussed cases. Means and standard deviations are given with superfluous significant figures to allow recalculation and comparison.

Case (b1):

			interpretation:
$\bar{x}_1 = 101.26$	$s_1 = 7.328$	$n_1 = 7$	
$\bar{x}_2 = 109.73$	$s_2 = 4.674$	$n_2 = 7$	H_0
$F = 2.46$	$F_c = 4.29$		
Eqs. (1.18), (1.19): $t = 2.58$			H_1
Eq. (1.20): $f = 12$, $t_c = 2.18$			

Comment: The two standard deviations are similar enough to pass the F test and the data are then pooled. The difference $\bar{x}_1 - \bar{x}_2$ is significant. Equation (1.14) gives the same results; if Eq. (1.16) had been applied, $t = 2.387$ would have been found.

Case (b3): interpretation:

$$\left.\begin{array}{lll} \bar{x}_1 = 101.26 & s_1 = 7.328 & n_1 = 6 \\ \bar{x}_2 = 109.73 & s_2 = 4.674 & n_2 = 8 \\ F = 2.46 & F_c = 3.97 & \end{array}\right\} \quad \mathrm{H}_0$$

Eq. (1.14): $t = 2.65$ Eq. (1.17): $f = 12$, $t_c = 2.18$ H_1

Eq. (1.15): $t = 2.48$ Eq. (1.17): $f = 12$, $t_c = 2.18$ H_1

Eq. (1.16): $t = 2.44$ Eq. (1.17): $f = 12$, $t_c = 2.18$ H_1

Comment: The same as for case (b1), except that n_1 and n_2 are not the same; the different models arrive at conflicting t values, but with identical interpretation: H_1 can thus be accepted.

Case (c): interpretation:

$$\left.\begin{array}{lll} \bar{x}_1 = 101.26 & s_1 = 8.328 & n_1 = 7 \\ \bar{x}_2 = 109.73 & s_2 = 3.974 & n_2 = 7 \\ F = 4.39 & F_c = 4.29 & \end{array}\right\} \quad \mathrm{H}_1$$

Eq. 1.21: $t = 2.43$ Eq. 1.22 +: $f = 9$, $t_c = 2.26$ H_1

Eq. 1.21: $t = 2.43$ Eq. 1.22 −: $f = 6$, $t_c = 2.45$ H_0

Eq. 1.21: $t = 2.43$ Eq. 1.23: $f = 8$, $t_c = 2.31$ H_1

Comment: All models yield the same t value but differ in the number of degrees of freedom to be used. The difference between the means is barely significant in two cases. Suggestion: Acquire more data to settle the case.

1.5.3. Extension of the t Test to More than Two Series of Measurements

The situation of having more than two series of measurements to compare is frequently encountered. One possibility resides in doing a t test as discussed above for every pairing of measurement series; this not only is inefficient, but also does not answer the question of whether all series belong to the same population. The technique that needs to be employed will be discussed in detail below (Section 1.5.4). The same how-to format of questions and instructions is used as above.

Data
Several groups of n_i replicate measurements of a given property on each of m different samples [for a total of $n = \Sigma(n_i)$]. The group sizes n_i need not be identical.

Question 1
Do all m groups have the same standard deviation?

To answer this, do the Bartlett test (Section 1.7.3).

If there is one group variance different from the rest, stop testing and concentrate on finding and eliminating the reason, if any, for

- systematic differences in the application of the measurement technique used, or
- inhomogeneities and improper sampling techniques, etc.

If the variances are indistinguishable, continue by pooling them (V_1, Tables 1.9 and 1.10).

Question 2
Do all m groups have the same mean $\bar{x}$?

Do the simple ANOVA test (Section 1.5.6) to detect variability between the group means in excess of what is expected due to chance alone.

If no excess between-group variance is found, stop testing and pool all values, because they probably all belong to the same population.

If significant excess variance is detected, continue testing.

Question 3
Which of the m groups is significantly different from the rest?

Do the multiple range test (Section 1.5.4) to find sets of averages $\bar{x}$ that are indistinguishable among themselves; it may occur that a given average $\bar{x}_i$ belongs to more than one such set of similar averages.

1.5.4. Multiple Range Test

A setting that turns up quite often is the following: A series of m batches of a given product have been produced, and a certain parameter, e.g., the content of a particular compound, was measured n_i times for each batch. The largest and the smallest means, x_{max} resp. x_{min}, appear to differ significantly: Which of the two is aberrant?

A simple t test cannot answer this question. The multiple range test combines several t tests into one simultaneous test.[38]

Provided that the m variances $V_j = s_j^2$ are roughly equal (Bartlett's test, see Section 1.7.3), the m means are ordered (cf. Program SORT). The smallest average has index 1; the largest has index m. A triangular matrix (see Tables 4.7 and 4.8) is then printed that gives the $m \cdot (m-1)/2$ differences $\Delta\bar{x}_{uv} = \bar{x}_u - \bar{x}_v$ for all possible pairings. Every element of the matrix is then transformed into a q value as

$$q = \Delta\bar{x}_{uv}\sqrt{2 \cdot n_u \cdot n_v/(D \cdot (n_u + n_v))} \tag{1.25}$$

with

$$D = \sum\sum(x_{iv} - \bar{x}_v)^2/f, \qquad i = 1 \cdots n, \quad v = 1 \cdots m,$$

$$f = \sum(n_v) - m, \qquad D \text{ is identical to } V_1 \text{ in Section 1.5.6.}$$

The calculated q value must be compared to a critical q that takes account of the distance that separates the two averages in the ordered set: If $\bar{x}_u$ and $\bar{x}_v$ are adjacent, the column labeled 2 in Table 1.6 must be used, and if $\bar{x}_u$ and $\bar{x}_v$ are separated by eight other averages, the column labeled 10 is used. An excerpt of the q table is given for reference purposes in Table 1.6.

Table 1.6. Critical q Values for Two Means with Index Number Differences $(u - v + 1)$ of 2, 3, 10, Resp. 20.

Degrees of Freedom	Difference $\lvert u - v\rvert + 1$			
	2	3	10	20
1	17.97	17.97	17.97	17.94
2	6.085	6.085	6.085	6.085
10	3.151	3.293	3.522	3.526
∞	2.772	2.918	3.294	3.466

Table 1.7. Reduced Critical q Values, Derived from Critical q Values for $p = 0.05$ by Division by $t(f, 0.05) \cdot \sqrt{2}$, see Data File QREDTBL.001

[With little risk of error, this table can also be used for $p = 0.025$ and 0.1 (divide q by $t(f, 0.025) \cdot \sqrt{2}$, respectively, $t(f, 0.1) \cdot \sqrt{2}$, as appropriate).]

Degrees of Freedom	Separation $\lvert u - v \rvert + 1$ (Difference between Index Numbers +1)														
	2	3	4	5	6	7	8	9	10	12	14	16	18	20	40
1	1.00	1.00	1.00	1.00	1.00	1.00	1.00	1.00	1.00	1.00	1.00	1.00	1.00	1.00	1.00
2	1.00	1.00	1.00	1.00	1.00	1.00	1.00	1.00	1.00	1.00	1.00	1.00	1.00	1.00	1.00
3	1.00	1.00	1.00	1.00	1.00	1.00	1.00	1.00	1.00	1.00	1.00	1.00	1.00	1.00	1.00
4	1.00	1.02	1.03	1.03	1.03	1.03	1.03	1.03	1.03	1.03	1.03	1.03	1.03	1.03	1.03
5	1.00	1.03	1.05	1.05	1.05	1.05	1.05	1.05	1.05	1.05	1.05	1.05	1.05	1.05	1.05
6	1.00	1.04	1.05	1.06	1.07	1.07	1.07	1.07	1.07	1.07	1.07	1.07	1.07	1.07	1.07
7	1.00	1.04	1.06	1.07	1.08	1.08	1.09	1.09	1.09	1.09	1.09	1.09	1.09	1.09	1.09
8	1.00	1.04	1.07	1.08	1.09	1.09	1.10	1.10	1.10	1.10	1.10	1.10	1.10	1.10	1.10
9	1.00	1.04	1.07	1.08	1.09	1.10	1.11	1.11	1.11	1.11	1.11	1.11	1.11	1.11	1.11
10	1.00	1.04	1.07	1.09	1.10	1.11	1.11	1.12	1.12	1.12	1.12	1.12	1.12	1.12	1.12
11	1.00	1.05	1.07	1.09	1.11	1.11	1.12	1.12	1.12	1.13	1.13	1.13	1.13	1.13	1.13
12	1.00	1.05	1.07	1.09	1.11	1.12	1.12	1.13	1.13	1.14	1.14	1.14	1.14	1.14	1.14
13	1.00	1.05	1.08	1.10	1.11	1.12	1.13	1.13	1.14	1.14	1.14	1.14	1.14	1.14	1.14
14	1.00	1.05	1.08	1.10	1.11	1.12	1.13	1.13	1.14	1.14	1.15	1.15	1.15	1.15	1.15
15	1.00	1.05	1.08	1.10	1.11	1.12	1.13	1.14	1.14	1.15	1.15	1.15	1.15	1.15	1.15
16	1.00	1.05	1.08	1.10	1.11	1.13	1.13	1.14	1.15	1.15	1.16	1.16	1.16	1.16	1.16
17	1.00	1.05	1.08	1.10	1.12	1.13	1.14	1.14	1.15	1.16	1.16	1.16	1.17	1.17	1.17
18	1.00	1.05	1.08	1.10	1.12	1.13	1.14	1.14	1.15	1.16	1.16	1.17	1.17	1.17	1.17
19	1.00	1.05	1.08	1.10	1.12	1.13	1.14	1.15	1.16	1.16	1.17	1.17	1.17	1.17	1.17
20	1.00	1.05	1.08	1.11	1.12	1.13	1.14	1.15	1.16	1.17	1.17	1.17	1.18	1.18	1.18
24	1.00	1.05	1.08	1.11	1.12	1.14	1.14	1.15	1.16	1.17	1.18	1.19	1.19	1.19	1.19
30	1.00	1.05	1.08	1.11	1.13	1.14	1.15	1.16	1.17	1.18	1.19	1.19	1.20	1.20	1.21
40	1.00	1.05	1.08	1.11	1.13	1.14	1.15	1.17	1.17	1.19	1.20	1.20	1.21	1.21	1.22
60	1.00	1.05	1.09	1.11	1.13	1.15	1.16	1.17	1.18	1.19	1.21	1.21	1.22	1.23	1.25
120	1.00	1.05	1.09	1.11	1.13	1.15	1.16	1.17	1.18	1.20	1.21	1.22	1.23	1.24	1.27
∞	1.00	1.05	1.09	1.11	1.14	1.15	1.17	1.18	1.19	1.20	1.22	1.23	1.24	1.25	1.30

Critical q values for $p = 0.05$ are available.[24,39] In lieu of using these tables, the calculated q values can be divided by the appropriate Student's $t(f, 0.05)$ and $\sqrt{2}$ and compared to the reduced critical q values; see Table 1.7 above.

A reduced q value that is smaller than the appropriate critical value signals that the tested means belong to the same population. A fully worked example is found in Chapter 4, Process Validation. Data file MOISTURE.001 used with program MULTI gives a good idea of how this concept is applied. MULTI uses a critical reduced q of 1.1 as a cutoff point.

1.5.5. Outlier Tests

The rejection of outliers is a deeply rooted habit; techniques range from the haughty "I know the truth" attitude, over "looks different from what we are used to, somebody must have made a mistake" optics, to attempts at objective proof of having observed an aberration.

The first attitude is not only unscientific, but downright fraudulent if "unacceptable" values are discarded and forgotten. The second perspective is closer to the mark: Measurements that apparently do not fit model or experience should always be investigated in the light of all available information. While there is the distinct possibility of a discovery about to be made, the other outcome of a sober analysis of the circumstances is more probable: a deviation from the experimental protocol. If this is documented, all the better: The probable outlier can, in good conscience, be rejected and replaced by a reliable repeat result.

Over the years an abundance of outlier tests have been proposed that have some theoretical rationale at their roots.[14] Such tests have to be carefully adjusted to the problem at hand because otherwise one would either not detect true outliers (false negatives) in every case, or then throw out up to 50% of the good measurements as well (false positives).[4, 14]

A well-known test is Dixon's: The data is first ordered according to size and a range (x_n–x_1) and a subrange (x_{n-i}–x_{1+j}) are compared. The ease of the calculations, which probably strongly contributed to the popularity of this test, is also its weakness: Since any out of a number of subrange ratio models (combinations of i and j) can be chosen, there is an arbitrary element involved. Obtaining numerically correct tables of critical quotients for convenient values of p is a problem; the use of this test is increasingly being discouraged.[40, 41] The Dixon tests build on and are subject to the stochastic nature of range measures; they use only a small portion of the available information and lack ruggedness.

As in the case of the detection limit (Section 2.2.7), one commonly used algorithm is based on the theory that any point outside $\pm z \cdot s_x$ is to be regarded as an outlier; if recalculation of $\bar{x}$ and s_x without this questionable point confirms the decision, the outlier is to be cast out. The coefficient z is often fixed at 2.0 or 3.0; the t function (see Fig. 1.20) and other functions have also been proposed. An obvious disadvantage of these approaches is that extreme values strongly affect s_x, and that more or less symmetrical outliers cannot be detected.

A wholly different approach is that of Huber,[15] who orders the values according to size and determines the median (cf. Section 1.1.1); then the absolute deviations $|x_i - x_m|$ are calculated and also ordered, the median absolute deviation (MAD) being found. MAD is then used as is s_x above,

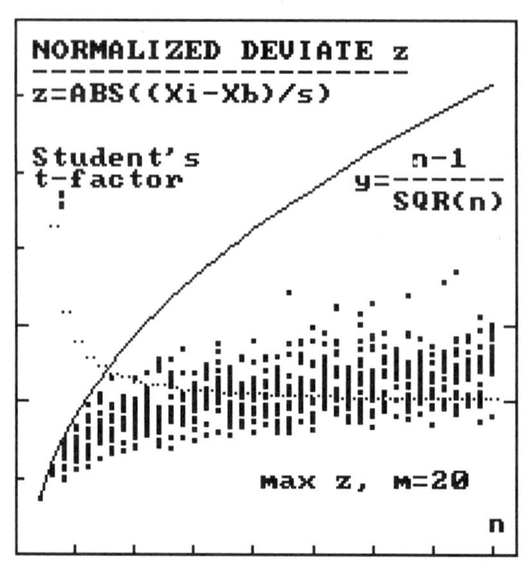

Figure 1.20. Rejection of suspected outliers. A series of normally distributed values was generated by the Monte Carlo technique; the mean and the standard deviation were calculated; the largest normalized absolute deviate (residual) $z = |x_i - \bar{x}|^{\max}/s_x$ is plotted versus n. The simulation was repeated $m = 20$ times for every $n = 2 \cdots 40$. The maximally possible z is given by the curve $y = (n - 1)/\sqrt{n}$. A rejection strategy denoted by "$\pm z \cdot s_x$" ($z = 2$ or 3) is implemented by drawing a horizontal line at $y = 2$ or 3 (large tic marks at left and right): Above $n = 5$ the probability of rejecting at least one value per n (false positive; cf. Section 1.9) increases dramatically if $z = 2$ is used. Repetitive application of this strategy can easily lead to the loss of 3–6 values out of $n = 20$. The dotted horizontals give the Student's $t(n - 1, p = 0.05)$ for comparison; this shows that the confidence limits cannot reliably be used for outlier rejection. Huber's rule with limits set at $x_m \pm k \cdot \text{MAD}$, $k = 3.5$, rejects only about 1 out of $n = 20$ values. The range R is not quite twice the largest residual, $R \approx 2 \cdot r^{\max}$; in this connection see Fig. 1.3.

the coefficient k being chosen between 3 and 5. This algorithm is much more robust than the ones described above.

Example: The $n = 19$ values in Table 1.14 yield a median of 2.37, absolute deviations ranging from 0.00 to 2.99, and a MAD of 0.98. The coefficient k can be as low as 3.05 before a single point is eliminated (-0.614): Use data file HISTO.001 in conjunction with program HUBER.

For the reasons described, no specific test will be advanced here as being superior, but Huber's model and the classical one for $z = 2$ and $z = 3$ are incorporated into program HUBER; the authors are of the opinion that the best recourse is to openly declare all values and do the analysis twice, once with the presumed outliers included, and once excluded from the statistical analysis; in the latter case the excluded points should nonetheless be included in tables (in parentheses) and in graphs (different symbol). Outliers should not be labeled as such solely on the basis of a fixed (statistical) rule; the decision should to the major part reflect scientific experience.[13] The justification must be explicitly stated in any report; cf. Sections 4.18 and 4.19. If the circumstances demand that a rule be set down, it is best to use a robust model such as Huber's; its sensitivity for the problem at hand, and the typical rate for false positives, should be investigated, e.g., by a Monte Carlo simulation. Program HUBER rejects points on the basis of Huber's rule and shows the results. For completeness, the means and standard deviations before and after the elimination are given, and the equivalent z values for the classical mean $\pm z \cdot \mathrm{SD}$ are calculated. The sensitivity of the elimination rules towards changes in the k resp. z values are graphically indicated.

An example, arrived at by numerical simulation, will be given here to illustrate the high probability of rejecting good data. Figure 1.20 shows that the largest residual of every series is around $1 \cdot s_x$ for $n = 3$, around $2 \cdot s_x$ for $n = 13$, and close to $3 \cdot s_x$ for $n = 30$. This makes it virtually certain that for a series of $n > 10$ measurements at least one will be rejected by a $\pm 2 \cdot s_x$ recipe, which only makes sense if the individual measurement is to be had at marginal cost. Rejection rates become higher still if the observation is taken into account that experimental distributions tend to be broader tailed than the Normal Distribution; a t Distribution for $f = 4$ is said to give the best fit.[14]

Because outlier elimination is something that is not to be taken lightly, the authors have decided to *not* provide on-line outlier deletion options in the programs. Instead, the user must first decide which points he regards as outliers, for example, by use of the program HUBER, then start program DATA and use options Edit or Delete Row, and finally create a modified data file with option Save. This approach was chosen to reinforce GLP-like procedures and documentation.

1.5.6. Analysis of Variance (ANOVA)

Analysis of Variance (ANOVA) tests whether one group of subjects (e.g., batch, method, laboratory, etc.) differs from the population of subjects investigated (several batches of one product, different methods for the same parameter; several laboratories participating in a round-robin test to validate a method; for examples see.[4,15,21] Multiple measurements are necessary to establish a benchmark variability (within group) typical for the type of subject. Whenever a difference significantly exceeds this benchmark, at least two populations of subjects are involved. A graphical analogue is the Youden plot (see Fig. 2.1). An additive model is assumed for ANOVA.

The type of problem to which ANOVA is applicable is exemplified as follows: Several groups of measurements are available pertaining to a certain product, several repeat measurements having been conducted on each batch. The same analytical method was used throughout (if not, between-group variance would be distorted by systematic differences between methods; this problem typically crops up when historical data series are compared with newer ones). Were one to do t tests, the data would have to be arranged according to the specific question asked; i.e, "do batches 1 and 2 differ among themselves?" Tests would be conducted according to cases (b1) or (b3).

	Batch					
	1	2	1	3	1	4...etc.
Measurement 1	□	□	□	□	□	□
2	□	□	□	□	□	□
·	□	□	□	□	□	□
n_j	□	□	□	□	□	□

The question to be answered here is not "do batches x and y differ?" but "are any individual batches different from the group as a whole?" The hypotheses thus have the form

$$H_0: \mu_1 = \mu_2 = \mu_3 = \cdots \mu_m$$

$$H_1: \mu_1 \neq \mu_2 = \mu_3 = \cdots \mu_m, \quad \text{or}$$

$$: \mu_1 = \mu_2 \neq \mu_3 \neq \cdots \mu_m.$$

Since a series of t tests is cumbersome to carry out, and does not answer all questions, all measurements will be simultaneously evaluated to find differences between means. The total variance (relative to the grand mean $\bar{\bar{x}}$) is broken down into a component V_1 "variance within groups," which corresponds to the residual variance, and component V_2 "variance between groups." If H_0 obtains, V_1 and V_2 should be similar, and all values can be pooled because they belong to the same population. When one or more means deviate from the rest, V_2 must be significantly larger than V_1. See Tables 1.8 and 1.9.

Table 1.8. Equations for Simple ANOVA.

The following variables are used:
m: number of groups, $j = 1 \cdots m$ (columns in example)
n_j: number of measurements in group j (column j)
n: total number of measurements (Σn_j)
x_{ij}: ith measurement in jth column

The following sums are calculated:

$u_j = \Sigma(x_{ij})$	Sum over all measurements in group j	(1.26)
$u_n = \Sigma(u_j)$	Sum over all n measurements (m groups)	(1.27)
$\bar{x}_j = u_j/n_j$	Average over group j	(1.28)
$\bar{\bar{x}} = u_n/n$	Grand average	(1.29)
$S_2 = \Sigma(n_j \cdot (\bar{x}_j - \bar{\bar{x}})^2)$	Sum of weighted squares	(1.30)
$S_1 = \Sigma(x_{ij} - \bar{x}_j)^2$	Sum of squares of residuals relative to the appropriate group	(1.31)
$S_1 = \Sigma(\Sigma(r_i)^2)$	Average, $\bar{x}_j$; $r_i = x_{ij} - \bar{x}_j$	(1.32)
$S_T = S_1 + S_2$	Total sum of squares	(1.33)
$f_1 = n - m$	Degrees of freedom within groups	(1.34)
$f_2 = m - 1$	Degrees of freedom between groups	(1.35)

Table 1.9. Presentation of ANOVA Results.

Sum of Squares	Degrees of Freedom	Variance	Comment
S_1	$f_1 = n - m$	$V_1 = S_1/f_1$	Variance within groups
$+S_2$	$f_2 = m - 1$	$V_2 = S_2/f_2$	Variance between groups
$= S_T$	$f_T = n - 1$	$V_T = S_T/f_T$	Total variance

Since the total number of degrees of freedom must be $n - 1$, and m groups are defined, there remain $n - 1 - (m - 1) = n - m$ degrees of freedom for within the groups.

For the data given in Table 1.10 an ANOVA calculation yields Table 1.11

V_1 and V_2 are subjected to an F test (see Section 1.6.1) to determine whether H_0 can be retained or not. Since V_2 must be larger than V_1 if H_1 is to obtain, $F = V_2/V_1$; should V_2 be smaller or equal to V_1, then H_1 could be rejected without an F test being performed.

If H_0 were to be retained, the individual averages x_j could not be distinguished from the grand average $\bar{\bar{x}}$; V_T would then be taken as the average variance associated with $\bar{\bar{x}}$ and $n - 1$ degrees of freedom.

Table 1.10. Raw Data and Intermediate Results of an ANOVA Test for Simulated Data [Eq. (1.30)].

	Example: Group j						
Index i	1	2	3	4	5		
1	7.87	6.35	4.65	8.74	5.29		
2	6.36	7.84	5.06	6.02	6.03		
3	5.73	5.31	6.52	6.69	6.06		
4	4.92	6.99	6.51	7.38	5.64		
5	4.60	8.54	8.28	5.82	4.33		
6	6.19	·	4.45	6.88	5.39		
7	5.98	·	·	8.65	5.85		
8	6.95	·	·	6.55	5.51		Sum
	6.08	7.01	5.91	7.09	5.51	Group average	$\bar{x}_j$: 31.60
	1.05	1.26	1.47	1.10	0.56	Group standard deviation	s_j
	8	5	6	8	8	Number of measurements	n_i: 35
	48.60	35.03	35.47	56.73	44.10	Sum	u_j: 219.9
	7.72	6.35	10.80	8.47	2.20	Sum of squared residuals	S_1: 35.52
	0.33	2.64	0.83	5.21	4.78	Sum of weighted squares	S_2: 13.76

Table 1.11. Results of an ANOVA Test.
(The "null" hypothesis is rejected because 2.91 > 2.69.)

Sum of Squares	Degrees of Freedom	Variance	Comment
35.520	30	1.184	Variance within groups
13.763	4	3.441	Variance between groups
49.283	34	1.450	Total variance

F test: $3.441/1.184 = 2.91$, $F_c(4, 30, 0.05) = 2.69$.

Interpretation of Table 1.11: Since V_2 is significantly larger than V_1, the groups cannot all belong to the same population. Therefore, the *grand average* $\bar{\bar{x}} = 219.93/35 = 6.28$ and the associated standard deviation $\sqrt{49.28/34} = \pm 1.2$ are irrelevant. The question of which means are indistinguishable among themselves and different from the rest is answered by the multiple range test.

Other forms of ANOVA: The simple ANOVA set out above tests for the presence of one unknown factor that produces (additive) differences between outwardly similar groups of measurements. Extensions of the concept allow one to simultaneously test for several factors (two-way ANOVA, etc.). The limit of ANOVA tests is quickly reached: They do not provide answers as to the type of functional relationship linking measurements and variables, but only indicate the probability of such factors being present. Thus, ANOVA is fine as long as there are no concrete hypotheses to test, as in explorative data analysis; cf. Section 2.1.

1.6. NUMBER OF DETERMINATIONS

Up to this point it was assumed that the number of determinations n was sufficient for a given statistical test. During the discussion of the t test [case (a)] an issue was skirted that demands more attention: Is $\bar{x}$ different from μ?

(1) If the question refers to the expectation $E(\bar{x}) = \mu$ (for $n \rightarrow \infty$), finding a difference could mean a deficiency in a theory.

(2) On the other hand, if $E(\bar{x}) < L$, where L is an inviolable limit, finding "$\bar{x}$ greater than L" could mean the infraction of a rule or an unsalable product. In other words, L would here have to be significantly larger than $\bar{x}$ for a high probability of acceptance of the product in question.

It is this second interpretation that will occupy us for a moment. Replacing μ in Eq. (1.13) by L yields

$$t_c < t = \frac{|E(\bar{x}) - L|}{s_{\bar{x}}} = \frac{|\bar{x} - L| \cdot \sqrt{n}}{s_x}. \tag{1.36}$$

Most likely, Student's t_c will be fixed by an agreed-upon confidence level, e.g., 95%, and L by a product specification; $E(\bar{x})$ is the true value of the parameter. The analyst has the option of reducing s_x or increasing n in order to augment the chances of obtaining a significant t. The standard deviation s_x is given by the choice of test method and instrumentation, and can be influenced to a certain extent by careful optimization and skillful working habits. The only real option left is to increase the number of determinations, n. A look at Table 1.4 and Eq. (1.36) reveals that the critical t value t_c is a function of n, and n is a function of t_c; thus a

Table 1.12. The Number of Determinations *n* Necessary to Achieve a Given Discrimination [Eq. (1.13), Table 1.5].

[If an experimental mean $\bar{x} = 90$ is found, how many measurements n are necessary to distinguish $\bar{x} = 90$ from $L = 100$ with an error probability of $p = 0.25 \cdots 0.00005$ (one sided) if the experimental standard deviation is $\pm 0.1 \cdots \pm 200$?]

	s_x										
p	0.1	0.2	0.5	1	2	5	10	20	50	100	200
0.25	2	2	2	2	2	2	2	3	13	47	183
0.1	2	2	2	2	2	3	4	9	43	166	659
0.05	2	2	2	2	2	3	5	13	70	273	1084
0.025	2	2	2	2	3	4	7	18	99	387	1538
0.01	2	2	3	3	3	5	9	25	139	546	2172
0.005	2	2	3	3	4	6	11	31	171	673	2679
0.001	3	3	3	4	5	7	15	44	244	960	3824
0.0005	3	3	3	4	5	8	17	50	277	1087	4330
0.00005	3	4	4	5	6	11	23	69	360	1392	5518

trial-and-error approach must be implemented that begins with a rough estimate based on experience, say, $n = 5$; t is calculated for given $\bar{x}$, L, and s_x and compared to the tabulated t_c. Depending on the outcome, n is increased or decreased. The search procedure can be automated by applying the polynomial $t = f(n - 1)$ from Section 5.1.2 and increasing n until $t > t_c$. The idea behind this is that t is proportional to $\sqrt{n}$, and t_c decreases from 12.7 ($n = 2$) to 1.96 ($n = \infty$) for a 97.5% one-sided confidence level. Note that a one-sided test is applied, because it is expected that $\bar{x} < L$. Equation (1.36) is rearranged to yield

$$t_c\sqrt{n} < |\bar{x} - L|/s_x = Q. \tag{1.37}$$

The quotient Q is fixed for a given situation; for estimates of the number n, see Table 1.12. The problem could also be solved graphically by drawing a horizontal at $y = Q$ in Fig. 1.15 or 1.17 and taking the intercept as an estimate of n. The necessary number of repeat determinations n is depicted in Fig. 1.21.

As an example, consider the following situation: The limit is given by $L = 100$, and the experimental average is $\bar{x} = 90$. How many measurements are necessary to find significant difference between L and $\bar{x}$ for various confidence levels? See Table 1.12.

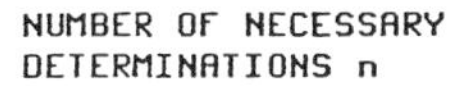

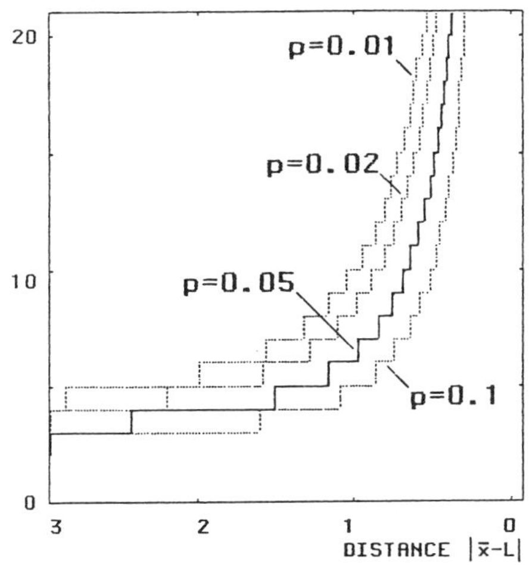

Figure 1.21. The number of measurements n that are necessary to obtain an error probability p of $\bar{x}$ exceeding L is given on the ordinate. The abscissa shows the independent variable Q in the range $L - 3 \cdot s_x \cdots L$ in units of s_x. It is quite evident that for a typical p of about 0.05, $\bar{x}$ must not be closer than about 0.5 standard deviations s_x from L in order that the necessary number of repeat measurements remains manageable. The enhanced line is for $p = 0.05$; the others are for 0.01 (left), 0.02, and 0.1 (right).

Example: For $s_x = \pm 5$ and an error probability of falsely accepting H_1 of $p = 0.005$ would require $n \geq 6$ because

$$\text{for } n = 6,\ t = 4.0321,\ \text{CL} = 90 + 5 \cdot 4.0321/\sqrt{6} = 98.23, \quad \text{and}$$

$$\text{for } n = 5,\ t = 4.6041,\ \text{CL} = 90 + 5 \cdot 4.6041/\sqrt{5} = 100.30.$$

Some equivalent statements:

- The upper confidence limit $\text{CL}_u(\bar{x} = 90)$ is less than 100.
- The true value μ is in the interval $\bar{x} = 80$ (of no interest here) $\cdots$ 100 with a probability of at least 99% $(100 - 2 \cdot 0.5 = 99)$.
- The null hypothesis H_0: $\mu < L$ has a probability of more than 99.5% of being correct.
- The probability of μ being larger than L is less than 0.5% or $p \leq 0.005$.

- Because for $n = 6$ ($f = 5$) the critical t factor is 4.0321, 99% of the individual measurements (two-sided) would be expected in the interval $69.8 \cdots 110.2$.
- With $n = 6$ and $f = 5$, $t = (100 - 90)/5 = 2$, and the critical $t(f = 5, p = 0.05$, one-sided$) = 2.0150$, thus about $1 - 0.05 = 0.95$ (95%) of all individual measurements would be expected to be below 100.0.

1.7. WIDTH OF A DISTRIBUTION

As was discussed in Section 1.1.2, there are several ways to characterize the width of a distribution:

(a) From a purely practical point of view the range or a quantile can serve as indicator. Quantiles are usually selected to encompass the central 60–90% of an ordered set; the influence of extreme values diminishes the smaller this % value is. No assumptions as to the underlying distribution are made.

(b) Theoretical models are a guide in setting up rules and conventions for defining a distribution's width, the standard deviation being a good example. Simply the fact of assuming a certain form of distribution, however, undermines whatever strength resides in the concept, unless theory and fact conform.

Measured distributions containing less than 100 events or so just as likely turn out to appear skewed as symmetrical; cf. Fig. 1.10. For this reason alone the tacit assumption of a Normal Distribution when contemplating analytical results is understandable, and excusable, if only because there is no practical alternative (alternative distribution models require more complex calculations, involve more stringent assumptions, or are more susceptible to violations of these basic assumptions than the relatively robust Normal Distribution).

1.7.1. The *F* Test

The F test is based on the Normal Distribution and serves to compare either

- an experimental result in the form of a standard deviation with a fixed limit on the distribution width, or
- such an experimental result with a second one in order to detect a difference.

Both cases are amenable to the same test, the distinction being a matter of the number of degrees of freedom f. The F test is used in connection with the t test, see program TTEST.

The test procedure is as follows:

	Distribution	
	Test Distribution	Reference Distribution
Standard deviation	$s_t = \sqrt{V_t}$	$s_r = \sqrt{V_r}$
Degrees of freedom	f_t	f_r
Null hypothesis H_0	The standard deviations s_t and s_r are indistinguishable	
Alternate hypothesis H_1	s_t and s_r are distinguishable on the confidence level given by p	
Test statistic	$F = V_t/V_r$ or $F = V_r/V_t \geq 1.00$ (1.38)	
Nomenclature	$F = V_1/V_2$ with f_1 resp. f_2 degrees of freedom	

The critical value F_c is taken from an F table or is approximated (cf. Section 5.1.3); if

$$\begin{aligned} F > F_c&: \text{accept } H_1, \text{reject} H_0, \\ F \leq F_c&: \text{retain } H_0, \text{reject } H_1. \end{aligned} \tag{1.39}$$

F_c depends on three parameters:

(a) The confidence level p must be chosen beforehand; most statistical works tabulate F_c for only a few levels of p, 0.05, 0.02, and 0.01 being typical.

(b) The number of degrees of freedom f_t is given by the test procedure, and can be influenced: Since the analyst for technical reasons ordinarily has only a limited set of analytical methods to choose from, each with its own characteristics, the wish to increase the confidence in a decision can normally only be achieved by drastically increasing the number of measurements.

(c) The number of degrees of freedom f_r is often an unalterable quantity, namely

- f_r is fixed if the reference measurements originate from outside one's laboratory, e.g., the literature,
- $f_r = \infty$ if the limit is theoretical or legal in nature.

Two **examples** will illustrate the concept ($p = 0.05$ will be used):

(a) A reference analytical method, for which much experimental evidence is at hand, shows a standard deviation $s_r = \pm 0.037$ with $n_r = 50$. A few tests with a purportedly better method yield $s_t = \pm 0.029$ for

$$F = (0.037/0.029)^2 = 1.28,$$
$$F_c = 5.7, \qquad \text{for } f_1 = 49,\ f_2 = 4.$$

On this evidence no improvement is detectable. If—for whatever reason—the reference method is to be proven inferior, for the same F ratio, over 45 measurements with the better method would be necessary (F_c is larger than 1.28 if $f_1 = 49$ and f_2 is less than 45), or, alternatively, s_t would have to be brought down to less than ± 0.0155 ($(0.037/0.0155)^2 = 5.7$) in order to achieve significance.

(b) A system suitability test prescribes a relative standard deviation of less than $\pm 1\%$ for the procedure to be valid; with $\bar{x} = 173.5$ this translates into $V_r = (\pm 1.735)^2$; because the limit is imposed, this is equivalent to having no uncertainty about the numerical value, or in other words, $f_r = \infty$. Since s_t was determined to be ± 2.43 for $n_t = 7$,

$$F = (2.43/1.735)^2 = 1.96 \quad \text{and} \quad F_c(0.05, 6, \infty) = 2.1.$$

Assuming the $\pm 1\%$ limit was set down in full cognizance of the statistics involved, the systems suitability test must be regarded as failed because $2.43 > 1.74$. Statistically it would have made more sense to select the criterion as $s_t \leq 0.01 \cdot \bar{x} \cdot \sqrt{F_c(0.05, f_t, \infty)}$ for acceptance and demanding, say, $n \geq 5$; in this particular case, s_t could have been as large as 3.6. Unfortunately, few chemists are aware of the large confidence interval a standard deviation carries, and thus are prone to setting down unreasonable specifications, such as the one above. The only thing that saves them from permanent frustration is the fact that if n is kept small enough, the chances for obtaining a few similar results in a row, and hence a small s_x, are relatively good; see Fig. 1.20. Assuming the commonly seen requirement $[n = 3]$ combined with either $[s_x < \pm 2]$ or $[R(n) < 2]$, for example, and measurements $x_1 = 99.5$, $x_3 = 101.5$, and x_2 anywhere in between, a range $R(3) \leq 2$ and an $s_x \leq \pm 1.15$ result; so far so good. Since the 95% CL(s_x) are $\pm 0.67 \cdots \pm 5.2$ for $2 \cdot p = 0.1$, the chances of obtaining an $s_x > 2$ for $\sigma \approx 1$ are very good, indeed! Use program MSD, options 4–6.

For identification, an excerpt from the F table for $p = 0.05$ is given in Table 1.13 and in program FVAL.

Table 1.13. F Table for $p = 0.05$.

	f_1						
f_2	1	2	4	8	16	32	∞
1	161.5	199.5	224.6	238.9	246.5	250.4	254.3
2	18.51	19.00	19.25	19.37	19.43	19.46	19.50
4	7.709	6.944	6.388	6.041	5.844	5.739	5.628
8	5.318	4.459	3.838	3.438	3.202	3.070	2.928
16	4.494	3.634	3.007	2.591	2.334	2.183	2.010
32	4.149	3.295	2.668	2.244	1.972	1.805	1.594
∞	3.842	2.996	2.372	1.938	1.644	1.444	1.000

1.7.2. Confidence Limits for a Standard Deviation

In Section 1.3.2 confidence limits are calculated to define a confidence interval within which the true value μ is expected with an error probability of p or less.

For standard deviations, an analogous confidence interval $\mathrm{CI}(s_x)$ can be derived via the F test. In contrast to $\mathrm{CI}(\bar{x})$, $\mathrm{CI}(s_x)$ is not symmetrical about the most probable value because s_x by definition can only be positive. The concept is as follows: An upper limit s_u on s_x is sought that has the quality of a very precise measurement; that is, its uncertainty must be very small and therefore its number of degrees of freedom f must be very large. The same logic applies to the lower limit s_l: Since

$$s_l < s_x, \qquad s_x < s_u,$$

$$F_l = V_x/V_l, \qquad F_u = V_u/V_x$$

$$F_l = F(n-1, \infty, p), \qquad F_u = F(\infty, n-1, p),$$

thus

$$V_l = V_x/F(n-1, \infty, p), \qquad V_u = V_x \cdot F(\infty, n-1, p), \tag{1.40}$$

and since

$$F(f, \infty, p) = \chi^2(f, p)/f, \quad F(\infty, f, p) = f/\chi^2(f, 1-p), \tag{1.41}$$

$$V_l = V_x \cdot f/\chi^2(f, p), \qquad V_u = V_x \cdot f/\chi^2(f, 1-p), \tag{1.42}$$

where $f = n - 1$.

The true standard deviation σ_x is expected inside the confidence interval $\mathrm{CI}(s_x) = \sqrt{V_l} \cdots \sqrt{V_u}$ with a total error probability $2 \cdot p$ (in connection with F and χ^2 p is taken to be one sided).

F values can be calculated according to Section 5.1.3 and χ^2 values according to Section 5.1.4; see also program MSD. Both could also be looked up in the appropriate tables, see programs FVAL and CHI.

Example: Given $s_x = 1.7$, $n = 8$, $f = 7$, $2 \cdot p = 0.1$, and $p = 0.05$:

Tabulated Value	Tabulated Value	Control [see Eq. (1.41)]
$F(7, \infty, 0.05) = 2.0096$	$\chi^2(7, 0.05) = 14.067$	$14.067/7 = 2.0096$
$F(\infty, 7, 0.05) = 3.2298$	$\chi^2(7, 0.95) = 2.167$	$7/2.167 = 3.2298$
$s_l = 1.7/\sqrt{2.0096} = 1.2$		
$s_u = 1.7 \cdot \sqrt{3.2298} = 3.0$		

Thus with a probability of 90% the true value $E(s_x) = \sigma_x$ is within the limits: $1.2 \le \sigma_x \le 3.0$.

For a depiction of the confidence limits on s_x for a given set of measurements and a range of probabilities $0.0002 \le p \le 0.2$, see program MSD, option 6. See Figure 1.22.

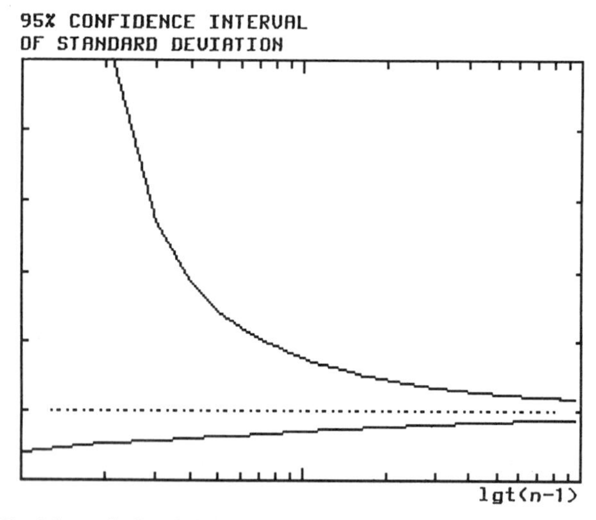

Figure 1.22. Confidence limits of the standard deviation for $p = 0.05$ and $f = 1 \cdots 100$. The f axis is logarithmically transformed for a better overview. For example, at $n = 11$, the true value σ_x is expected between 0.7 and 1.8 times the experimental s_x. The ordinate is scaled in units of s_x.

1.7.3. Bartlett Test

Several series of measurements are to be compared as regards the standard deviation. It is of interest to know whether the standard deviations could all be traced to one population characterized by σ (H_0: no deviation observed), and any differences versus σ would only reflect stochastic effects, or whether one or more standard deviations belong to a different population (H_1: difference observed):

$$H_0: \sigma_1 = \sigma_2 = \sigma_3 = \cdots = \sigma_m,$$

$$H_1: \sigma_1 \neq \sigma_2 = \sigma_3 \cdots = \sigma_m, \text{ etc.}$$

For example, a certain analytical procedure could have been repeatedly performed by m different technicians. Do they all work at the same level of proficiency, or do they differ in their skills?

The observed standard deviations are $s_1, s_2, s_3, \ldots, s_m$, and the corresponding number of degrees of freedom are $f_1 = n_1 - 1, \ldots, f_m = n_m - 1$, with f_j larger than 2. The interpretation is formulated in analogy to Section 1.5.6 above.

The following sums have to be calculated:

$$A = \sum(f_j \cdot V_j), \text{ sum of all squared residuals; } A/D \text{ is the total variance,} \tag{1.43}$$

$$B = \sum(f_j \cdot \ln(V_j)), \tag{1.44}$$

$$C = \sum(1/f_j), \tag{1.45}$$

$$D = \sum(f_j). \tag{1.46}$$

Then

$$E = D \cdot \ln(A/D) - B, \qquad \chi^2, \tag{1.47}$$

$$F = (C - 1/D)/(3 \cdot m - 3) + 1, \qquad \text{correction term}, \tag{1.48}$$

$$G = E/F, \qquad \text{corrected } \chi^2. \tag{1.49}$$

If the found G exceeds the tabulated critical value $\chi^2(p, m-1)$, the null hypothesis H_0 must be rejected in favor of H_1: The standard deviations do not all belong to the same population; that is, there is at least one that is larger than the rest. The correction factor F is necessary because Eq. (1.47) overestimates χ^2.

For a completely worked example, see Section 4.4, Process Validation, and data file MOISTURE.001 in connection with program MULTI.

1.8. CHARTING A DISTRIBUTION

In explorative data analysis, an important clue is the form of the population's distribution (cf. Fig. 1.9); after accounting for artifacts due to the analysis method (usually an increase in distribution width) and sampling procedure (often a truncation), width and shape can offer insight into the system from which the samples were drawn. Distribution width and shape can be judged using a histogram, see program HISTO. The probability chart (program HISTO, option NPP = Normal Probability Paper) tests for normality and the χ^2 test can be used for any shape.

1.8.1. Histograms

When the need arises to depict the frequency distribution of a number of observations, a histogram (also known as a bar chart), is plotted. The data is assumed to be available in a vector $x(\ )$; the individual values occur within a certain range given by x_{min} and x_{max}, and these two limits can be identical with the limits of the graph, or different. How are the values grouped? From a theoretical point of view, $\mathbf{c} = \sqrt{n}$ classes (or bins) of equal width would be optimal. It is suggested that, starting from this initial estimate $\mathbf{c}$, a combination of an integer $\mathbf{c}$, and lower resp. upper bounds on the graph ($\mathbf{a} \leq x_{min}$) and ($\mathbf{b} \geq x_{max}$) be chosen so that the class boundaries are defined by $\mathbf{a} + i \cdot \mathbf{w}$, with $\mathbf{w} = (\mathbf{b} - \mathbf{a})/\mathbf{c}$ and $i = 0, 1, 2, \ldots, \mathbf{c}$, assume numerical values appropriate to the problem (not all software packages are capable of satisfying this simple requirement). An example: $x_{min} =$ 0.327, $x_{max} = 0.952$, $n = 21$

$$\mathbf{a} = 0.20,\quad \mathbf{b} = 1.20,\quad \mathbf{c} = 5,\quad \mathbf{w} = 0.20,\quad \text{or}$$
$$\mathbf{a} = 0.15,\quad \mathbf{b} = 1.35,\quad \mathbf{c} = 8,\quad \mathbf{w} = 0.15,$$
$$\mathbf{a} = 0.20,\quad \mathbf{b} = 1.20,\quad \mathbf{c} = 10,\quad \mathbf{w} = 0.10.$$

This results in sensible numbers for the class boundaries and allows for comparisons with other series of observations that have different extrema, for example (0.375/0.892) or (0.25/1.11). Strict adherence to the theory-inspired rule would have yielded class boundaries 0.327, 0.4833, 0.6395, 0.7958, and 0.952, and the extreme values exactly on the edges of the graph. That this is impractical is obvious, mainly because the class boundaries depend on the stochastic element innate in x_{min} resp. x_{max}. Program

HISTO, option Scale, allows for an independent setting of a subdivided range R with C bins of width R/C, and lower/upper limits on the graph.

Next, observations $x_1 \cdots x_n$ are classed by first subtracting the x value of the lower boundary of bin 1, **a**, and then dividing the difference by the class width **w**. The integer $\mathrm{INT}((x_i - \mathbf{a})/\mathbf{w} + 1)$ gives the class number (index) j. The number of events per class is counted and expressed in form of an absolute, a relative, and a cumulative result (E, $100 \cdot E/n$, resp. $100 \cdot \Sigma(E/n)$).

Before plotting the histogram, the vertical scale (absolute or relative frequency) has to be set; as above, practical considerations like comparability among histograms should serve as guide. The frequencies are then plotted by drawing a horizontal from one class boundary to the other, and dropping verticals along the boundaries.

Artistic license should be limited by the fact that the eye perceives the area of the bar as the significant piece of information. The ordinate should start at zero. A violation of these two rules—through purpose or ignorance—can skew the issue or even misrepresent the data. Hair-raising

Table 1.14. Data and Results of a Histogram Analysis.
(Class 0 covers the events below the lower limit of the graph, i.e., from $-\infty$ to -0.5.)

Values:					
	3.351	1.971	2.681	2.309	0.706
	2.973	2.184	−0.614	1.848	3.749
	3.431	−0.304	4.631	3.010	3.770
	2.996	0.290	1.302	2.371	

Results:				
Number of values	n: 19	Number of classes	:6
Left boundary	a: −0.5	Right boundary	b: 5.499
Smallest value	:− 0.61	Largest value	:4.63
To left of class	1: 1 event	To right of class	6: 0 event
Mean	$\bar{x}$: 2.245	Std. deviation	s_x: ±1.430

Class	Events	% Events	%	lcb	ucb
0	1	5.26	5.26	$-\infty$	−0.501
1	2	10.53	15.79	−0.500	0.499
2	2	10.53	26.32	0.500	1.499
3	5	26.32	52.63	1.500	2.499
4	6	31.58	84.21	2.500	3.499
5	2	10.53	94.74	3.500	4.499
6	1	5.26	100.0	4.500	5.499
	19	100.00			

With lcb: lowers class boundary; ucb: upper class boundary.

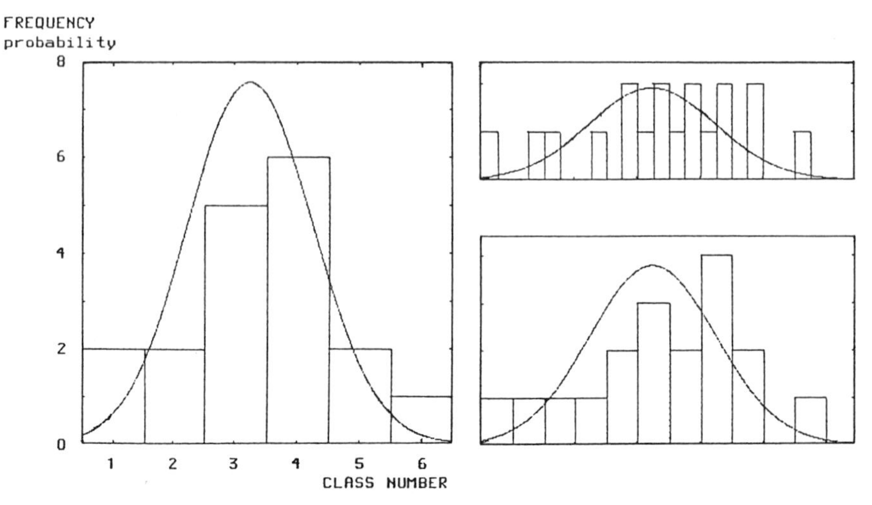

Figure 1.23. Histograms of data set containing $n = 19$ values (see above); the x range depicted is in all cases -0.5 to 5.5 inclusive, leaving one event to the left of the lower boundary. The number of classes is 6, 12, resp. 24; in the latter case the observed frequency never exceeds 2 per class, which is clearly insufficient for a χ^2 test (see text). The superimposed Normal Distribution has the same area as the sum of events n times the bin width, namely, 19, 8.5, respectively, 4.25.

examples can be found in the economics sections even of major newspapers.

A numerical example (see data file HISTO.001) can be found in Table 1.14. Note that the single event to the left of class 1, equivalent to 5.6%, must be added to the first number in the % Events column to obtain the correct $\Sigma\%$, unless a separate class is established (here Class 0).

The appropriate theoretical probability distribution can be superimposed on the histogram; ideally, scaling should be chosen to yield an area under the curve (integral from $-\infty$ to $+\infty$) equal to $n \cdot w$ (see program HISTO, option ND).

In Fig. 1.23 the above data is plotted using 6, 12, resp. 24 classes within the x boundaries $a = -0.5$, $b = 5.5$. The left panel gives a sensible subdivision of this x range. If, for comparison purposes, more classes are needed, the two other panels depict the consequences: Many bins contain only one or two events.

1.8.2. χ^2 Test

This test is used to judge both the similarity of two distributions and the fit of a model to data. The distribution of a data set in the form of a

Table 1.15. Intermediate and Final Results of a χ^2 Test.
[The experimental (observed) frequencies are compared with the theoretical (expected) number. The critical χ^2 value for $p = 0.05$ and $f = 4$ is 9.49; thus no difference in distribution function is detected. Note that the first and last classes extend to infinity; it might even be advisable to eliminate these poorly defined classes by merging them with the neighboring ones: χ^2 is found as 1.6648 in this case; the critical χ^2 is 5.991. Read 0.9999 as 1.000. Columns z_l, z_r, ΔCP, expected events, and χ^2 are provided in program HISTO, option table.]

Observed	z Values		Probability			Expected	
Events	z_l	z_r	CP_l	CP_r	ΔCP	Events	χ^2
1	$-\infty$	−1.920	0.0000	0.0274	0.0274	0.521	0.4415
2	−1.920	−1.221	0.0274	0.1111	0.0837	1.590	0.1056
2	−1.221	−0.521	0.1111	0.3011	0.1900	3.610	0.7184
5	−0.521	0.178	0.3011	0.5708	0.2696	5.123	0.0030
6	0.178	0.878	0.5708	0.8100	0.2392	4.545	0.4658
2	0.878	1.577	0.8100	0.9426	0.1327	2.520	0.1074
1	1.577	$+\infty$	0.9426	0.9999	0.0574	1.091	0.0075
19					0.9999	19.000	1.8492

histogram can always be plotted, without reference to theories and hypotheses. Once enough data has accumulated, there is the natural urge to see whether it fits an expected distribution function. To this end, both the experimental frequencies and the theoretical probabilities must be brought to a common scale: A very convenient one is that which sets the number of events n equal to certainty (probability $p = 1.00$) = area under the distribution function. The x scale must also be made identical. Next, the probability corresponding to each experimental class must be determined from tables or algorithms. Last, a class-by-class comparison is made, and the sum of all lack-of-fit figures is recorded. In the case of the χ^2 test the weighing model is

$$\chi^2 = \sum \frac{\left[\begin{pmatrix}\text{number of}\\ \text{observations/class}\end{pmatrix} - \begin{pmatrix}\text{expected number of}\\ \text{observations/class}\end{pmatrix}\right]^2}{\begin{matrix}\text{expected number of}\\ \text{observations/class}\end{matrix}}. \quad (1.50)$$

Consider as an example the set of 19 Monte Carlo generated, normally distributed values with a mean = 2.25 and a standard deviation = ± 1.43

used in Section 1.8.1. Table 1.15 is constructed in six steps:

(1) Calculate normalized deviations $z_l = (\text{lcb} - \bar{x})/s_x$ resp. $z_r = (\text{ucb} - \bar{x})/s_x$ for each class; e.g., $(1.5 - 2.25)/1.43 = -0.521$. The number 1.5 is from Table 1.14, column lcb.
(2) The cumulative probabilities are either looked up or are calculated according to Section 5.1.1: $z = -0.521$ yields CP = 0.3011.
(3) Subtract the CP of the left from that of the right class boundary to obtain the ΔCP for this class (this corresponds to the hatched area in Fig. 1.9); e.g., 0.5708 − 0.3011 = 0.2696.
(4) ΔCP is multiplied by the total number of events to obtain the number of events expected in this class, e.g., $0.2696 \cdot 19 = 5.123$.
(5) From columns 1 and 7, χ^2 is calculated, $\chi^2 = (\text{Obs} - \text{Exp})^2/\text{Exp}$; e.g., $(5.0 - 5.123)^2/5.123 = 0.0030$.
(6) The degrees of freedom f is calculated as $f = m - 3$ (m: number of classes).

In terms of the weighing model (see Section 3.1) the χ^2 function is intermediate between a least-square fit using

absolute deviations $\left(\text{measure} = \sum (x_i - E(x_i))^2\right)$, and one using

relative deviations $\left(\text{measure} = \sum ((x_i - E(x_i))/E(x_i))^2\right)$.

The effect is twofold: As with the least-squares model, both positive and negative deviations are treated equally, and large deviations contribute the most. Second, there is a weighing component with $w_i = 1/E(x_i)$ that prevents a moderate (absolute) error of fit on top of a large expected $E(x_i)$ from skewing the χ^2 measure, and at the same time, deviations from small $E(x_i)$ carry some weight.

The critical χ^2 values are tabulated in most statistical handbooks. An excerpt from the table for various levels of p is given in Table 1.16. This table is used for the two-sided test; that is, one simply asks "are the two distributions different?" Approximations to tabulated χ^2 values for different confidence levels can be made by using the algorithm and the coefficients given in the Section 5.1.4.

Because of the convenient mathematical characteristics of the χ^2 value (it is additive), it is also used to monitor the fit of a model to experimental data; in this application the fitted model $Y = ABS(f(x, \ldots))$ replaces the expected probability increment ΔCP [see Eq. (1.7)] and the measured value y_i replaces the observed frequency. Comparisons are only carried out between successive iterations of the optimization routine (e.g., a

Table 1.16. Critical χ^2 Values for $p = 0.975, 0.95, 0.9, 0.05$, and 0.025.

	p					
f	0.975	0.95	0.90	0.10	0.05	0.025
1	0.000982	0.00393	0.0158	2.706	3.841	5.024
2	0.0506	0.103	0.211	4.605	5.991	7.378
3	0.216	0.352	0.584	6.251	7.815	9.348
10	3.247	3.940	4.865	15.987	18.307	20.307
20	9.591	10.851	12.443	28.412	31.410	34.170
50	32.357	34.764	37.689	63.167	57.505	71.420
100	74.222	77.929	82.358	118.498	124.342	129.561

Table 1.17. Results of χ^2 Test for Testing the Goodness of Fit of a Model. [The y column shows the values actually measured, while the Y columns give the model estimates for the coefficients A1, A2, and A3 below. The χ^2 columns are calculated as $(y - Y)^2/Y$. The fact that the sums over these terms, 23.03, 15.48, and 0.065 decrease for successive approximations means that the coefficient set 6.499 ··· yields a better approximation than either the initial or the first proposed set. If the χ^2 sum, e.g., 0.065, is smaller than a benchmark figure, the optimization can be stopped; the reason for this is that there is no point in fitting a function to a particular set of measurements and thereby reducing the residuals to values that are much smaller than the repeatability of the measurements. The benchmark is obtained by repeating the measurements and comparing the two series as if one of them were the model Y. For example, for -7.2, -4.4, $+0.3$, $+5.4$, and $+11.1$ and the above set, a figure of χ^2 of 0.11 is found. It is even better to use the averages of a series of measurements for the model Y, and then compare the individual sets against this.]

Data		Initial		Proposal 1		Proposal 2	
x_i	y_i	$Y1$	χ^2	$Y2$	χ^2	$Y3$	χ^2
1	−7.4	−9.206	0.06	−6.641	0.01	−7.456	0.000
2	−4.6	−6.338	0.14	−2.771	0.16	−4.193	0.008
3	0.5	−1.884	22.73	2.427	14.85	0.545	0.008
4	5.2	3.743	0.08	8.394	0.38	6.254	0.041
5	11.4	9.985	0.02	14.709	0.08	12.436	0.008
			23.03		15.48		0.065
Coefficients		A1:	6.672		6.539		6.499
		A2:	0.173		0.192		0.197
		A3:	3.543		2.771		3.116

Simplex program), so that critical χ^2 values need not be used. For example, a mixed logarithmic/exponential function $Y = A1^* \text{LOG}(A2 + \text{EXP}(X - A3))$ is to be fitted to the data tabulated above. Do the proposed sets of coefficients improve the fit? The conclusion is that the new coefficients are indeed better. See Table 1.17.

As noted above, the χ^2 test for goodness of fit gives a more balanced view of the concept of "fit" than does the pure least-squares model; however, there is no direct comparison between χ^2 and the reproducibility of an analytical method. A simplex-optimization program that incorporates this scheme is used in the example "Nonlinear Fitting" (Section 4.2).

1.8.3. Probability Charts

The χ^2 test discussed above needs a graphical counterpart for_a fast, visual check of data. A useful method exists for normally distributed data

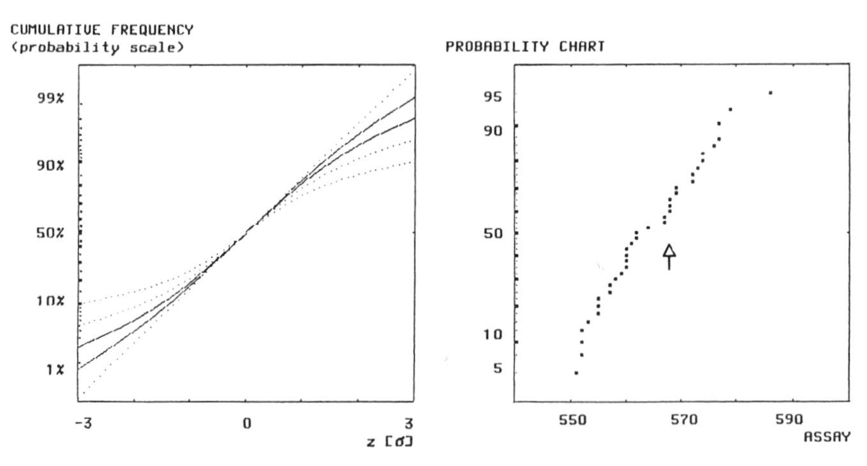

Figure 1.24. Probability-scaled chart for the testing of a distribution for normality. The abscissa is in the same units as the measurements, the ordinate is nonlinear in terms of cumulative probability. The left figure shows a straight line and four curves corresponding to t distributions with $f = 9$ and 4 (enhanced), and $f = 2$ and 3 degrees of freedom. The straight line corresponding to a Normal Distribution evolved from the CP function familiar from Section 1.2.1 through nonlinear stretching in the vertical direction. The median and the standard deviation can be graphically estimated (regression line, interpolation at 50, respectively, 17 and 83%). The right figure depicts 40 assay values for the active principle content in a cream (three values are off scale, namely, 2.5, 97.5, and 100%). The data by and large conforms to a normal distribution. Note the vertical runs of 2–4 points (arrow!) at the same x value; this can be attributed to a quantization raster that is impressed on the measurement because only certain absorbance values are possible due to the three-digit display on the employed instrument.

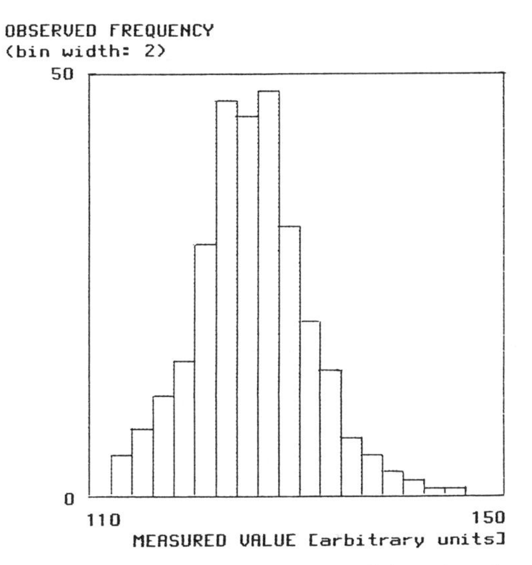

Figure 1.25. Mechanical measurements (readings to 0.1 units) conducted on 298 samples of a packaging component are grouped into classes of width 2.0.

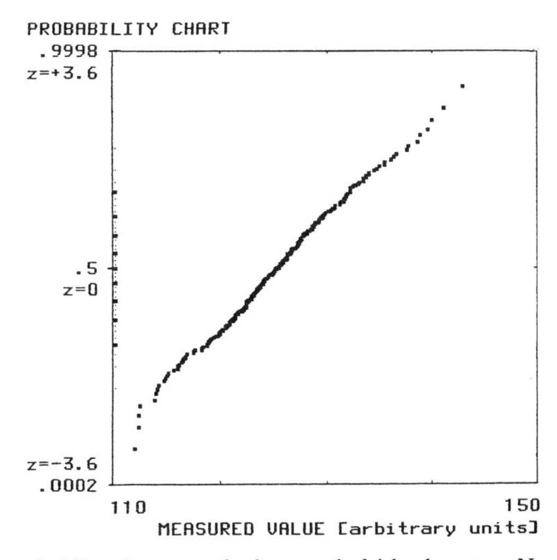

Figure 1.26. The probability chart reveals that one is fairly close to a Normal Distribution; the units with the smallest measured values had been selectively eliminated. The data is the same as that in Fig. 1.25.

that could also be adapted to other distributions. The idea is to first order the observations and then to assign each one an index value $1/n$, $2/n, \ldots, n/n$. These index values are then entered into the $z = f(\mathrm{CP})$ algorithm (also used in the MCSIM program) and the result is plotted versus the observed value. If a perfect Normal Distribution obtains, a straight line will result. A practical application is shown in Section 4.1. This technique effectively replaces the probability-scaled paper. A numerical example will illustrate the concept: If 27 measurements are available, the fourth smallest one corresponds to a cumulative probability $\mathrm{CP} = 4/27 = 0.148$ and the z value is -1.045; a symbol would be plotted at the coordinates (fourth smallest x value$/-1.045$). The last value is always off scale because the z value corresponding to $\mathrm{CP} = 1.000$ is $+\infty$ (use program HISTO, option NPP). See Figures 1.24, 1.25, and 1.26.

If non-Normal Distributions were to be tested for, the $z = f(\mathrm{CP})$ algorithm in Section 5.1.1 would have to be replaced by one that linearizes the cumulative CP for the distribution in question.

1.8.4. Conventional Control Charts (Shewhart Charts)

In a production environment, the quality control department does not ordinarily concern itself with single applications of analytical methods, that being a typical issue for analytical R&D, but concentrates on redoing the same tests over and over in order to release products for sale or reject them. When data of a single type accumulates, new forms of statistical analysis become possible. In the following, conventional control and cumsum charts will be presented. In the authors' opinion, newer developments in the form of tight (multiple) specifications and the proliferation of PCs have increased the value of control charts; especially in the case of on-line in-process controlling, monitors depicting several stacked charts allow the floor supervisors to recognize trends before they become dangerous. V templates (a graphical device to be superimposed on a CUMSUM chart that consists of a V-shaped confidence limit[22]) can be incorporated for warning purposes, but are not deemed a necessity because the supervisor's experience with the process gives him the advantage of being able to apply several (even poorly defined) rules of thumb simultaneously, while the computer must have sharp alarm limits set. The supervisor can often associate a pattern of deviations with the particular readjustments that are necessary to bring the process back in line. A computer would have to run a fuzzy-logic Expert System to achieve the same effect, but these are expensive to install and need retraining every time a new phenomenon is observed and assigned to a cause. Such refinements are rather common during the initial phases of a product's life cycle; in today's global markets, though, a process barely has a chance of reaching maturity before it is scaled up or replaced.

Although a control or a CUMSUM chart of first glance resembles an ordinary two-dimensional graph, one important aspect is different: The abscissa values are an ordered set graphed according to the rank (index), and not according to the true abscissa values (in this case time) on a continuous scale. Because of this, curve fitting to obtain a "trend" is not permissible. There is one exception, namely, when the samples are taken from a continuous process at strictly constant intervals. A good example is a refinery running around the clock for weeks or months at a time with automatic sampling of a process stream at, say, 10-minute intervals. Consider another process run on an 8- or 16-hour shift schedule with a shutdown every other weekend: Curve fitting would here only be permissible on data taken on the same day (e.g., at 1-hour intervals), or on data points taken once a day between shutdowns (e.g., always at 10 a.m.). Sample scheduling is often done differently, though, for good reasons, viz., (a) every time a new drum of a critical reagent is fed into the process, (b) every time a receiving or transport container is filled, (c) every time a batch is finished, etc.

The statistical techniques applicable to control charts are thus restricted to those of Section 1.5, that is, detecting deviations from the long-term mean respectively crossing of the specified limits.

The conventional control chart is a graph having a "time" axis (abscissa) consisting of a simple raster, such as that provided by graph or ruled stationary paper, and a measurement axis (ordinate) scaled to provide 6–8 standard deviations centered on the process mean. Overall standard deviations are used that include the variability of the process and the analytical uncertainty, see Figure 1.8. Two limits are incorporated: The outer set of limits corresponds to the process specifications and the inner one to "warning" or "action" levels for in-house use. Control charts are plotted for two types of data:

(1) A standard sample is incorporated into every series, e.g., daily, to detect changes in the analytical method and to keep it under statistical control.
(2) The data for the actual unknown samples are plotted to detect changes in the product and/or violation of specification limits.

The specification limits [case (2)] can either be imposed by regulatory agencies, agreed-upon standards, or market pressure, or can be set at will. In the latter case, given the monetary value associated with a certain risk of rejection, the limits can be calculated by statistical means. More likely, limits are imposed: In pharmaceutical technology +2, ±5, or ±10% about the nominal assay value are commonly encountered; some uses to which chemical intermediates are put dictate asymmetrical limits, reflecting perhaps solubility, purity, or kinetic aspects; a reputation for high quality demands that tight limits be offered to customers, which introduces a nontechnical side to the discussion. The classical $\pm 2\sigma$ and $\pm 3\sigma$ SL are dangerous because for purely mathematical reasons, a normalized deviate $z = r_i/s_x \geq 3$ can only be obtained for $n \geq 11$[40]; cf. Fig. 1.20.

Action limits must be derived from both the specification limits and the characteristics of the process: They must provide the operators with ample leeway (time, concentration, temperature, etc.) to react and bring the process back to nominal conditions without danger of the specification limits being exceeded. An important factor, especially in continuous production, is the time constant: A slow-to-react process demands narrow action limits relative to the specification limits, while many a tightly feedback-controlled process can be run with action limits close to the specification limits. If the product conforms to the specifications, but not to the action limits, the technical staff is alerted to look into the potential problem and come up with improvements before a rejection occurs. Generally, action limits will be moved closer to the specification limits as experience accrues.

For an example of a control chart see Fig. 1.27 and Sections 4.1 and 4.8. Control charts have a grave weakness: The number of available data points must be relatively high in order to be able to claim "statistical control." As is often the case in this age of shorter and shorter product life

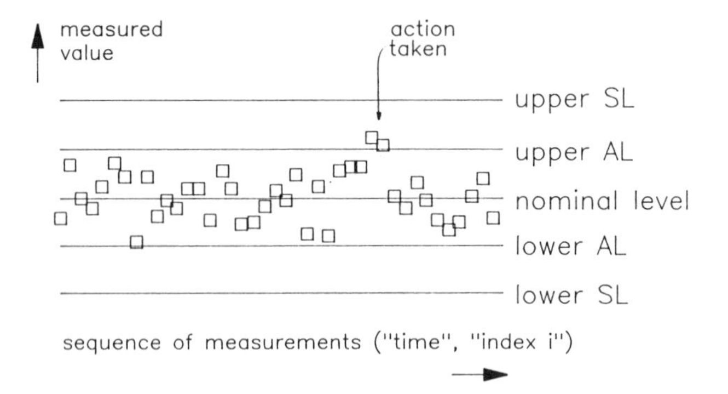

Figure 1.27. Typical control chart showing the specification and action limits. The four limits can be arranged asymmetrically relative to the nominal level, depending on the process characteristics and experience.

cycles, decisions will have to be made on the basis of a few batch release measurements, unless the number of in-process controls is appropriately increased.

1.8.5. CUMSUM Charts

A disadvantage of the conventional control charts is that a small or gradual shift in the observed process parameter is only confirmed long after it has occurred because the shift is swamped in statistical (analytical) noise. A simple way out is the CUMSUM chart (cumulated sum of residuals; see program SMOOTH), because changes in a parameter's average quickly show up; see Fig. 1.28. The basic idea is to integrate (sum) the individual values x_i. In practice, many consecutive results are electronically stored; a reference period is chosen, such as the previous month, and the corresponding monthly average **a** is calculated. This average is then subtracted from all values in the combined reference and observation period; thus $r_i = x_i - \mathbf{a}$. On the CUMSUM chart, the sum $\Sigma(r_i)$, $i = 1 \cdots j$, is plotted at time i. Different types of graphs can be distinguished[42]; see Fig. 1.29.

Scales should optimally be chosen so that the same distance (in mm) that separates two points horizontally vertically corresponds to about $2 \cdot s_y/\sqrt{m}$, that is twice the standard deviation of the mean found for m repeat measurements.[22] A V-formed mask can be used as a local confidence limit.[22] This approach, however, is of statistical nature; the combi-

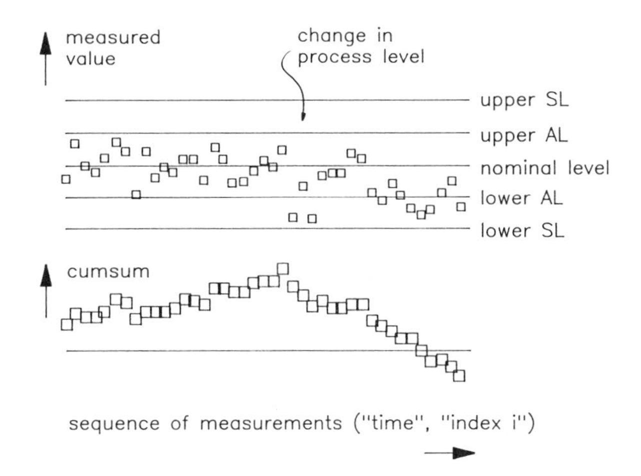

Figure 1.28. Typical CUMSUM chart showing a change in process mean.

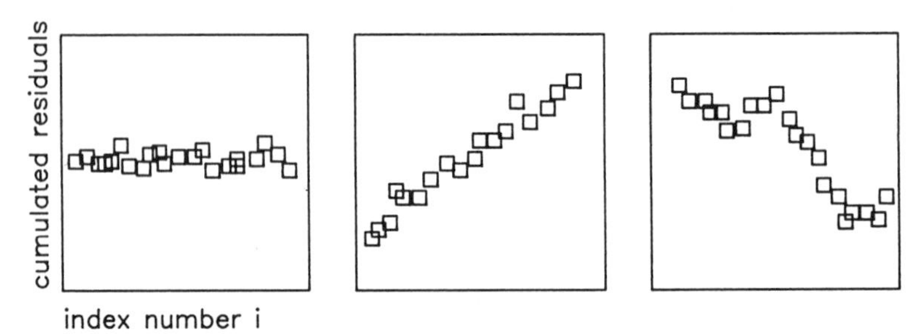

Figure 1.29. CUMSUM charts (schematic). Type 1, a horizontal line, indicates that the system comprising the observed process and the analytical method is stable and that results stochastically vary about the average. Type 2, a sloping straight line, indicates a nonoptimal choice of the average **a**, e.g., by using too few decimal places. Type 3, a series of straight-line segments of changing slope, shows when a change in any of the system variables resulted in a new average **a**′ different from the previous average **a**. Due to the summation process the system noise is damped out and changes in the trend rapidly become obvious.

nation of intuition, experience, and insider's knowledge is much more revealing. For examples, see Ref. 43 and Section 4.8.

1.9. ERRORS OF THE FIRST AND SECOND KIND

The two error types mentioned in the title are also designated with the Roman numerals I and II; the associated error probabilities are termed alpha (α) and beta (β).

When one attempts to estimate some parameter, the possibility of error is implicitly assumed. What sort of errors are possible? Why is it necessary to distinguish between two types of error? After a decision is passed, the route taken can be "true" or "false," and by the same token the theoretically correct route (as hindsight would show later on, but unknown at the time) could be "true" or "false."[13] This gives four outcomes that can be depicted as in Table 1.18.

The different statistical tests discussed in this book are all defined by the left column; that is, the initial situation H_0 is known and circumscribed, whereas H_1 is not (accordingly one should use the error probability α). In this light, the type II error is a hypothetical entity, but very useful. A graphical presentation of the situation, as in Fig. 1.30, will help.

Assume the ensemble of all results A (e.g., reference Method A) to be fixed and the ensemble B (e.g., test Method B) to be movable in the horizontal direction; this corresponds to the assumption of a variable bias $\Delta x = \bar{x}_B - \bar{x}_A$, to be determined, and usually, to be proven noncritical.

Table 1.18. Null and Alternative Hypotheses.
(Calculations involving the error probability beta are demonstrated in Section 4.1. The expression "X = Y" is to be read as "X is indistinguishable from Y.")

		real situation	
		X = Y	X ≠ Y
decision taken, answer	null hypothesis H_0 X = Y	correctly retain H_0 correctly reject H_1 probability $1 - \alpha$	**"false negative"** falsely retain H_0 falsely reject H_1 probability β
given, or situation assumed	alternative hypothesis H_1 X ≠ Y	**"false positive"** falsely reject H_0 falsely accept H_1 probability α	correctly reject H_0 correctly accept H_1 "power" of a test: probability $1 - \beta$

The critical x value, x_c, is normally defined by choosing an error probability α in connection with H_0 (that is, $\bar{x}_B$ is assumed to be equal to $\bar{x}_A$). In the case presented here, a one-sided test is constructed under the hypothesis "H_1: $\bar{x}_B$ larger than $\bar{x}_A$." The shaded area in Fig. 1.30(a) gives the probability of erroneously rejecting H_0 if a result $\bar{x} > x_c$ is found (false positive). In Fig. 1.30(b) the area corresponding to an erroneous retention of H_0, β (false negative) under assumption "B different from A" is shaded. Obviously, for different exact hypotheses H_1, i.e., clearly defined position of $\bar{x}_B$, β varies while α stays the same. Thus, the power of the test, $1 - \beta$, to discriminate between two given exact hypotheses H_0 and H_1 can for all intents and purposes only be influenced by the number of samples n by narrowing the distribution functions for the observed mean; see Fig. 1.14. If there is any reason to believe that an outcome of a decision would either carry a great risk (e.g., in the sense of safety), or would have immense impact (for example, medication X better than Y), H_0 and H_1 will have to be appropriately assigned (H_0: X = Y and H_1: X ≠ Y) or (H_0: X ≠ Y and H_1: X = Y), and α will have to be set to reflect the concern associated with not obtaining decision H_0 when H_0 is deemed to be the right decision. For example, if a chemical's properties

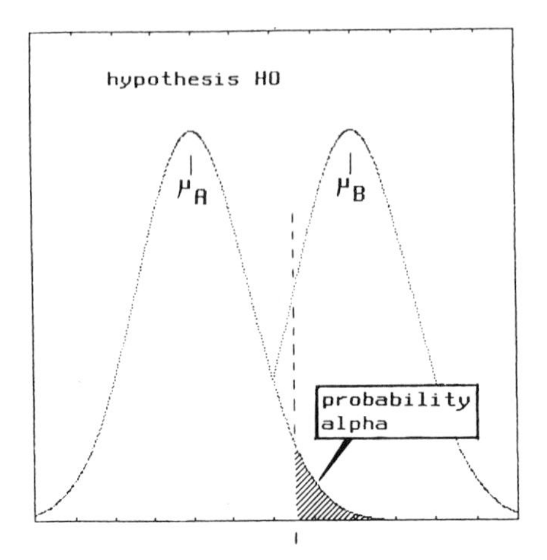

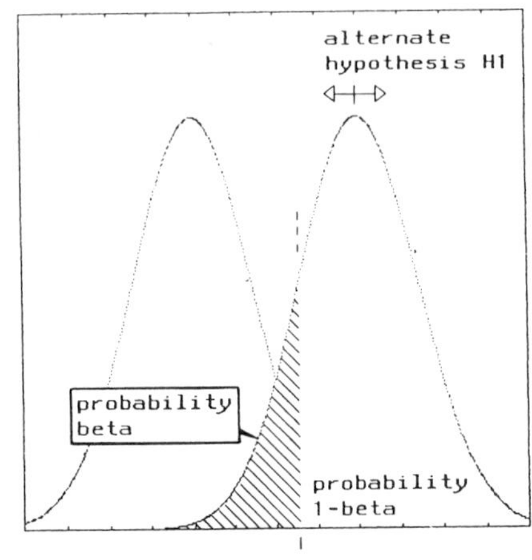

Figure 1.30. Alternative hypothesis and the power of a t test. Alpha (α) is the probability of rejecting an event.that belongs to the population associated with H_0; it is normally in the range $0.05 \cdots 0.01$. Beta (β) is the probability that an event effectively to be associated with H_1 is accepted as belonging to the population associated with H_0. Note that the power of the test to discriminate between hypotheses increases with the distance between μ_A and μ_B. μ_A is fixed either by theory or by previous measurements, while μ_B can be adjusted (shifted along the x axis); for examples, see H_1–H_4, Section 4.1, and program HYPOTHES.

Table 1.19. Interpretation of a Null / Alternative Hypothesis Situation.

		real situation	
		our product is *the same*	our product is *better*
decision taken, answer	null hypothesis H_0: our product is the same	have R & D come up with new ideas	**"false negative"** loss of a good marketing argument; hopefully the customer will appreciate the difference in quality
given, or situation assumed	alternative hypothesis H_1: our product is better	**"false positive"** the Marketing Department launches an expensive advertising campaign; the company is perceived to be unserious	capture the market, you have a good argument

are to be used for promotional purposes, there had better be hard evidence for its superiority over the competitor's product; otherwise, the company's standing as a serious partner will soon be tarnished. Thus, instead of postulating "H_0: our product is better" and proving this with a few measurements, one should assign "H_0: our product is the same." The reason is that in the former case one would, by choosing a small α, make it hard to reject H_0 and herewith throw a pet notion overboard, and at the same time, because of an unknown and possibly large β, provoke the retention of H_0 despite the lack of substantiation. In the case "H_0: ours is the same," the tables are turned: "H_1: superiority" must be proven by hard facts, and if found, is essentially true (α small), while the worst that could happen due to a large β would be the retention of an unearned "the same," that is, the loss of a convenient marketing argument. See Table 1.19.

CHAPTER 2

BI- AND MULTIVARIATE DATA

Data sets become two dimensional in any of the following situations:

- One-dimensional data are plotted versus an experimental variable; a prime example is the Lambert-Beer plot of absorbance vs. concentration, as in a calibration run.
- More than one dimension, i.e., parameter, of the experimental system is measured, say absorbance and pH of an indicator solution; the correlation behavior is tested to find connections between parameters.
- The same parameter is tested on at least two samples by each of several laboratories using the same method (round-robin test).
- At least two parameters are tested by the same laboratory on many nominally similar samples or repeatedly on the same sample (e.g., stability tests).

The last two situations are amenable to a graphical technique known as the Youden plot[22] that complements ANOVA analysis. Figures 2.1(a) and 2.1(b) give examples. Essentially, the interpretation is as follows: If a round patch of points is observed, no participating laboratory (or sample or point in time) is exceptional, i.e., H_1 must be rejected. On the other hand, an elliptical patch, especially if the slope deviates from what is expected, shows that some effects are at work that need further investigation.

2.1. CORRELATION

Two statistical measures found in most software packages are the *correlation coefficient* r and the *coefficient of determination* r^2. The range of r is bounded by $-1 \leq r \leq +1$; $|r| = 1$ is interpreted as perfect correlation, and $r = 0$ as no correlation whatsoever:

$$r = \frac{S_{xy}}{\sqrt{S_{xx}} \cdot \sqrt{S_{yy}}}. \tag{2.1}$$

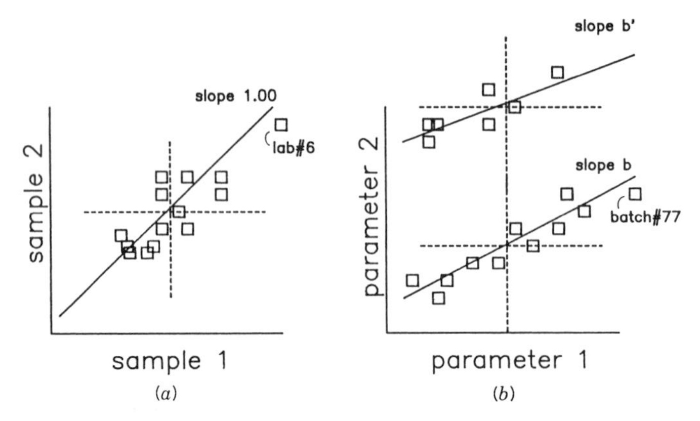

Figure 2.1. (a) Dotted line, expected averages; solid line: diagonal if the same parameter is tested on two samples, or slope b if two parameters are tested on the same sample (modification of Youden's plot). As an example, two exceptional points are marked: laboratory No. 6 could have trouble calibrating its methods (both samples have the same bias). (b) Tablets containing fixed ratios of two active components (slopes b and b') either have a large dispersion in weight (e.g., the batch No. 77 sample might have been improperly weighed) and/or the injection into the HPLC could be out of control (perhaps no internal standard, see Section 4.14).

For the definition of S_{xx}, S_{xy}, and S_{yy}, see Eqs. (2.4)–(2.6). Under what circumstances is r to be used? When hidden relations between different parameters are sought, as in explorative data analysis, r is a useful measure. Any pairing of parameters that results in a statistically significant r is best reviewed in graphical form, so that a decision about further data treatment can be reached. Even here caution must be exercised, because r is a summary measure that can suggest effects where none are present, and vice versa, i.e., because of an outlier or inhomogeneous data. Program CORREL provides the possibility of finding and inspecting the correlations in a large data set. For most other applications, and calibration curves in particular, the correlation coefficient must be viewed as a relic of the past[44]: Many important statistical theories were developed in sociological or psychological settings, along with appropriate measures such as r^2. There the correlation coefficient indicates the degree of parallelism between one parameter and another, say, reading skills and mathematical aptitude in IQ tests. The question of cause and effect, so important in the physical sciences, is not touched upon: One parameter might be the (partial) cause and the other the effect; or, more likely, both are effects of several interacting and unquantifiable factors, say intelligence, upbringing, and heredity.

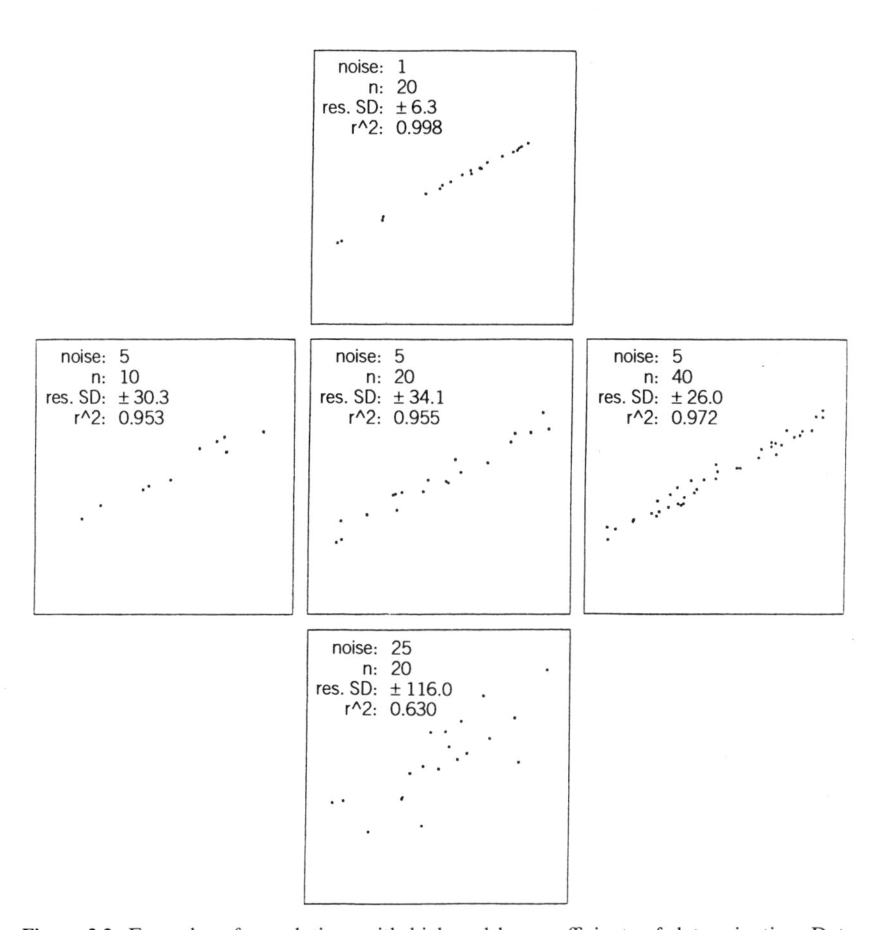

Figure 2.2. Examples of correlations with high and low coefficients of determination. Data were simulated for combinations of various levels of noise ($\sigma = \pm 1$, ± 5, ± 25, top to bottom) and sample size (n = 10, 20, 40, left to right). The residual standard deviation follows the noise level (± 6, ± 30, ± 116, from top to bottom). Note that the coefficient 0.998 in the top panel is on the low side for many analytical calibrations where the points so exactly fit the theoretical line that $r^2 > 0.999$ even for low n and small calibration ranges. The coefficient of determination r^2 is a dimensionless number, the significance of which resides in the last, non-9, digits; the calculation has to be performed with high precision and is subject to truncation effects similar to those discussed in connection with the standard deviation; see Table 1.1.

The situation in analytical chemistry is wholly different: Cause and effect relationships are well characterized; no one is obliged to rediscover the well-documented hypothesis that runs under the name Lambert-Beer law. What is in demand, though, is a proof of adherence. With today's high-precision instrumentation, correlation coefficients larger than 0.999 are commonplace. The absurdity is that the numerical precision of the algorithms used in some soft-/firmware packages limit the number of reliable digits in r^2 to two or three (but the authors are not aware of anyone having qualms about which type of calculator to use). Furthermore, it is hard to develop a feeling for numbers when all the difference resides in the fourth or fifth digit. As an alternative goodness-of-fit measure, the residual standard deviation is proposed because it has a dimension the chemist is used to (same as ordinate) and can be directly compared to instrument performance [cf. Eqs. (1.26) and (2.13)].

A numerical **example**: Using file VALID.003 ($r = 0.99991038\ldots$) and a suitably modified program VALID, depending on whether the means are subtracted as in Eqs. (2.4)–(2.6), or not, as in Eqs. (2.7)–(2.9), whether single or double precision is used, and the sequence of the mathematical operations, the last four digits can assume the values 1030, 1041, 0792, 0802, or 0797. Plainly, there is no point in quoting **r** to more than 3, or at most 4, decimal places. See Fig. 2.2.

2.2. LINEAR REGRESSION

Whenever a property is measured in function of another the question of which model should be chosen to relate the two crops up. By far the most common model function is the linear one; i.e., the dependent variable y is defined as a linear combination containing two adjustable coefficients and x, the independent variable, namely,

$$Y = a + b \cdot x. \tag{2.2}$$

A good example is the absorption of a dyestuff at a given wavelength λ (lambda) for different concentrations, as expressed by the well-known Lambert-Beer law:

$$\text{Absorbance} = A_{\text{blank}} + \text{path length} \cdot \text{absorptivity} \cdot \text{concentration},$$

with the identifications $Y =$ absorbance, $a = A_{\text{blank}}$, $x =$ concentration, and $b =$ path length · absorptivity. If the measurements do not support the assumption of a linear relationship,[12] one often tries transformations to linearize it. One does not have to go far for good reasons:

(a) Because only two parameters need to be estimated, the equation of the straight line is far easier to calculate than that of most curves.
(b) The function is transparent and robust, and lends itself to manipulations like inversion ($X = f(y)$).
(c) Relatively few measurements suffice to establish a regression line.
(d) A simple ruler is all one needs for making or checking graphs. A linear relationship inherently appeals to the mind and is simple to explain.
(e) Before the advent of the digital computer high-order and nonlinear functions were impractical at best, and without a graphics plotter much time is needed to draw a curve. Interpolation, particularly in the form $X = f(y)$, is neither transparent nor straightforward if confidence limits are requested.

Thus the linear model is undoubtedly the most important one in the treatment of two-dimensional data and will therefore be discussed in detail.

Overdetermination of the system of equations is at the heart of regression analysis; that is, one determines more than the absolute minimum of two coordinate pairs (x_1/y_1) and (x_2/y_2) necessary to calculate a and b by classical algebra. The unknown coefficients are then estimated by invoking a further model. Just as with the univariate data treated in Chapter 1, the least-squares model is chosen, which yields an unbiased best-fit line subject to the restriction

$$\sum (r_i)^2 = \text{minimum}, \tag{2.3}$$

where r_i is the residual associated with the ith measurement. The question is now how this residual is to be geometrically defined. The seemingly logical thing to do is to drop a perpendicular from the coordinate (x_i/y_i) onto the regression line, as shown by Fig. 2.3 (left).

While it is perfectly permissible to estimate a and b on this basis, the calculation can only be done in an iterative fashion; that is, both a and b are varied in smaller and smaller steps (see Optimization Techniques, Section 3.1) and each time the squared residuals are calculated and summed. The combination of a and b that yields the smallest of such sums represents the solution. Despite digital computers, Adcock's solution, a special case of the maximum likelihood method,[46] is not widely used; the additional computational effort and the more complicated soft-

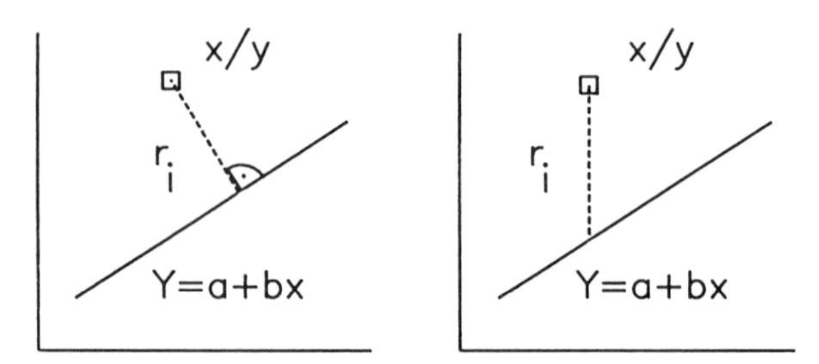

Figure 2.3. The definition of the residual. The sum of the squared residuals is to be minimized.

ware are not justified by the "improved" (a debatable notion) results, and the process is not at all transparent, i.e., not amenable to manual verification.

2.2.1. Standard Approach

The standard approach today[24, 34, 45] is shown in Fig. 2.3 (right): The vertical residuals are minimized according to $r_i = y_i - Y = y_i - (a + b \cdot x_i)$. A closed (noniterative) solution is obtained that is easily verifiable by manual calculations. There are three assumptions that must be kept in mind, though:

(a) The uncertainty inherent in the individual measurements of the property Y must be much larger than that for property X, or, in other words, the repeatability s_y relative to the range of the y values must be much larger than the repeatability s_x relative to the range of the x values. Thus, if the graph $(x_{min}, x_{max}, y_{min}, y_{max})$ is scaled so as to be approximately square (the regression line is nearly identical with a 45-degree diagonal), the confidence intervals are related as $\text{CI}(y) \gg \text{CI}(x)$ if measured in millimeters. Maximum likelihood is the technique to use if this assumption is grossly violated.[46]

(b) Homoscedacity must obtain; i.e., the reproducibilities of y measured for small, medium, and large x values must be approximately the same, so the uniform weighing scheme can be implemented.

(c) The distribution of errors must be Gaussian; regression under conditions of *Poisson*-distributed noise is dealt with in Ref. 47.

Restrictions (a) and (b) are of no practical relevance, at least as far as the

Table 2.1. Equations for Linear Regression.

Correct Formulation (sums Σ are taken over all measurements, $i = 1 \cdots n$):

Equation	Description	Eq.
$\bar{x} = (\Sigma x_i)/n \qquad \bar{y} = (\Sigma y_i)/n$	Averages	(1.1)
$S_{xx} = \Sigma(x_i - \bar{x})^2$	Sums of	(2.4), (1.3)
$S_{yy} = \Sigma(y_i - \bar{y})^2$	squares	(2.5)
$S_{xy} = \Sigma(y_i - \bar{y}) \cdot (x_i - \bar{x})$		(2.6)

Algebraically Equivalent Formulation as Used in Pocket Calculators (beware of numerical artifacts when using Eqs. 2.7–2.9; cf. Table 1.1):

Equation	Description	Eq.
$S_{xx} = \Sigma(x_i^2) - (\Sigma x_i)^2/n$	Sums of	(2.7)
$S_{yy} = \Sigma(y_i^2) - (\Sigma y_i)^2/n$	squares	(2.8)
$S_{xy} = \Sigma(x_i \cdot y_i) - (\Sigma x_i) \cdot (\Sigma y_i)/n$		(2.9)
$Y = a + b \cdot x$	Regression model	(2.10)
$b = S_{xy}/S_{xx} \qquad a = \bar{y} - b \cdot \bar{x}$	Estimates b, a	(2.11), (2.12)
$V_{\text{res}} = s_{\text{res}}^2 = \dfrac{\Sigma(y_i - Y)^2}{n-2} = \dfrac{S_{yy} - b \cdot S_{xy}}{n-2}$	Residual variance	(2.13)
$V_b = V_{\text{res}}/S_{xx}$	Variance of slope b	(2.14)
$V_a = V_{\text{res}} \cdot \left\{\dfrac{1}{n} + \dfrac{\bar{x}^2}{S_{xx}}\right\}$	Variance of intercept a	(2.15)
$V_Y = V_{\text{res}} \cdot \left\{\dfrac{1}{n} + \dfrac{(x - \bar{x})^2}{S_{xx}}\right\}$	Variance of estimate $Y(x)$	(2.16)
$\text{CL}(Y) = a + b \cdot x \pm t(f, p) \cdot \sqrt{V_Y}$	Estimate Y	(2.17)
$V_X = \dfrac{V_{\text{res}}}{b^2} \cdot \left\{\dfrac{1}{n} + \dfrac{1}{m} + \dfrac{(y^* - \bar{y})^2}{b^2 \cdot S_{xx}}\right\}$	Variance of estimate X	(2.18)
$\text{CL}(X) = (y^* - a)/b \pm t(f, p) \cdot \sqrt{V_x}$	Estimate X	(2.19)

Notation

- n: number of calibration points
- x_i: known concentration (or other independent variable)
- y_i: measured signal at x_i
- $\bar{x}$: mean of all x_i
- $\bar{y}$: mean of all y_i
- S_{xx}: sum of squares over Δx
- S_{yy}: sum of squares over Δy
- S_{xy}: sum of cross-product $\Delta x \cdot \Delta y$
- a: intercept
- b: slope

Table 2.1 ***(Continued)***

r_i:	ith residual $r_i = y_i - (a + b \cdot x_i)$
s_{res}:	residual standard deviation
V_{res}:	residual variance
$t(f, p)$:	Student's t factor for $f = n - 2$ degrees of freedom, probability p
m:	number of replicates y_k^* on unknown
y^*:	mean of several y_k^*, $k = 1 \cdots m$, $y^* = (\Sigma y_k^*)/m$
Y:	expected signal value (at given x)
X:	estimated concentration for mean signal y^*

slope and the intercept are concerned, when all data points closely fit a straight line; the other statistical indicators are influenced, however.

In the following the standard unweighed linear regression model is introduced. All necessary equations are found in the following section (Table 2.1) and are used in program LR. In a later section (2.2.10) nonuniform weighing will be dealt with. The equations given in the box were derived by combining Eqs. (2.2) and (2.3), forcing the regression line through the center of gravity ($\bar{x}/\bar{y}$), and setting the partial derivative $\partial(\Sigma r^2)/\partial b = 0$ to find the minimum.

2.2.2. Slope and Intercept

Estimating the slope b is formally equivalent to the calculation of a mean from a distribution (Section 1.1.1) in that the slope also represents a statistical mean (best estimate). The slope is calculated as the quotient of the sums S_{xy} and S_{xx}. See Table 2.1. Since the regression line must pass through the center of mass ($\bar{x}/\bar{y}$), the intercept is given by extrapolating from this point to $x = 0$. The question of whether to force the regression line through the origin ($a \equiv 0$) has been discussed at length.[48–51]

The confidence interval CI(b) serves the same purpose as CI($\bar{x}$) in Section 1.3.2; the quality of these average values is described in a manner that is graphic and allows meaningful comparisons to be made. An example from photometry, see Table 2.2, is used to illustrate the calculations (see also data file UV.001); further calculations, comments, and interpretations are found in the appropriate sections. Results in Table 2.3 are tabulated with more significant digits than is warranted, but this allows the reader to check recalculations and programs. Figure 2.4 gives the corresponding graphical output. If the calibration procedure is repeated

Table 2.2. Calibration Measurements.
(X is the calibration (or standard) sample's concentration in percent of the product's nominal concentration, y is the measured absorbance, and AU refers to absorbance units.)

x	y
$x = 50\%$	$y = 0.220$ AU
75	0.325
100	0.428
125	0.537
150	0.632
Degrees of freedom: $n = 5$, thus $f = n - 2 = 3$	Critical Student's t: $t(f = 3, p = 0.05) = 3.1824$

Table 2.3. Intermediate and Final Results of a Linear Regression.

[Interpretation (see Section 2.2.1): The slope is found as 0.004144 ± 0.000046 (mean b, $\pm$SD s_b). Using the Student's t for $f = 3$ degrees of freedom and an error probability of $p = 0.05$, the confidence limits are 0.00400, resp. 0.00429, and the confidence interval is the difference 0.00029; note that these last three figures, while being accurate estimates, provide no "feeling." The relative confidence interval, given by $\pm 100 \cdot t \cdot s_b/b$, is a convenient measure: $b = 0.00414 \pm 3.5\%$. This figure of merit indicates that the slope is relatively well defined. The corresponding calculation for the intercept a yields $a = 0.014 \pm 111\%$; in this case a large figure of merit, that is, one above 100%, indicates that the intercept is not significantly different from zero. In such a case the argument could be advanced to force the regression line through the origin, i.e., $a \equiv 0$ and $b = \bar{y}/\bar{x}$; while this might be justifiable from a purely scientific point of view, the practical consequence is to enlarge V_{res}, and therefore to diminish the power to accurately predict $X(y^*)$ or $Y(x)$; cf. Refs. 48–51 and Section 4.6.]

Intermediate Results			Final Results		
Item	Value	Equation	Item	Value	Equation
$\bar{x}$	= 100	(1.1)	b	= 0.004144	(2.11)
$\bar{y}$	= 0.4284	(1.1)	a	= 0.01400	(2.12)
S_{xx}	= 6250	(2.4)	V_{res}	= 0.00001320	(2.13)
S_{xy}	= 25.90	(2.5)	s_b	= ±0.0000459	(2.14)
S_{yy}	= 0.10737	(2.6)	s_a	= ±0.00487	(2.15)
			r^2	= 0.99963	(2.1)
			$t(3, 0.05)$	= 3.18	
			s_{res}	= ±0.00363	(2.13)
			$t \cdot s_b$	= ±0.000146	
			$t \cdot s_a$	= ±0.0155	
$t \cdot s_b$ amounts to 3.5% of b			CL(b): 0.0040, and 0.0043		CI(b): 0.0003
$t \cdot s_a$ amounts to 111% of a			CL(a): −0.0015 and 0.029		CI(a): 0.03

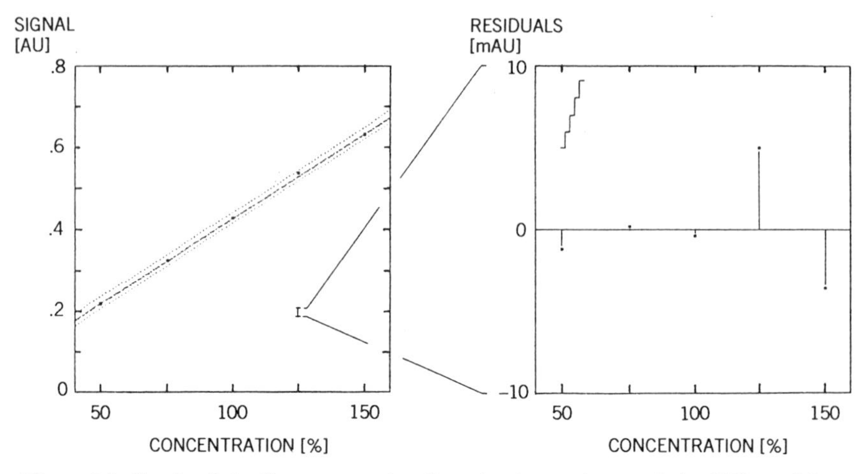

Figure 2.4. Graph of the linear regression line, the data points, and the 95% confidence limits (left), and the residuals (right). The 40-fold magnification of the right panel is indicated; the digital resolution ± 1 mAU of a typical UV spectrophotometer is illustrated by the steps.

and a number of linear regression slopes b are available, these can be compared as are means $\bar{x}$ (see Section 1.5.1, but also Section 2.2.4).

2.2.3. Residual Variance

The residual variance V_{res} summarizes the vertical residuals from Fig. 2.4; it is composed of

$$V_{res} = V_{reprod} + V_{nonlin} + V_{misc}, \tag{2.20}$$

where V_{reprod} is that variance due to repetitive sampling and measuring of an average sample; V_{nonlin} stands for the apparent increase if the linear model is applied to an inherently curved set of data; V_{misc} contains all other variance components, e.g., that stemming from the x-dependent s_y (heteroscedacity), see Fig. 2.5.

Under controlled experimental conditions the first term will dominate. The second term can be assessed by using a plot of the residuals (see next section); for a correctly validated GLP/GMP-compatible method, curvature must be avoided and checked for during the R&D phase (justification for the linear model!) by testing a concentration range that goes far beyond the 80–120%-of-nominal that is often used under routine QC conditions. Thus the residual variance should not be much larger than the

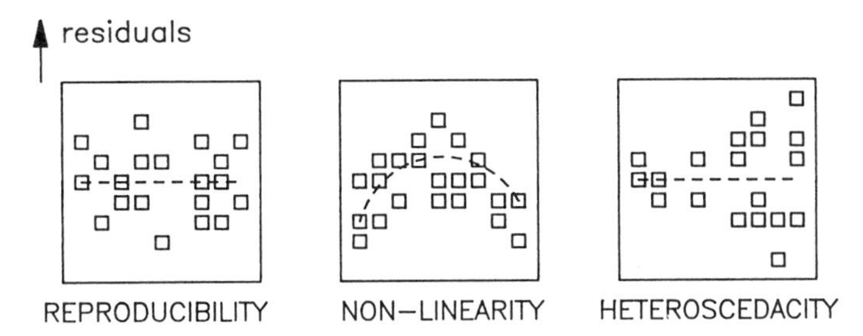

Figure 2.5. Three important components of residual variance. The residuals are graphed versus the independent variable x.

first term. Depending on the circumstances, a factor of 2–4 could already be an indication of noncontrol.

Taking the square root of V_{res}, one obtains the residual standard deviation, s_{res}, a most useful measure:

- s_{res} has the same dimension as the reproducibility and the repeatability, namely, the dimension of the measurement; those are the units the analyst is most familiar with, such as absorbance units, milligrams, etc.
- s_{res} is nearly independent of the number of calibration points and their concentration values x; cf. Fig. 2.8.
- s_{res} is easy to calculate, and, since the relevant information resides in the first significant digits, its calculation places no particular demands on the soft- or hardware (cf. Section 3.3) if the definition of r_i in Table 2.1 and Eqs. (1.3a)–(1.3d) is used.
- s_{res} is necessary to obtain other error-propagation information and can be used in optimization schemes (cf. Section 3.3).

Example: (Table 2.1, Section 2.2.1.). A residual standard deviation of less than ± 0.004 relative to $y = 0.428$ indicates that the experiment is relatively well controlled: On the typical UV/VIS spectrometer in use today, three decimal places are displayed, the least significant giving milliabsorbance units; noise is usually ± 1–2 mAU. If the residual standard deviation is felt to be too large, one ought to look at the individual residuals to find the contributions and any trend: The residuals are -0.0012, 0.0002, -0.0004, 0.005, and -0.0036. No trend is observable. The relative contributions are obtained by expressing the square of each residual as a percentage of $V_{res} \cdot (n - 2)$, i.e., $100 \cdot (0.0012) \cdot 2/0.0000132/3$, etc., which yields 3.6, 0.1, 0.4, 63, resp. 33%. Since the last two residuals make up over 96% of the total variance, bringing these two down to about 0.002 by more careful experimentation would result in a residual standard deviation of about ± 0.003, an improvement of 25%.

2.2.4. Testing Linearity and Slope

The test for the significance of a slope b is formally the same as a t test (Section 1.5.2): If the confidence interval CI(b) includes zero, b cannot significantly differ from zero; thus $b = 0$. If a horizontal line can be fitted between the plotted CL, the same interpretation applies; cf. Figs. 2.6a and 2.6b. Note that s_b corresponds to $s_{\bar{x}}$; that is the standard deviation of a mean. In the above example the confidence interval clearly does not include zero; this remains so even if a higher confidence level with $t(f = 3, p = 0.001) = 12.92$ is used.

Two slopes are compared in a similar manner as are two means: The simplest case obtains when both calibrations were carried out using identical calibration concentrations (as is usual when standard operating procedures, SOPs, are followed); the average variance V_b' is used in a t test:

$$V_b' = (V_{b,1} + V_{b,2})/2, \tag{2.21}$$

$$t = |b_1 - b_2|/\sqrt{V_b'}, \qquad \text{with } f = 2 \cdot n - 4 \text{ degrees of freedom.} \tag{2.22}$$

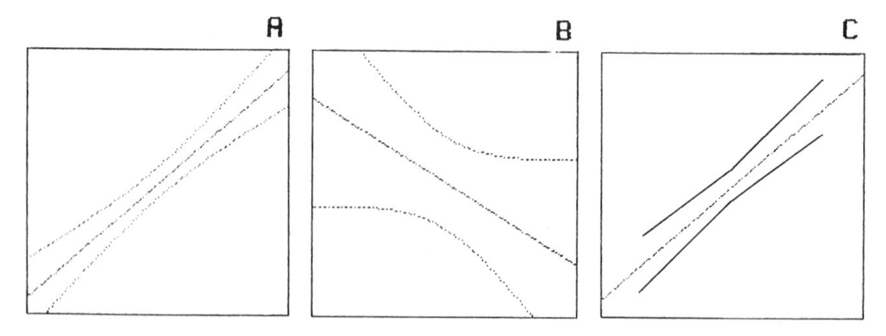

Figure 2.6. A graphical depiction of a significant and a nonsignificant slope [slopes $\pm s_b$ = 4.3 ± 0.5 (a) resp. -0.75 ± 1.3 (b)]. If a horizontal line can be fitted between the confidence limits an interpretation $X = f(y^*)$ is impossible. The confidence limits can, for many applications, be approximated by straight segments (c).

Example: Two calibrations are carried out with the results $b = 0.004144 \pm 0.000046$ and $b' = 0.003986 \pm 0.000073$; V' is thus ± 0.000061 and $t = 0.0000158/0.000061 = 2.6$; since $n = n' = 5$, $f = 6$, and $t(6, 0.05) = 2.45$ so that a significant difference in the slopes is found. The reader is reminded to round only final results; if already rounded results had been used here, different conclusions would have been reached: the use of five or four decimal places results in $t = 2.47$, respectively, $t = 1.4$, the latter of which is clearly insignificant relative to the tabulated $t_c = 2.45$.

A graphical test can be applied in all cases: The regression lines and their confidence limits for the chosen confidence level are plotted as given by Eq. (2.24) (next section); if one regression line is wholly within the area defined by the confidence limits of the other, the two cannot be distinguished. As an alternative to plotting the CL(Y) point by point over the whole x interval of interest, an approximation by straight-line segments as shown in Fig. 2.6c will suffice in most cases: The CL(Y) are plotted for x_{min}, $\bar{x}$, and x_{max}.

Eight combinations are possible with the true/false answers to the following three questions: (1) Is $s_{res,1} = s_{res,2}$? (2) Is $b_1 = b_2$? (3) Is $\bar{y}_1 = \bar{y}_2$? A rigorous treatment is given in Ref. 24. First, question 1 must be answered: if H_0 is retained, question 2 must be answered. Only if this also leads to a positive result can question 3 be posed.

There are several ways to test the linearity of a calibration line; one can devise theory-based tests, or use common sense. The latter approach is suggested here because if only a few calibration points are available on which to rest one's judgement, a graph of the residuals will reveal a trend,

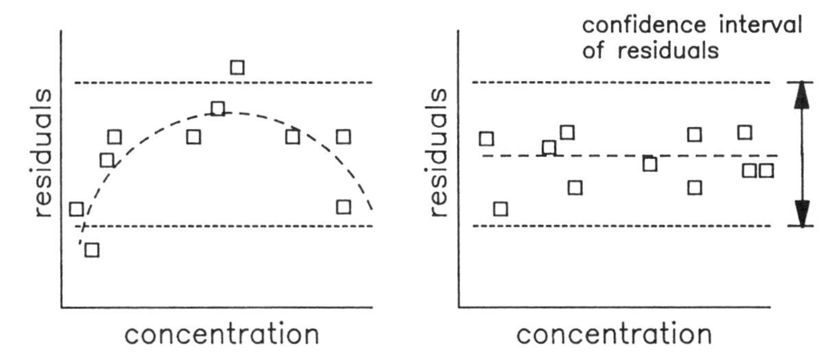

Figure 2.7. Using residuals to judge linearity. Dotted lines, Accepted variation of a single point, e.g., $\pm 2 \cdot s_{res}$; dashed line, perceived trend; note that in the middle and near the ends there is a tendency for the residuals to be near or beyond the accepted limits; that is, the model does not fit the data. For an example, see Section 4.13.

if any is present,[52] while numerical tests need to be adjusted to have the proper sensitivity. It is advisable to add two horizontal lines offset by $\pm$ the measure of repeatability accepted for the method; unless the apparent curvature is such that points near the middle, respectively, the end of the x range are clearly outside this reproducibility band, no action need be taken. See Fig. 2.7.

On regarding the residuals, many an investigator would be tempted to cast out outliers; the reader is advised to consult Section 1.5.5. If values are grouped (i.e., several values y_i are measured at the same x), outlier tests can be applied to the individual group; however, blind reliance on a rule, such as $\bar{y} \pm 2 \cdot s_y$, is strongly discouraged. A test of linearity as applied to instrument validation is given in Ref. 53.

2.2.5. Interpolating $Y(x)$

The estimate of Y for a given x is an operation formally equivalent to the calculation of a mean, as in Table 2.4. The expression for $V_Y = s_y^2 = V_{res}/n$ is true if $x = \bar{x}$; however, if x is different from $\bar{x}$, an extrapolation penalty is paid that is proportional to the square of the deviation, namely

$$V_Y = V_{res} \cdot \left\{ \frac{1}{n} + \frac{(x - \bar{x})^2}{S_{xx}} \right\}; \qquad \text{see (2.16)}.$$

This results in the characteristic "trumpet" shape observed in Figs. 2.4 and 2.6.

The $CL(Y_x)$ correspond to the y values of the two dotted curves at x:

$$CL(Y_x) = a + b \cdot x \pm t \cdot \sqrt{V_Y}; \qquad \text{see (2.17)}.$$

Table 2.4. Comparing the Confidence Intervals of the Estimates $\bar{x}$ and $Y(x)$.

	Univariate Data (Sections 1.1.1 and 1.1.2)	Linear Regression (Section 2.2.1)	Equation
Model	$\bar{x} = \mu$	$Y = a + b \cdot \bar{x}$	(2.2)
Estimated variance	$V_{\bar{x}} = \frac{V_x}{n}$	$V_Y = \frac{V_{res}}{n}$	(2.23)
Estimated confidence limits	$CL(X) = \bar{x} \pm t \cdot \sqrt{V_{\bar{x}}}$	$CL(Y) = Y(\bar{x}) \pm t \cdot \sqrt{V_Y}$	(2.24)

Table 2.5. Estimation of $Y(x)$.
[See Program LR, Option $Y(x)$, cf. Tables 2.2 and 2.3 for raw data.]

Example		
	$a = 0.0140$,	$n = 5$
	$b = 0.004144$,	$S_{xx} = 6250$
	$s_{res} = \pm 0.00363$,	$t = 3.18$
	$x = 120.0$,	$\bar{x} = 100.0$
	$Y = 0.0140 + 120 \cdot 0.004144 = 0.511$	
	$s_y = \pm 0.00363\sqrt{1/5 + (120 - 100)^2/6250} = \pm 0.00187$	
Result	$Y = 0.511 \pm 0.006\ (\pm 1.2\%)$;	see (2.17)
Thus	$0.505 \le Y(x) \le 0.517$;	see (2.17)
	$0.498 \le y(x) \le 0.524$;	see (2.25)

The confidence limits thus established indicate the y interval within which $Y(x)$ is expected to fall; the probability that this is an erroneous assumption is $100 \cdot p\%$; in other words, if the measurements were to be repeated and slightly differing values for a and b were obtained, the chances would only be $100 \cdot p\%$ that a Y is found outside the confidence limits CL(Y). Use option $Y(x)$ in program LR. See Table 2.5.

The CL(Y) obviously refer to the expected average ordinate Y at the given abcissa x; if one were interested in knowing within which interval one would have to expect individual measurements y, the CL(y) apply (Y refers to an estimate, y to a measurement!). Equation (2.16) above for V_Y is expanded to read

$$Vy = V_{res} \cdot \left\{ \frac{1}{n} + 1 + \frac{(x - \bar{x})^2}{S_{xx}} \right\}. \tag{2.25}$$

The additional term "+1" is explained in Fig. 2.9 and in the following. See also Fig. 2.8.

If it is assumed that a given individual measurement y_i at x_i is part of the same population from which the calibration points were drawn (same chemical and statistical properties), the reproducibility s_y associated with this measurement should be well represented by s_{res}. Thus, the highest y value still tolerated for y_i could be modeled by superimposing CI($Y(x_i)$) on CI(y_i) as shown by Fig. 2.9 (left). A much easier approach is to integrate the uncertainty in y_i into the calculation of CL(Y); because

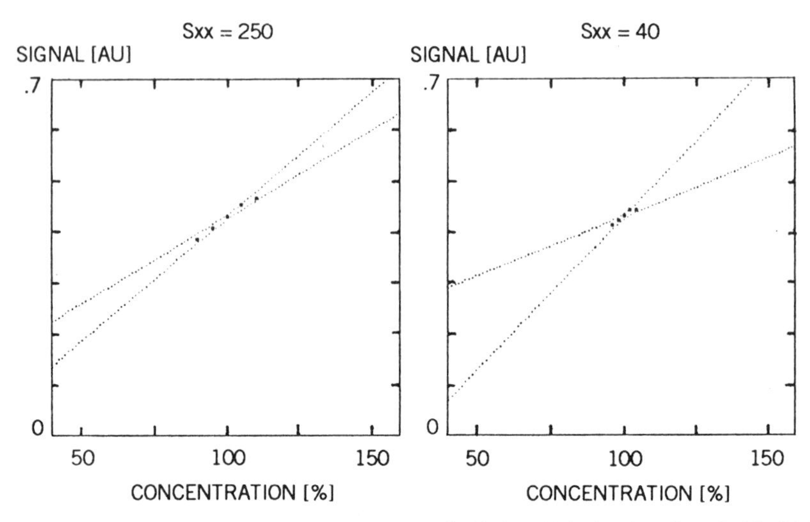

Figure 2.8. Compare these two figures to Fig. 2.4 (left), in which the data from Table 2.1 is depicted. In the above figures, the slopes, intercepts, and residuals are the same as in Fig. 2.4, but the x values are more densely clustered: 90, 95, 100, 105, and 110% of nominal (left), respectively 96, 98, 100, 102, and 104% of nominal (right). The following figures of merit are found for the sequence Figs. 2.4, 2.8 (left), 2.8 (right): the residual standard deviations: ± 0.00363 in all cases; the coefficients of determination: 0.9996, 0.9909, 0.9455; the relative confidence intervals of b: $\pm 3.5\%$, $\pm 17.6\%$, $\pm 44.1\%$. Obviously the extrapolation penalty increases with decreasing S_{xx}, and can be readily influenced by the choice of the calibration concentrations. The difference in S_{xx} (6250, Fig. 2.4; 250 resp. 40, above) exerts a very large influence on the estimated confidence limits associated with a, b, $Y(x)$, and $X(y^*)$.

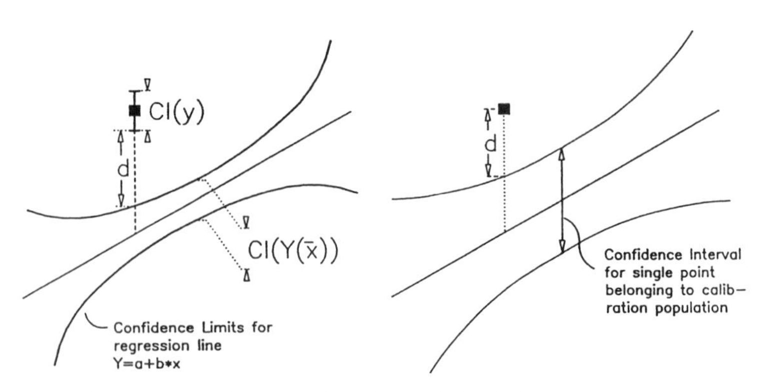

Figure 2.9. The confidence interval CI(y) for an individual value y is derived from the confidence interval of the regression line CI(Y) using the reproducibility. An outlier could be recognized by the lack of overlap of the corresponding confidence intervals (distance d); cf. Section 1.5.5.

variances (not standard deviations) are additive, this is done by adding V_{res} outside, respectively "+1" inside the parentheses of Eq. (2.16) to obtain Eq. (2.25).

In the example above, the term in the parentheses is increased from 0.264 to 1.264; this increase by a factor of 4.8 translates into CL(y) being 2.2 times larger than CI(Y). The corresponding test at $x = 125$ ($0.517 \leq y(x) \leq 0.547$) shows the measured value 0.537 to be well within the tolerated limits.

A test for outliers can be based on this concept, for instance, by using an appropriate t value or by making use of a special table[54], see Section 5.1.2, but as with all outlier tests, restraint is advised: Data points should never be suppressed on statistical reasoning alone. A good practice is to run through all calculations twice, once with and once without the suspected outlier, and to present a conservative interpretation, e.g., "the concentration of the unknown is estimated at 16.3 ± 0.8 mM using all seven calibration points. If the suspected outlier (marked x in the graph) were left out, a concentration of 16.7 ± 0.6 mM with $n = 6$ would be found." The reader can then draw his own conclusions. See Fig. 2.10.

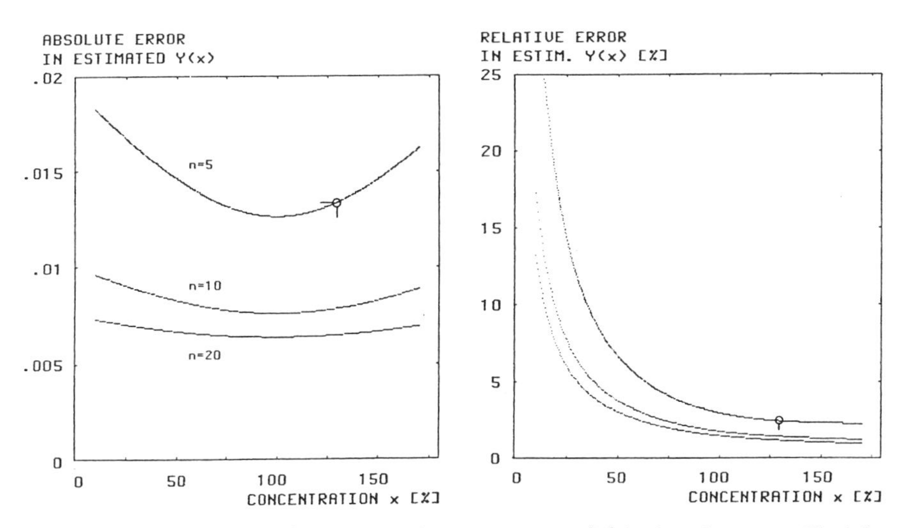

Figure 2.10. For several n (**5**, 10, resp. 20) the estimated CI(Y) is plotted versus x. The left figure shows the absolute values $|t \cdot s_Y|$, while the right figure depicts the relative ones, namely, $100 \cdot t \cdot s_Y/Y$ in percent. At $x = 130$ one finds $Y = 0.553$ with a CI of ± 0.013 ($\pm 2.4\%$, circles). It is obvious that it would be inopportune to operate in the region below about 90% of nominal if relative precision were an issue. There are two remedies in such a case: Increase n (and costs) or reduce all calibration concentrations by an appropriate factor, say 10%.

2.2.6. Interpolating $X(y)$

The quintessential statistical operation in analytical chemistry consists in estimating, from a calibration curve, the concentration of an analyte in an unknown sample. If the regression parameters a and b and the unknown's analytical response y^* are known, the most likely concentration is

$$X(y^*) = (y^* - a)/b; \qquad \text{see (2.19)}.$$

While it is useful to know $X(y^*)$, knowing the CL(X) or, alternatively, whether X is within the preordained limits, given a certain confidence level, is a prerequisite to interpretation (see Fig. 2.11). The variance and confidence intervals are

$$V_x = \frac{V_{\text{res}}}{b^2} \cdot \left\{1/n + 1/m + \frac{(y^* - \bar{y})^2}{b^2 \cdot S_{xx}}\right\}, \qquad \text{see (2.18)},$$

$$\text{CI}(X) = X \pm t(f, p) \cdot \sqrt{V_x}.$$

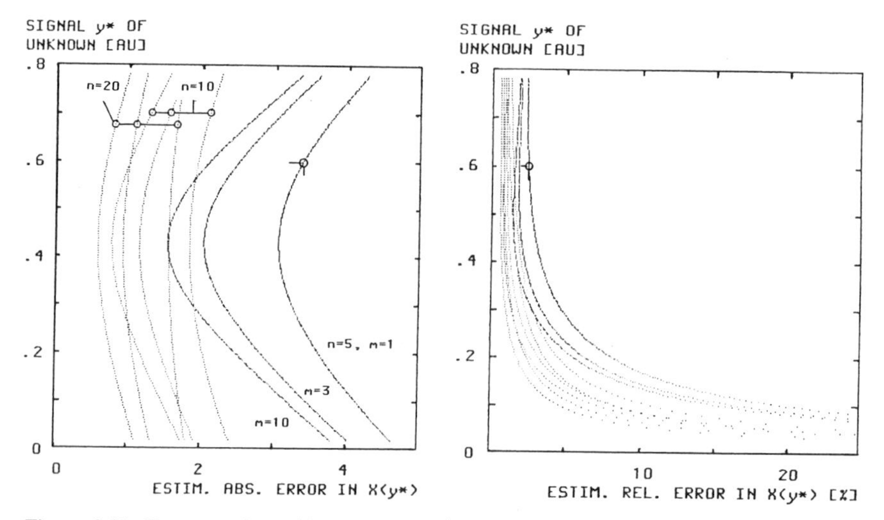

Figure 2.11. For several combinations of n (5, 10, resp. 20) and m (1, 3, resp. 10) the estimated CI(X) is plotted versus y^*. The left figure shows the absolute values $|t \cdot s_X|$, while the right figure depicts the relative ones, namely, $100 \cdot t \cdot s_X/X$ in percent. It is obvious that it would be inopportune to operate in the region below about 90% of nominal (in this particular case below $y = 0.36$) if relative precision were an issue. There are three remedies: Increase n or m (and costs) or reduce the calibration concentrations to shift the center of mass $(\bar{x}, \bar{y})$ below 100/0.42, respectively, increase the concentration of the test sample. At $y^* = 0.6$ and $m = 1$ (no replicates!) one finds $X = 141.4$ with a CI of ± 3.39 ($\pm 2.4\%$, circles).

Table 2.6. Estimation of $X(y)$.
[See also program LR, option $X(y)$. Beware: using rounded intermediate values can yield slightly different final results. See Tables 2.2 and 2.3 for raw data.]

$y^* = 0.445$ $\quad \bar{y} = 0.4284$ $\quad S_{xx} = 6250$ $\quad s_{res} = 0.00363$
$a = 0.0140$ $\quad b = 0.004144$ $\quad n = 5$ $\quad m = 1$

$$X(y^*) = (0.445 - 0.0140)/0.004144 = 104.0\% \text{ of nominal}$$

$$V_x = \frac{(0.003630)^2}{(0.004144)^2} \cdot \left\{ \frac{1}{5} + \frac{1}{1} + \frac{(0.445 - 0.4284)^2}{(0.004144)^2 \cdot 6250} \right\} = 0.924, \qquad \text{see (2.18)}$$

$$t \cdot s_x = 3.182 \cdot \pm \sqrt{0.924} = \pm 3.06$$

Confidence limits CL($X(y^*)$):
$t \cdot s_x = \pm 3.06$ for $m = 1$ independent measurement: 100.9 ⋯ 107.1
$t \cdot s_x = \pm 2.04$ for $m = 3$ independent measurements: 102.0 ⋯ 106.0
$t \cdot s_x = \pm 1.53$ for $m = 10$ independent measurements: 102.5 ⋯ 105.5

Example: (see Section 2.2.1): Assume that the measurement of a test article yields an absorbance of 0.445; what is the probable assay value? See Table 2.6.

The true value of $X(y^*)$ thus is found within the confidence limits 100.9/107.1 ($m = 1$), 102.0/106.0 ($m = 3$), resp. 102.5/105.5 ($m = 10$). These large confidence intervals imply that the result cannot be quoted other than 104%. The situation can be improved as follows: In the above equation the three terms in the parentheses come to 1/5, $1/m$, resp. 0.0026, that is 16.6, 83.2, and 0.2% of the total for $m = 1$. The third term is insignificant owing to the small difference between y^* and $\bar{y}$ and the large S_{xx}, so that an optimization strategy would have to aim at reducing $(1/n + 1/m)$, which is best achieved by increasing m to, say, 5. Thus the contributions would come to 49.7, 49.7, resp. 0.65% of the new total. Assuming $n = m$ is the chosen strategy, about $n = 16$ would be necessary to define $X(y^*)$ to ± 1 ($n = 60$: ± 0.5). Clearly, a practical limit becomes visible here that can only be overcome by improving s_{res} (better instrumentation and/or more skillful work). Evidently, knowing $X(y^*)$ but ignoring CL(X) creates the illusion of precision that is simply not there.

The CI(X) yields information as to which digit the result should be rounded to. As discussed in Sections 1.1.5 and 1.6, there is little point in quoting $X(y^*)$ to four significant digits and drawing the corresponding conclusions, unless the CI(X) is very small indeed; in the above example, one barely manages to demonstrate a difference between $X(y^*) = 104$ and the nominal value $X_n = 100$, so that it would be irresponsible to quote a single decimal place, and out of the question to write down whatever the calculator display indicates, say, 104.005792.

The CL(X) are calculated as given in Eqs. (2.18) and (2.19); a comparison with Eq. (2.16) reveals the formal equivalence: The expression $(y^* - \bar{y})/b$ is identical to $(x - \bar{x})$ and dividing s_{res} by b converts a measure of uncertainty in y to one in x.

The estimation of the intersection of two regression lines, as used in titrimetry, is discussed in Ref. 55.

A sin that is casually committed under routine conditions is to once and for all validate an analytical method at its introduction, and then to assume $a \equiv 0$, thus calculating $X(y^*)$ from the measurement of a reference y_R, and that of the sample y_S by means of a simple proportionality. The possible consequences of this are discussed in Ref. 56. The only excuse for this shortcut is the nonavailability of a PC; this approach will have to be abandoned with the increasing emphasis regulatory agencies are putting on statements of precision.

Formalizing a Strategy

What are the options the analyst has to increase the probability of a correct decision? V_{res} will be more or less given by the available instrumentation and the analytical method; an improvement would in most cases entail investments, a careful study to reduce sample workup related errors, and operator training. Also, specification limits are often fixed.

A general course of action could look like this:

(1) Assuming the specification limits SL are given (regulations, market, etc): postulate a tentative confidence interval CI(X) no larger than about SI/4; SI = specification interval; see Fig. 2.12.

(2) Draw up a list of all analytical methodologies that are in principle capable of satisfying this condition.

(3) Eliminate all methodologies that do not stand up in terms of selectivity, accuracy, linearity, etc.

(4) For all methodologies that survive step 3, assemble typical data, such as V_{res}, costs per test, etc.; for examples, see Refs. 13 and 57.

(5) For every methodology set up reasonable scenarios, that is, tentative analytical protocols with realistic n, m, S_{xx}, estimated time, and costs.

(6) Play with the numbers to improve CI(X) and/or cut costs.

(7) Drop all methodologies that impose impractical demands on human and capital resources: Many analytical techniques, while perfectly sound, will be eliminated at this stage because manpower, instru-

mentation, and/or scheduling requirements make them noncompetitive.

(8) If necessary, repeat steps 5–7 to find an acceptable compromise.

[Note concerning point 7: In the medium to long run it is counterproductive to install methodologies that work reliably only if the laboratory environment is controlled to unreasonable tolerances and operators have to acquire work habits that go against the grain. While work habits can be improved up to a certain point by good training (there is a cultural component in this), automation might be the answer if one does not want to run into GLP/GMP compliance problems. In the pharmaceutical industry, if compliance cannot be demonstrated, a product license might be revoked or a factory closed, at enormous cost.]

Step 6 above can be broken down as given in Table 2.7 below. If the hardware and its operation is under control, and some experience with

Table 2.7. Tactics for Improving a Calibration.

Tactic	Target	Costs	Example
Shift calibration points to reduce $(y^* - \bar{y})$, or increase S_{xx}	Reduce third term under parentheses in Eq. (2.18)	Organizational	Figs. 2.8 and 2.11
Increase n or m	Reduce terms 1 or 2 under parentheses in Eq. (2.18)	Time, material	Figs. 2.10 and 2.11
Improve skills	Decrease one component of V_{res}	Training, organizational, time	Reduce weighing errors, Fig. 4.8
Buy better hardware	Decrease other component of V_{res}	Capital investment	Better balances mechanical dilutors, better detector
Shift calibration points so the y range within which the interpolation will take place does not include values below about $0.9 \cdot \bar{y}$	Reduce interpolation error for low y^*	Organizational	Figs. 2.10 and 2.11
Do each test analysis a first time to obtain a rough estimate and repeat it with the optimal sample dilution	$y^* \approx \bar{y}$	Run a repeat analysis using nonstandard dilution scheme	Section 4.13

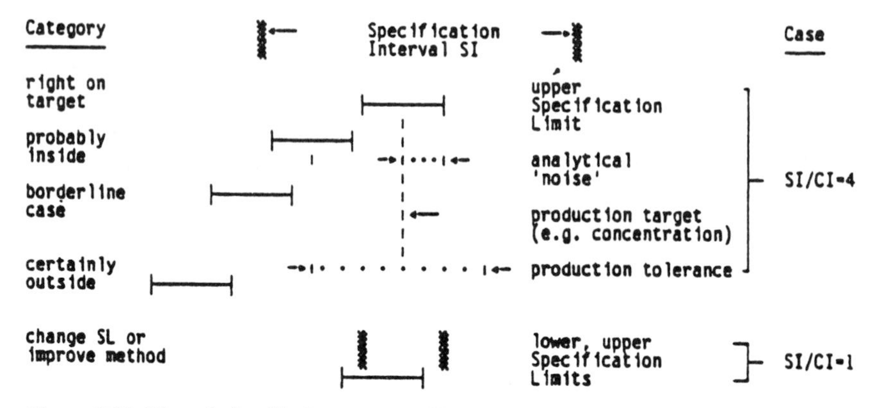

Figure 2.12. The relationship between specification interval SI and confidence intervals of the test results CI for a constant CI.

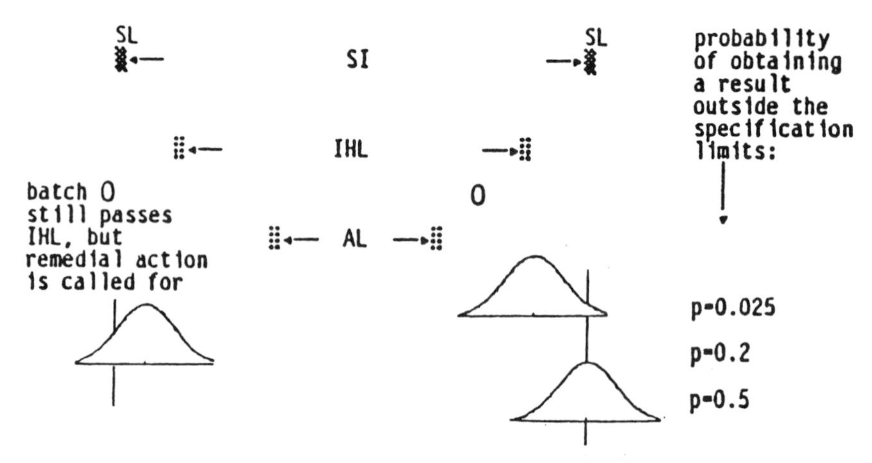

Figure 2.13. The distance between the upper in-house limit (IHL) and the upper specification (SL) limits is defined by the CL(X). The risk of the real value being outside the SL grows from negligible (measured value far inside IHL) to 50% (measured value on SL). Note that the definition of the upper in-house limit takes into account the details of the analytical method (n, x_i, m, V_{res}); this is the minimal separation between IHL and SL for an error probability for (μ > SL) of less than $p = 0.025$. The alarm limits (AL), as drawn here, are very conservative; when there is more confidence in man and machine, the AL will be placed closer to the IHL. IHL and AL need not be symmetrical relative to the SL.

similar problems is available, experiments need only be carried out late in the selection process to prove/disprove the viability of a tentative protocol. Laboratory work will earnestly begin with the optimization of instrumental parameters, and will continue with validation. In following such a simulation procedure, days and weeks of costly lab work can be replaced by hours or days of desk work.

As shown in Fig. 2.12, the specification/confidence interval ratio SI/CI is crucial to interpretation: While SI/CI ≥ 4 allows for distinctions, with SI/CI $= 1$ doubts will always remain. SI/CI $= 4$ is the minimum one should strive for when setting up specifications (provided one is free to choose) or when selecting the analytical method, because otherwise the production department will have to work under close to zero tolerance conditions as regards composition and homogeneity. Once a target CI(X) is given, optimization of experimental parameters can be effected as shown in Section 2.2.8. See Fig. 2.13.

Depending on the circumstances, the risk and the associated monetary value of being out of specifications might well be very high, so in-house limits IHL for X could be set that would guarantee that the risk of a deviation would be less than a given level.

It is evident that the distance between the in-house and the specification limits is influenced by the quality of the calibration/measurement procedure; a fixed relation, such as $2\sigma, 3\sigma$, as has been proposed for control charts, might well be too optimistic or too pessimistic (for a test result exactly on the 2σ in-house limit, the true value μ would have a 15% chance of being outside the 3σ SL). Also, it takes at least $n = 6$ (resp. $n = 11$) values to make a $z = 2$ ($z = 3$) even theoretically possible. Action limits (AL) can be identical with the in-house limits IHL, or even tighter. IHL are a quality assurance concept that reflects a mandated policy "we will play on the safe side," while AL are a production/engineering concept traceable to process validation and a concern to prevent downtime and failed batches. The three sets of limits need not be symmetrically placed with respect to the nominal value, or each other.

2.2.7. Limit of Detection

Analytical measurements are commonly performed in one of two ways:

- When sufficient amounts of sample are available one tries to exploit the central part of the dynamical range because the signal-to-noise ratio is high and saturation effects need not be feared (cf. Figs. 2.11 and 2.19). Assays of a major component are mostly done in this manner.

- Smaller concentrations (or amounts) are chosen for various reasons, for example, to get by with less of an expensive sample, or to reduce overloading an analytical system in order to improve resolution.

In the second case the limit of detection sets a lower boundary on the accessible concentration range.

Different concepts of limit of detection (LOD) have been advanced over the years,[58] the most well-known one defining the LOD as that concentration (or sample amount) for which the signal-to-noise ratio, SNR, $S/N = z$, with $z = 3 \cdots 6$ (Refs. 59–64). Evidently this LOD is

- Dependent only on baseline noise, N, and the signal, S;
- Independent of any calibration schemes; and
- Independent of heteroscedacity.

While the concept as such has profited the analytical community, the proposal of one or the other z value as being the most appropriate for some situation has raised the question of arbitrariness. This concept affords protection against Type I errors (concluding that an analyte is present when it is not), providing z is set large enough, but not against Type II errors (false negatives); cf. Sections 1.9 and 1.55.

There are a number of misconceptions about this popular index of quality of measurement; the correct use of the SNR is discussed in Ref. 65. The determination of the LOD in connection with transient signals is discussed in Ref. 66.

Another approach[67,68] is endorsed here for reasons of logical consistency, response to optimization endeavors, and easy implementation: The use of the confidence limit function. Figure 2.14 gives an (highly schematic) overview: The procedure is as follows:

- A horizontal is drawn through the upper confidence limit for the estimated intercept a, marked with a circle in Fig. 2.14.
- The intercept of the horizontal with the regression line defines the limit of detection, LOD, any value below which would be reported as "less than X_{LOD}."
- The intercept of the horizontal with the lower confidence limit function of the regression line defines the limit of quantitation, LOQ, any value above which would be quoted as "$X(y^*) \pm t \cdot s_x$."
- X values between LOD and LOQ should be reported as "$X(y^*)$."

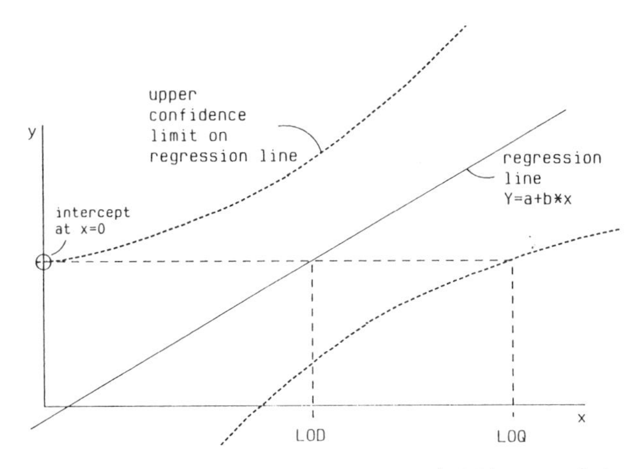

Figure 2.14. The definition of the limits of detection (LOD), respectively, quantitation (LOQ).

A linear regression program according to Sections 2.2 and 5.2.5 (menu item LOD) needs only the following additions:

- CL_u is obtained from Eqs. (2.16) and (2.17) for $x = 0$ using the add sign (+) (line 700).
- CL_u is inserted in Eq. (2.10) ($CL_u = a + b \cdot X_{LOD}$) to obtain X_{LOD} (line 2000).
- CL_u is inserted in Eqs. (2.18) and (2.19) with $M = \infty$, and using the + sign, to obtain X_{LOQ} (lines 2010, 2020, 730). See Table 2.8.

Table 2.8. Interpretation of Interpolations Near the LOD, as Implemented in Program LR, Option $X(y)$.

Example (see Section 2.2.2): $CL_u(a) = 0.0140 +$ $0.0155 = 0.0295$	$X_{LOD} = (0.0155)/0.004144 = 3.7$ X_{LOQ} is estimated as $3.7 + 3.6 = 7.3$ [Eq. (2.19) with $1/m = 0$ and $y^* = 0.0295$]

The results for three unknown samples ($x = 2.5, 5.2$, resp. 15.6) would be reported as

	Limits		
	LOD 3.7	LOQ 7.3	
Estimated result	2.5	5.2	15.6
Reported as	'< 3.7'	'5.2'	'15.6 ± 4.3 ($m = 1$)'

Pathological situations arise only when

(a) Slope b is close to zero and/or s_{res} is large, which in effect means the horizontal will not intercept the lower confidence limit function, and
(b) The horizontal intercepts the lower confidence-limit function twice; i.e., if n is small, s_{res} is large, and all calibration points are close together; this can be guarded against by accepting X_{LOQ} only if it is smaller than $\bar{x}$.

How stringent is this model in terms of the necessary signal height relative to the baseline noise? First, some redefinition of terms is necessary:

- "Signal" is replaced by the calculated analyte concentration X_{LOD} at LOD resp. LOQ.
- "Noise" is understood to mean the residual standard deviation expressed in abscissa units, $N_X = s_{res}/b$.

Since for this model a calibration scheme is part of the definition, the following practical case will be evaluated: For $n = 3 \cdots 50$ the lowest

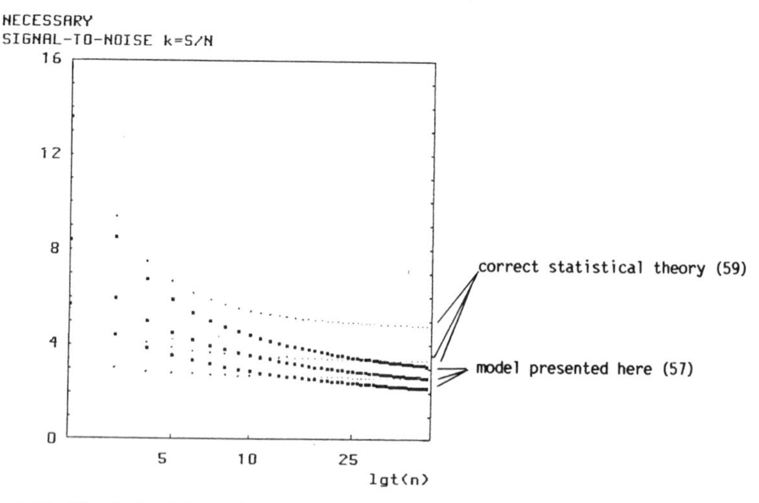

Figure 2.15. The limit of detection LOD: The minimum signal/noise ratio necessary according to two models (ordinate) is plotted against $\log_{10}(n)$ under the assumption of evenly spaced calibration points. The three sets of curves are for $p = 0.02$ (top), 0.05, and 0.1 (bottom). The correct statistical theory is given by the dotted curves,[69] while the model presented here is depicted with coarser dots.[67] The widely used $S/N = 3 \cdots 6$ models would be represented by horizontals at $y = 3 \cdots 6$.

concentration value will be at 50 and the highest at 150% of nominal (the expected concentration of the sample); all other points will be evenly spaced in between. In Fig. 2.15 the quotient signal/noise X_{LOD}/N_X is plotted versus the logarithm of n. It is quite evident that the limit of detection defined by the above quotient becomes smaller as n grows for constant repeatability s_{res}. The quotient for $p = 0.05$ is around 2.0 for the largest n,[59] and rises to over 7 for $n = 3$. These results are nearly the same if the x range is shifted or compressed. The obvious value of the model is to demonstrate the necessity of a thoughtful calibration scheme (cf. Section 4.13) and careful measurements when it comes to defining the LOQ/LOD pair. At large n the model is a bit more lenient than a very involved, correct statistical theory,[69] and for small n it is more demanding; see Fig. 2.15. This stringency is alleviated by redistributing the calibration points closer to the LOD. A comparison of various definitions of LOD/LOQ and the practical consequences thereof is given in Fig. 4.29 and Table 4.25.

2.2.8. Minimizing the Cost of Calibration

The traditional analyst depended on a few general rules of thumb for guidance while he coped with technical intricacies; his modern counterpart has a multitude of easy-to-use high-precision instruments at his disposal[70] and is under constant pressure to justify the high costs of the laboratory. With a few program lines it is possible to juggle the variables to often obtain an unexpected improvement in precision,[71–75] organization, or instrument utilization; cf. Table 2.7. A simple example will be provided here; a more extensive one is found in Section 4.3.

Assume that a simple measurement costs 20 monetary units; n measurements are performed for calibration and m for replicates of each of five unknown samples. Furthermore, a calibration series of n measurements must be paid for by the unknowns to be analyzed. The slope of the calibration line is $b = 1.00$ and the residual standard deviation is $s_{\mathrm{res}} = \pm 3$; cf. Refs. 31 and 57. The n calibration concentrations will be evenly spaced between 50 and 150% of nominal, e.g., $n = 4$: x_i: 50, 83, 117, 150. For an unknown corresponding to 130% of nominal, s_X should be below ± 3.3 units. What combination of n and m will provide the most economical solution? Use Eq. (2.18) for V_X:

$$S_{xx} = \sum (x_i - \bar{x})^2, \qquad \text{see (2.4)},$$

$$V_X = \frac{9}{1} \cdot \left\{ \frac{1}{n} + \frac{1}{m} + \frac{30^2}{S_{xx}} \right\} \leq 3.3^2, \qquad \text{see (2.18)}$$

$$c = 20 \cdot (n + 5 \cdot m), \qquad \text{costs for five unknowns + calibration.}$$

Table 2.9. Left: Costs in Monetary Units per Unknown Analyzed (c / 5) for Different Combinations of n and m. Right: Variances According to Eq. (2.18).

[The figures between the vertical bars pertain to conditions that do not satisfy the $V_X \leq 3.3^2 = 10.89$ criterion; the **bold** figures are for the most favorable case.]

<table>
<tr><th></th><th></th><th colspan="4">Monetary Units c</th><th colspan="4">Variances V_X</th></tr>
<tr><th>n</th><th>S_{xx}</th><th>$m = 1$</th><th>2</th><th>3</th><th>4</th><th>$m = 1$</th><th>2</th><th>3</th><th>4</th></tr>
<tr><td>3</td><td>5000</td><td>|32|</td><td>52</td><td>72</td><td>92</td><td>|13.6|</td><td>9.1</td><td>7.6</td><td>6.9</td></tr>
<tr><td>4</td><td>5578</td><td>|36|</td><td>56</td><td>76</td><td>96</td><td>|12.7|</td><td>8.2</td><td>6.7</td><td>6.0</td></tr>
<tr><td>5</td><td>6250</td><td>|40|</td><td>60</td><td>80</td><td>100</td><td>|12.1|</td><td>7.6</td><td>6.1</td><td>5.3</td></tr>
<tr><td>6</td><td>7000</td><td>|44|</td><td>64</td><td>84</td><td>104</td><td>|11.7|</td><td>7.2</td><td>5.7</td><td>4.9</td></tr>
<tr><td>7</td><td>7756</td><td>|48|</td><td>68</td><td>88</td><td>108</td><td>|11.3|</td><td>6.8</td><td>5.3</td><td>4.6</td></tr>
<tr><td>8</td><td>8572</td><td>|52|</td><td>72</td><td>92</td><td>112</td><td>|11.1|</td><td>6.6</td><td>5.1</td><td>4.3</td></tr>
<tr><td>9</td><td>9276</td><td>56</td><td>76</td><td>96</td><td>116</td><td>10.9</td><td>6.4</td><td>4.9</td><td>4.1</td></tr>
<tr><td>10</td><td>10260</td><td>60</td><td>80</td><td>100</td><td>120</td><td>10.7</td><td>6.2</td><td>4.7</td><td>3.9</td></tr>
</table>

Solution: Since S_{xx} is a function of the x values, and thus a function of n (e.g., $n = 4$: $S_{xx} = 5578$), solve the three equations in the given order for various combinations of n and m and tabulate the costs per result, i.e., $c/5$; then select the combination that offers the lowest cost at an acceptable variance, see Table 2.9.

Similar simulations were run for more or less than five unknown samples per calibration run. The conclusions are fairly simple: Costs per analysis range from 32 to 120 units; if no constraints as to precision were imposed, ($n = 3/m = 1$, underlined) would be the most favorable combination. In terms of the cost the combination ($n = 3/m = 2$) is better than ($n = 10/m = 1$) for up to six unknowns per calibration run.

This sort of calculation should serve as a rough guide only; nonfinancial reasoning must be taken into account, such as an additional safety margin in terms of achievable precision, or double determinations as a principle of GLP. Obviously, using the wrong combination of calibration points n and replicates m can enormously drive up costs. An alternate method is to incorporate previous calibrations with the most recent one to draw upon a broader data base and thus reduce estimation errors CI(X); one way of weighing old and new data is given in Ref.[76] (Bayesian calibration).

2.2.9. Standard Addition

A frequently encountered situation is that of no blank matrix being available for spiking at levels below the expected (nominal) level. The only recourse is to modify the recovery experiments above in the sense that the sample to be tested itself is used as a kind of blank, to which further analyte is spiked. This results in at least two measurements, namely, untreated sample and spiked sample, which can then be used to establish a calibration line from which the amount of analyte in the untreated sample is estimated. It is unnecessary to emphasize that linearity is a prerequisite for accurate results. This point has to be validated by repeatedly spiking the sample. The results can be summarized in two different ways.

The traditional manner of graphing standard addition results is shown in Fig. 2.16 (left): The raw (observed) signal is plotted vs. the amount of analyte spiked into the test sample; a straight line is drawn through the two measurements (if the sample was repeatedly spiked, more points will be available, so that a linear regression can be applied); the line is extrapolated to zero signal. A glance at Section 2.2.6 makes apparent that extrapolation, while perfectly legitimate, widens the confidence interval

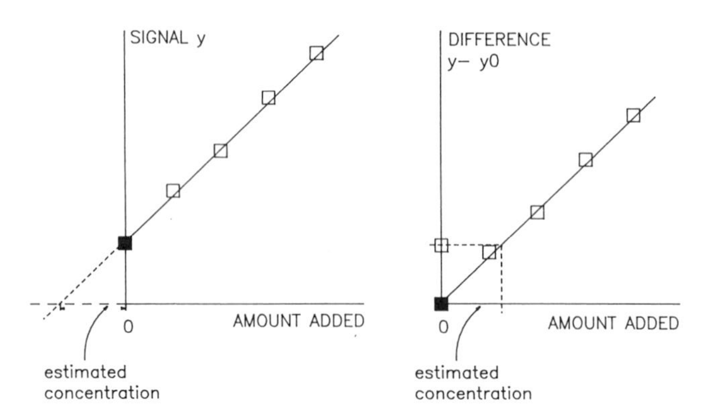

Figure 2.16. Depiction of the standard addition method: extrapolation (left), interpolation (right). The data and the numerical results are given in the example below.

around conc. = $-X(0)$, the sought result. This effect can be countered somewhat by higher spiking levels and thereby increasing S_{xx}: Instead of roughly estimating x and spiking the sample to $2 \cdot X(0)$, the sample is spiked to a multiple of concentration level $X(0)$. This strategy is successful only if spiking to such large levels does not increase the total analyte concentration beyond the saturation level (see Fig. 2.19).

Furthermore, there is the problem that the signal level to which one extrapolates need not necessarily be $y = 0$; if there is any interference by a matrix component, one would have to extrapolate to a level $y > 0$. This uncertainty can only be cleared if the standard addition line perfectly coincides with the calibration line obtained for the pure analyte in absence of the matrix, i.e., sample slope and 100% recovery; see also Fig. 2.20. This problem is extensively treated in Ref. 78–82. A modification is presented in Ref. 83.

Another approach to graphing standard addition results is shown in Fig. 2.16 (right): The signal for the unspiked test sample is marked off on the ordinate as before; at the same time, this value is subtracted from all spiked-sample measurements. The same standard addition line is obtained as in the left figure, with the difference that it passes through the origin $x = 0/y = 0$. This trick enables one to carry out an interpolation instead of an extrapolation, which improves precision without demanding a single additional measurement.

The trade-offs between direct calibration and standard addition are treated in Ref. 77. The same recovery as is found for the native analyte has to obtain for the spiked analyte; see Section 2.2.11 and Table 2.10.

Table 2.10. Alternative Interpretation of Standard Addition Data. [The minus signs in column 4 indicate that by extrapolation the negative concentration is obtained. The improvements possible by increasing the number of measurements and by changing from extrapolation to interpolation are evident. The application of spiking in potentiometry is reviewed in Refs. 84 and 85.]

Example: The test sample is estimated, from a conventional calibration, to contain the analyte in question at a level of about 0.8 mg/ml; the measured GC signal is 58376 area units.

Amount Spiked	Signal	Levels Used	Conventional Estimate	Alternative Estimate
0 mg/ml	58376			
1	132600	0-1*	(−)0.787*	0.787*
2	203670	0-2	(−)0.811 ± 48%	0.796 ± 33%
3	276410	0-3	(−)0.813 ± 10%	0.797 ± 8%

*Note that if only these two data points are used, confidence limits cannot be calculated.

2.2.10. Weighed Regression

In the previous sections of chapter 2 it was assumed that the standard deviation s_y obtained for a series of repeat measurements at a concentration x would be the same no matter which x was chosen; this concept is termed homoscedacity (homogeneous scatter across the observed range). Under certain circumstances the above assumption is untenable: One has to do with heteroscedacity, a very common form of which is presented in Fig. 2.17

The reasons for heteroscedacity can be manifold, for example:

(a) The relative standard deviation RSD (or c.o.v., coefficient of variation) is constant over the whole range, such as in many GC methods, that is, the standard deviation s_y is proportional to y.

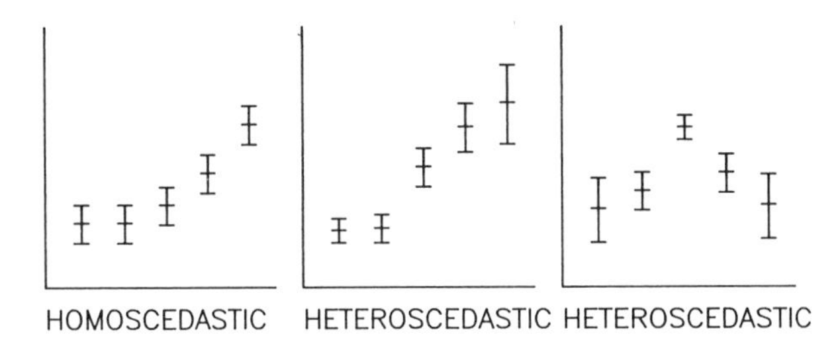

Figure 2.17. Schematic depiction of homo- (left) and heteroscedacity (right).

(b) The RSD is a simple function of y, as in isotope labeling work (RSD proportional to $1/\sqrt{y}$).
(c) The RSD is small in the middle and large near the ends of the linear range, as in photometry.

Curve fitting need not be abandoned in this case, but some modifications are necessary so that precisely measured points to a greater degree influences the form of the curve than a similar number of less precisely measured ones. Thus, a weighing scheme is introduced. There are different ways of doing this; the most accepted model makes use of the experimental standard deviation,[86,87] namely,

$$w_i = k \cdot \left\{ \frac{1}{s_y(x_i)} \right\}^2, \qquad \text{with} \qquad \sum w_i = n. \tag{2.26}$$

How is one to obtain the necessary s_y values? There are two ways:

(1) One performs so many repeat measurements at each concentration point that standard deviations can be reasonably calculated, e.g., as in validation work; the statistical weights w_i are then taken to be inversely proportional to the local variance, as given above.
(2) One roughly models the variance as a function of x using the data that are available[87–89]. The standard deviations are plotted versus the concentrations and if any trend is apparent, a simple curve is fitted, for example,

$$s_y = a_s + b_s \cdot x \qquad \text{see (2.27)}$$

As more experience is accumulated, this relation can be modified.[90,91] It is important to realize that for the typical analytical application (with relatively few measurements well characterized by a straight line) a weighing scheme has little influence on slope and intercept, but appreciable influence on the confidence limits of interpolated $X(y)$ resp. $Y(x)$.

The step-by-step procedure for option (2) is nearly the same as for the standard approach above [for option (1) only Eqs. (2.27) and (2.29) have to be appropriately modified to include the experimental values]:

(1) $$s(x_i) = a_s + b_s \cdot x_i \quad \text{(model of SD)}, \qquad (2.27)$$

(2) $$k = n \Big/ \sum V(x_i)^{-1} \quad \text{(normalization factor)}, \qquad (2.28)$$

(3) $$w_i = k/V(x_i), \quad \text{(statistical weight)}, \qquad (2.29)$$

(4) $$\bar{x}_w = \sum (w_i \cdot x_i)/n \quad \text{(weighed summation)}, \qquad (2.30)$$

(5) $$\bar{y}_w = \sum (w_i \cdot y_i)/n \quad \text{(weighed summation)}, \qquad (2.31)$$

(6) $$S_{xx,w} = \sum \left(w_i \cdot (x_i - \bar{x}_w)^2 \right) \quad \text{(weighed } S_{xx}\text{)}, \qquad (2.32)$$

(7) $$S_{yy,w} = \sum \left(w_i \cdot (y_i - \bar{y}_w)^2 \right) \quad \text{(weighed } S_{yy}\text{)}, \qquad (2.33)$$

(8) $$S_{xy,w} = \sum (w_i \cdot (x_i - \bar{x}_w) \cdot (y_i - \bar{y}_w)) \quad \text{(weighed } S_{xy}\text{)}. \qquad (2.34)$$

All further calculations proceed as under Section 2.2.1, Standard Approach.

The following numerical **example** demonstrates the difference between a weighed and an unweighed regression: four concentrations, two measurements per concentration. Data: $(x_i/y_1/y_2)$:

10	462.7	571.3
20	1011	1201
30	1419	1988
40	2239	2060

For the weighed regression the standard deviation was modeled as $s(x) = 100 + 5 \cdot x$; this information stems from experience with the analytical technique. See Table 2.11.

Conclusions: The residual standard deviation is somewhat improved by the weighing scheme; note that the coefficient of determination gives no clue as to the improvements discussed below. In this specific case, weighing actually worsens the relative confidence interval associated with the slope b. However, because the smallest absolute standard deviations $s(x)$

Table 2.11. Raw Data and Intermediate Results, Weighed Regression Analysis. (Because of the round numbers, the statistical sums for unweighed regression are unburdened by decimals, whereas at right, the introduction of the weighing factor k leads to additional significant digits.)

Comparison of Results			
Variable	Unweighed Regression	Weighed Regression	Eq./Fig.
k	1	41'427	(2.28)
n	8	8	
$\bar{x}$	25.0	19.36	(2.30)
$\bar{y}$	1369.0	1053.975	(2.31)
S_{xx}	1000.0	865.723	(2.32)
S_{yy}	3'234'135	2'875'773.6	(2.33)
S_{xy}	54'950	48'554.81	(2.34)
V_{res}	35'772	25'422.53	(2.13)
s_{res}	± 189	± 159.44	(2.13)
r^2	0.9336	0.9470	(2.1)
a	-4.75	-31.6	(2.12)
rel CI(a)	$\pm 8436\%$	$\pm 922\%$	(2.15)
b	54.95	56.09	(2.11)
rel (CI(b)	26.6%	23.6%	(2.14)
LOD	7.3	5.19	Fig. 2.14
LOQ	13.0	9.35	Fig. 2.14

are found near the origin, the center of mass $\bar{x}/\bar{y}$ moves towards the origin and the estimated limits of detection resp. quantitation, LOD resp. LOQ, are improved. The interpolation $Y = f(x)$ is improved for the smaller x values, and worse for the largest x values. The interpolation $X = f(y^*)$, here given for $m = 1$, is similarly influenced, with an overall improvement. The largest residual has hardly been changed, and the contributions at small x have increased. This example shows that weighing is justified, particularly when the poorly defined measurements at $x = 30$ and 40 were just added to better define the slope, and interpolations are planned at low y^* levels. Note that the CL of the interpolation $X = f(y^*)$ are much larger ($m = 1$) than what one would expect from Fig. 2.18; increasing m to, say, 10 already brings an improvement by about a factor of 2. The same y^* values were chosen for the weighed as well as for the unweighed regression to show up the effect of the weighing scheme on the interpolation. The weighed regression, especially the CL(X), give the best indication of how to dilute the samples; although the relative CL(X) for

Table 2.11 (Continued)
The Sum of the Weighed Residuals is 0.

Interpolations:	Unweighed	Weighed
$x = 10$	$Y = 545 \pm 274$ (50%)	529 ± 185 (35%)
$x = 20$	$Y = 1094 \pm 179$ (16%)	1090 ± 138 (13%)
$x = 30$	$Y = 1644 \pm 179$ (11%)	1651 ± 197 (12%)
$x = 40$	$Y = 2193 \pm 274$ (12%)	2212 ± 306 (14%)
$y = 544.8$, $M = 1$	$X = 10.0 \pm 9.8$ (98%)	10.3 ± 7.7 (75%)
$y = 1094$, $M = 1$	$X = 20.0 \pm 9.0$ (45%)	20.1 ± 7.4 (37%)
$y = 1644$, $M = 1$	$X = 30.0 \pm 9.0$ (30%)	29.9 ± 7.8 (26%)
$y = 21.93$, $M = 1$	$X = 40.0 \pm 9.8$ (24%)	39.7 ± 8.8 (22%)

Contribution of the Points towards the Total Variance:

Unweighed			Weighed		
w_i	r_i	%V	w_i	r_i	%V
1	−82.0	3.1	1.84	−66.6	5.4
1	26.5	0.3	1.84	42.0	2.1
1	−83.2	3.2	1.04	−79.1	4.3
1	107	5.3	1.04	111	8.4
1	−225	23.5	0.66	−232	23.4
1	344	55.2	0.66	337	49.4
1	45.8	1.0	0.46	27.2	0.2
1	−133	8.3	0.46	−152	7.0
8	0.0	100.0	8	= 0.0	100.0

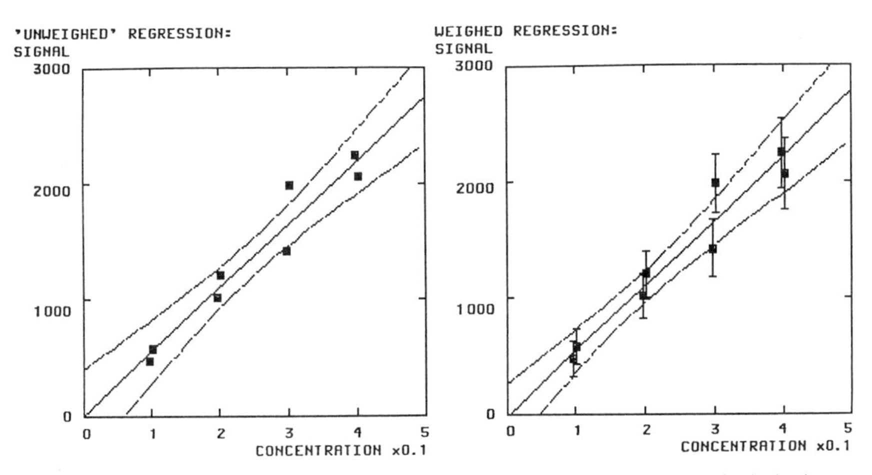

Figure 2.18. The data from the above example is analyzed according to Eqs. (2.4)–(2.6) and (2.10)–(2.17), left, resp. Eqs. (2.27)–(2.34) and (2.10)–(2.17) right; the effect of the weighing scheme on the center of mass and the confidence limits is clearly visible.

the unweighed regression are smallest for large signals y^*, account must be taken that it is exactly these signals that have the largest uncertainty: $s_y = 100 \pm 5 \cdot x$; if there is marked heteroscedacity, the unweighed regression is a poor model.

2.2.11. Validating an Analytical Method

By far the majority of all analytical methods in one way or another make use of the linear relationship presented in Section 2.2. For this reason a section covering the various aspects of method validation and their statistical implications is inserted here.

Demonstrating the validity of an analytical procedure is a GLP/GMP concern[2, 13, 25–27, 92–95]; some of the key terms encountered are calibration and control, selectivity, ruggedness, system suitability tests, and error recovery protocol. Those that relate to statistical data treatment are discussed in more detail.

Calibration and *control* means appropriate standards as well as positive and negative control samples are available, and their use to secure control over the methodology established. Control samples are blinded, if possible, and pains are taken to achieve stability over months to years so that identical control samples can be run with calibrations that are far removed in time or location from the initial method development site.

Selectivity means that only that species is measured which the analyst is looking for. A corollary is the absence of interferences. A lack of selectivity is often the cause of nonlinearity of the calibration curve.

Accuracy is the term used to describe the degree of deviation (bias) between the (often unknown) true value and what is found by means of a given analytical method. Accuracy cannot be determined by statistical means; the test protocol must be devised to include the necessary comparisons (blanks, other methods).

Ruggedness is achieved when small deviations from the official procedure, e.g., a different make of stirrer or similar-quality chemicals from another supplier are tolerable and when the stochastic variation of instrument readings does not change the overall interpretation upon duplication of an experiment. A nonrugged method leaves no margin for error and thus is of questionable value for routine use in that it introduces undue scatter and bias in the results. Ruggedness is particularly important when the method must be transferred to other hands, instruments, or locations.

System suitability tests are to be conducted, if necessary, every time an analysis method is installed, in order to ensure that meaningful results are generated. Example: Make certain that aging and contamination have not degraded an HPLC column to the point where the required resolution is no longer attained; it is futile to apply statistical methods to results of doubtful quality.

Error recovery protocols are standard operating procedures that tell the analyst what measures are to be taken in case of unexpected results. Example: "if the above test did not yield a significant difference, continue with the measurements according to Procedure A, else recalibrate using Procedure B."

Linearity appeals to mind and eye and makes for easy comparisons. Fortunately, it is a characteristic of most analytical techniques to have a linear signal-to-concentration relationship over at least a certain concentration range[96, 97], in some instances a transformation might be necessary for one axis (e.g., logarithm of ion activity) to obtain linearity. At both ends there is a region where the calibration line gradually merges into a horizontal section; at the low-concentration end one normally finds a baseline given by background interference and measurement noise. The upper end of the calibration curve can be abrupt (some electronic component reaches its cutoff voltage, higher signals will be clipped), or gradual (various physical processes begin to interfere). As long as there is no disadvantage associated with it, an analyst will tend toward using the central part of the linear portion for quantitative work. This strategy serves well for routine methods: A few calibration points will do and interpolation is straightforward. See Figure 2.19.

It should be noted that nonlinear calibration curves are not forbidden, but they do complicate things quite a bit: More calibration points are necessary, and interpolation from signal to concentration is often tedious. It would be improper to apply a regression of concentration x on signal y to ease the calculational load; cf. Section 2.2.1, because all error would be assigned to the concentrations, and the measured signals would be regarded as relatively free of uncertainty. The authors are aware of AAS

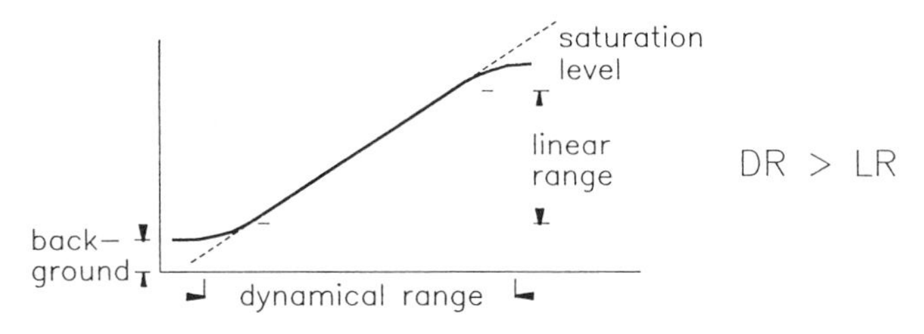

Figure 2.19. Definition of linear (LR) and dynamic (DR) ranges. The DR is often given as a proportion, i.e., 1 : 75, which means the largest and the smallest concentrations that could be run under identical conditions would be different by a factor of 75.

equipment that offers the user the benefit of so-called direct concentration readout of unknown samples, at the price of improper statistical procedures and/or curve-fitting models that allow for infinite or negative concentrations under certain numerical constellations.[55] Unless there is good experimental evidence for or sound theoretical reasoning behind the assumption of a particular nonlinear model, justification is not easy to come by. A good example for the use of the nonlinear portion near the detection limit is found in an interference-limited technique, namely, the ion-selective electrode. An interpolation is here only possible by the time-honored graphical method, or then by first fitting a moderately complex nonlinear theoretical model to the calibration data and then iteratively finding a numerical solution. In both cases, in linear regression, the preferred option where appropriate, and in curve fitting, the model is best justified by plotting the residuals $r_i = y_i - Y(x_i)$ versus x_i (Section 2.2.3) and discussing the evidence.

The *reproducibility* characterizes the degree of control exerted over the analytical method. A well-designed standard operating procedure permits one to repeat the sampling, sample workup, and measurement process and repeatedly obtain very similar results. As discussed in Sections 1.1.3 and 1.1.4, the absolute or relative standard deviation calculated from experimental data is influenced by a variety of factors, some of them beyond the control of the analyst.[57,71,75]. Thus there is no one agreed-upon relative standard deviation to judge all methods and techniques against (example: $\pm 0.5\%$ is very good, $\pm 3\%$ is acceptable, $\pm 5\%$ is insufficient); the analyst will have to assemble from different sources a notion of what constitutes an acceptable RSD for his or her particular problem. The residual curvature will also have to be judged against this value.

Recovery is a measure of the efficiency of an analytical method, especially the sampling and sample workup steps, to recover and measure the analyte spiked into a blank matrix (also called a placebo in the case of a pharmaceutical dosage form containing no active principle). Note, even an otherwise perfect method suffers from lack of credibility if the recovered amount is much below what was added; the rigid connection between content and signal is then severed.

Recovery is best measured by adding equal and known amounts of analyte to (a) the solvent and (b) the blank matrix from which the analyte is to be isolated and determined. The former samples establish a reference calibration curve. The other samples are taken through the whole sample workup procedure and yield a second calibration curve. If both curves coincide recovery is 100% and interference is negligible. A difference in slope shows the lack of extractive efficiency or selectivity. Problems crop up, however, if the added standard does not behave the same way as the

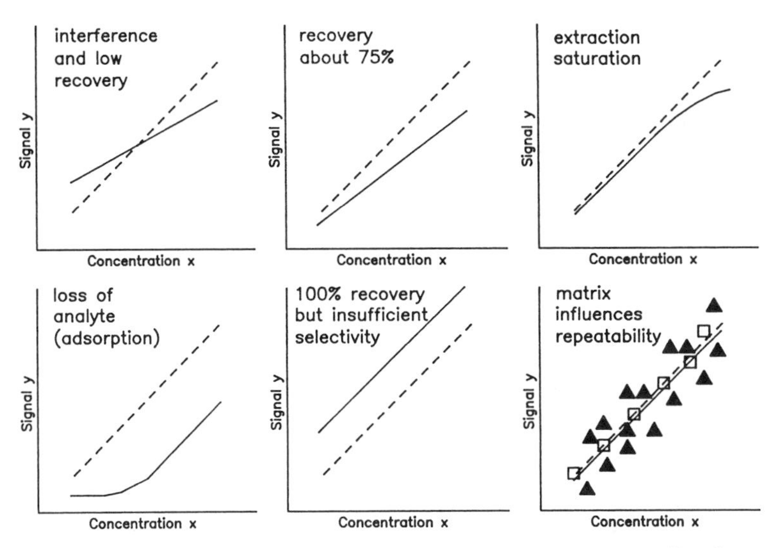

Figure 2.20. Schematic representation of some of the effects on the calibration curve observed during validation.[98, 99] The dotted line represents the reference method or laboratory, the solid line the test method or new laboratory.

analyte that was added at the formulation stage of the product. Examples of such unexpected matrix effects are the slow adsorption of the analyte into packaging material or the complexation of the analyte with a matrix component under specific conditions during the manufacturing process that then leave the analyte in a frozen equilibrium from which it cannot be recovered. A series of schematic graphs shows some of the effects that are commonly observed (combinations are possible); see Fig. 2.20.

Data acquisition is not treated in this book. The most common technique is to convert a physico-chemical signal into a voltage by means of a sensor, and feed the electrical signal into a digital voltmeter or a chart recorder. With today's instrumentation this is no longer the problem it used to be.

Data treatment is quite a different subject, though. While operations such as the calculation of the area under a peak are widely used, the availability of an integrator is no guarantee for correct results. One has to recognize that the algorithms and the hardware built into such machines were designed around a number of more or less implicit assumptions, such as the form of chromatographic signals.[9, 11] It should not come as a surprise then, that analytical methods that were perfectly well behaved on one instrument configuration will not work properly if only one element, such as an integrator, is replaced by one of a different make. The same is true, of course, for software packages installed on PCs (e.g., use of single- or double-precision math as a hidden default option). The signal path from detector up to the hard copy output of the final results must be perceived as a chain of error-prone components: There are errors due to conception,[55] construction, installation, calibration, and (mis)use. Method validation checks into these aspects, linear regression being an important link.

2.2.12. Intersection of Two Linear Regression Lines

There are a number of analytical techniques that rely on finding the point on the abscissa where two straight-line segments of an analytical signal intersect, e.g., titration curves. The signal can be any function, $y = f(x)$, such as electrical potential versus amount added, that changes slope when, e.g., a species is consumed. The evaluation proceeds by defining a series of measurements $y_{i1} \cdots y_{i2}$ before and another series $y_{j1} \cdots y_{j2}$ after the break point, and fitting linear regression lines to the two segments. Finding the intersection of the regression lines is a straightforward exercise in algebra. There are several models for finding the confidence limits on the intersection.[100]

2.3. NONLINEAR REGRESSION

Whenever a linear relationship between dependent and independent variables (ordinate resp. abscissa values) obtains, the straightforward linear regression technique is used: The equations make for a simple implementation, even on programmable calculators. The situation changes drastically when curvature is observed[17]:

(1) Many more measurements are necessary, and these have to be carefully distributed over the x range to ensure optimally estimated coefficients.

(2) The right model has to be chosen; this is trivial only when a well-proven theory foresees a certain function $y = f(x)$. Constraints add a further dimension.[101] In all other cases a choice must be made between approaches (3) and (4).

(3) If the graph y vs. x suggests a certain functional relation, there are often several alternative mathematical formulations that might apply, e.g., $y = \sqrt{x}$, $y = a \cdot (1 - \exp(b \cdot (x + c)))$, and $y = a \cdot (1 - 1/(x + b))$: Choosing one over the others on sparse data may mean faulty interpretation of results later on.[102] An interesting example is presented in Ref. 55 (cf. 2.3.1). An important aspect is whether a function lends itself to linearization (see 2.3.1), to direct least-squares estimation of the coefficients, or whether iterative techniques need to be used.

(4) In some instances all one is interested in is an accurate numerical representation of data, without any intent of physico-chemical interpretation of the estimated coefficients: A simple polynomial might suffice, e.g., the approximation to tabulated values, Chapter 5.

(5) An interpretation of the inverse, $x = f^{-1}(y)$, of the chosen function $y = f(x)$ might not yield more than X; without confidence limits such a result is nearly worthless.[102]

The linearization technique mentioned under (3) is treated in the next section.

2.3.1. Linearization

Linearization is here defined as one or more transformations applied to the x and/or y coordinates in order to obtain a linear y' vs. x' relationship for easier statistical treatment. One of the more common transformations is the logarithmic one; it will nicely serve to illustrate some pitfalls.

Two aspects—wanted or unwanted—will determine the usefulness of a transformation:

(a) Individual coordinates (x_i/y_i) are affected so as to eliminate or change a curvature observed in the original graph.

(b) Error bars defined by the confidence limits $CL(y_i)$ will shrink or expand, most likely in an asymmetric manner. Since we here presuppose near absence of error from the abscissa values, this point applies only to y transformations. A numerical example: 17 ± 1 ($\pm 5.9\%$, symmetric CL), upon logarithmic transformation becomes $1.23045 - 0.02633/+0.02482$ (-2.1, $+2.0\%$, asymmetric).

Both aspects are combined in Fig. 2.21; see also Table 2.12.

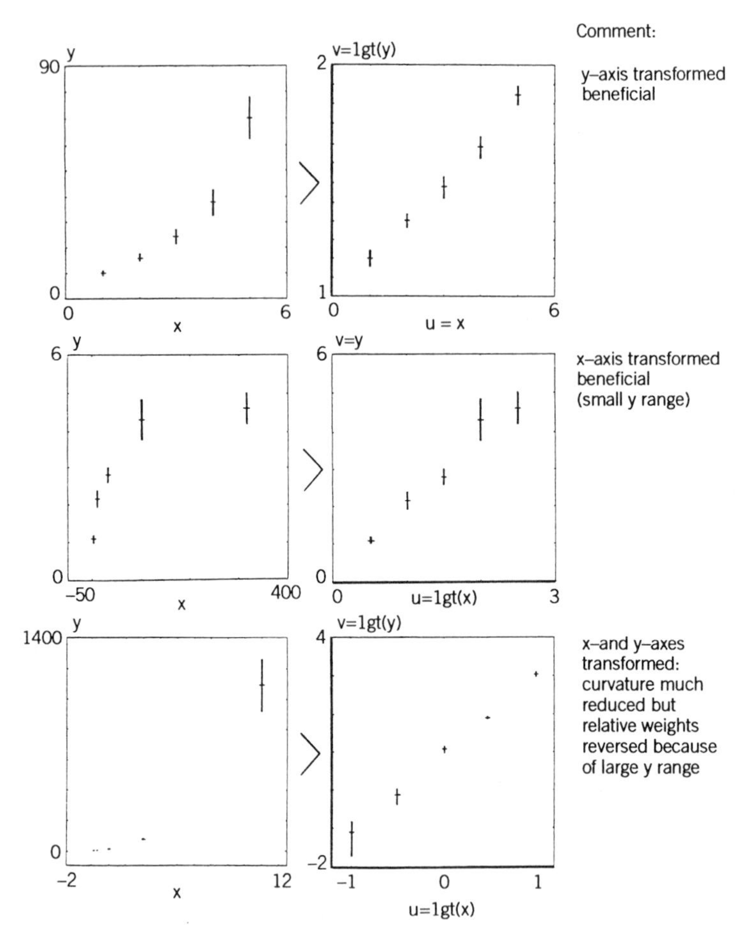

Figure 2.21. Logarithmic transformations on x- or y-axes as used to linearize data. Notice how the confidence limits change in an asymmetric fashion. In the top row, the y axis is transformed; in the middle row, the x axis is transformed; in the bottom row, both axes are transformed simultaneously.

2.3.2. Nonlinear Regression and Modeling

Many functional relationships do not lend themselves to linearization; the user has to choose either option (3) or (4) above to continue.

Option (4) (multilinear regression, polynomials[102, 103]) is uncomplicated, insofar as clear-cut procedures for finding the equation's coefficients exist. Today, linear algebra (use of matrix inversion) is most commonly em-

Table 2.12. Logarithmic Transformations.

[The linear coordinates are **x** resp. **y**, the logarithmic ones **u**, resp. **v**. Regression coefficients established for the lin/lin plot are **a**, **b**, whereas those for the transformed coordinates are **p**, **q**. Note that the intercept **p** in the y transformed graph becomes a multiplicative preexponential factor in the original non- (resp. back-) transformed graph and that functions always intersect the ordinate at $y = 10^p$. A straight line in logarithmic coordinates, if the intercept **p** is not exactly zero, will become an exponential function with intercept 10^p after back-transformation. Since double-logarithmic transformations are often employed to compress data, cf. GC-FID response over a 1 : 1000 dynamic range, statistical indistinguishability of two such transformed response functions must not be interpreted as an indication of identity: For one, any straight line in a lin/lin plot takes on a slope of 1.000 in a log/log plot, and any difference between intercepts **p**, however small, translates into two different slopes in the original plot, while the intercept **a** is always zero. The logarithmic transformations take on the following algebraic form: $u = \text{lgt}(x)$, $v = \text{lgt}(y)$.]

Transformed Axis	lin/lin Presentation	Transformation	Logarithmized Presentation
x axis	$y = a + b \cdot x$	$u = \text{lgt}(x) \Rightarrow$	$y = a + b \cdot 10^u$
	$y = p + q \cdot \text{lgt}(x)$	$\Leftarrow x = 10^u$	$y = p + q \cdot u$
y axis	$y = a + b \cdot x$	$v = \text{lgt}(y) \Rightarrow$	$10^v = a + b \cdot x$
	$y = 10^p \cdot (10^x)^q$	$\Leftarrow y = 10^v$	$v = p + q \cdot x$
Both	$y = a + b \cdot x$	$u = \text{lgt}(x) \Rightarrow$	$10^v = a + b \cdot 10^u$
axes		$v = \text{lgt}(y) \Rightarrow$	
	$y = 10^p \cdot x^q$	$\Leftarrow x = 10^u$	$v = p + q \cdot u$
		$\Leftarrow y = 10^v$	

ployed; there even exist cheap pocket calculators that are capable of solving these problems. Things do become quite involved and much less clear when one begins to ponder the question of the correct model to use. First, there is the weighing model: The least-squares approach, see Eq. (2.3)[104], is implicit in most commercially available software. Weighing can be intentionally introduced in some programs, and in others there is automatic calculation of weights for grouped data (more than one calibration point for a given concentration x). Second, there is the fitted model: Is one to choose a polynomial of order 5, 8, or even 15? Or, using a multilinear regression routine, should the terms x^2, y^2, xy, or the terms x^2y and x^2y^2 be introduced[102]? If one tries several alternative fitting models, how is one to determine the optimal one? From a purely statistical point of view, the question can be answered. There exist powerful program packages that automatically fit hundreds of models to a data set and rank-order them according to goodness of fit, a sort of hit parade of mathematical functions. So far, so good, but what is one to say if for a series of similar experiments the proposed best models belong to completely different mathematical functionalities, or functions are proposed that are at odds with the observed processes (example: diffusion is an exponential process and should not be described by fractional functions)? Practicability demands that the model be kept as simple as possible. The more terms (coefficients) a model includes, the larger the danger that a perfect fit will entice the user into perceiving more than is justified. Numerical simulation is a valuable tool for shattering these illusions: The hopefully good model and some sensible assumptions concerning residual variance are used to construct several sets of synthetic data that are similar to the experimental data (Monte Carlo technique; see programs SIMGAUSS and SIMILAR); these data sets are analyzed according to the chosen fitting model(s). If the same model, say a polynomial of order 3, is consistently found to best represent the data, the probability of a wise choice increases. However, if the model's order of one or more of the coefficients are unstable, then the simplest model that does the job should be picked. For a calculated example see Section 3.4.

Option (3) (arbitrary models) must be viewed in a similar light as option (4), with the difference that more often than not no direct procedure for estimating the coefficients exist. Here, formulation of the model and the initial conditions for a sequential simplex search of the parameter space are a delicate matter. The simplex procedure[105–107] improves on a random or systematic trial-and-error search by estimating, with a minimal set of vertices (points, experiments), the direction of steepest decent (toward lower residual variances), and going in that direction for a fixed (classical simplex) or variable (Fletcher-Powell, or similar algorithms) distance, retaining the best vertices (e.g., sets of estimated coefficients), and repeating until a constant variance, or one below a cutoff criterion is found. The system, see Fig. 2.23, is not foolproof; plausibility checks and graphics are an essential aspect; see Section 4.2. Even if an arbitrary model is devised that permits direct calculation of its coefficients, this is no guarantee that such a model will not break down under certain conditions and produce nonsense; this can even happen to the unsuspecting user of built-in, unalterable firmware.[55] The limit of quantitation in the nonlinear case is discussed in Ref. 108.

2.4. MULTIDIMENSIONAL DATA

When confronted with multidimensional data it is easy to plug the figures into a statistical package and have nice tables printed that purportedly accurately analyze and represent the underlying factors. Have the questions been asked:

- Does the model conform to the problem?
- Is the number of factors meaningful?
- Is their algebraic connection appropriate?

The point made here is that the data can be forced to fit nearly any model; the more factors (parameters) and the higher the order of the independent variable(s) (x, x^2, x^3, etc.) the better the chance of obtaining near-zero residuals and a perfect fit. Does this make sense, statistically or chemically?

It is proposed to visualize the data before any number crunching is applied, and to carefully ponder whether an increase in model complexity is necessary or justified. The authors abstain from introducing more complicated statistics such as multiple linear regression, principal component analysis, or partial least squares[109] because much experience is necessary to correctly use them. Since the day the computer invaded the laboratory, publications have appeared that feature elegant, and increasingly abstract, recipes for extracting results from a heap of numbers; only time will tell whether many of these concepts are generally useful.

2.4.1. Visualizing Data

The reader may have guessed from previous sections that graphical display contributes much toward understanding the data and the statistical analysis. This notion is correct, and graphics become more important as the dimensionality of the data rises, especially to three and more dimensions. Bear in mind:

- The higher the dimensionality, the more acute the need for a visual check before statistical programs are indiscriminately applied.
- The higher the dimensionality, the harder it becomes for humans to grasp the situation; with color coding and pseudo-three-dimensional displays, four dimensions can just be managed.
- Fancy graphics programs help, but most are oriented toward presenting marketing or financial information.

The reader is urged to try graphics before using mathematics for reasons that will become evident in the example of Table 2.13.

A bar chart plotted with SYMPHONY™ is shown in Fig. 2.22 (left); the line with the dot on top of each bar depicts one standard deviation.

Table 2.13. Multidimensional Data.

[The data are for batches A, B, C, and D, and the calculated means respectively standard deviations ($\times 100$) are for methods 1, 2, 3, and 4. The averages for both rows and columns are given. The lowest mean in each row is given in **bold**. The overall mean is 99.25; relative to this, the overall SD is ± 0.245, which includes the average repeatability (± 0.095) and a between-group component of ± 0.226. Because $n = 6$, $t(p = 0.05, f = 5)/\sqrt{6} = 1.04$, which means the standard deviation is nearly equal to the confidence interval of the mean.]

Four batches A, B, C, and D of an amino acid hydrochloride were investigated; four different titration techniques were applied to every sample:
(1) Direct titration of $^{+}H\text{-}OOC\text{-}R\text{-}N^{+}R_1R_2R_3 \cdot Cl^{-}$ with NaOH
(2) Indirect titration using an excess of NaOH and back-titration to the first,
(3) respectively second equivalence point, and
(4) titration of the chloride.
Six repeat titrations went into every mean and standard deviation listed (for a total of $4 \times 4 \times 6 = 96$ measurements):

	Calculated Means					Calculated SDs × **100**				
Method	1	2	3	4	av.	1	2	3	4	av.
Batch A	**99.03**	99.16	99.27	99.41	99.22	±7	±2	±12	±2	±6.6
B	99.45	99.34	**99.06**	99.35	99.42	±2	±1	±17	±15	±10.6
C	**99.17**	99.55	99.42	99.55	99.42	±11	±7	±10	±7	±8.3
D	99.04	99.36	**98.58**	99.20	99.30	±10	±7	±7	±14	±9.3
av.	99.17	99.35	99.33	99.38	**99.25**	±7.7	±4.7	±11.2	±10.2	±**9.5**

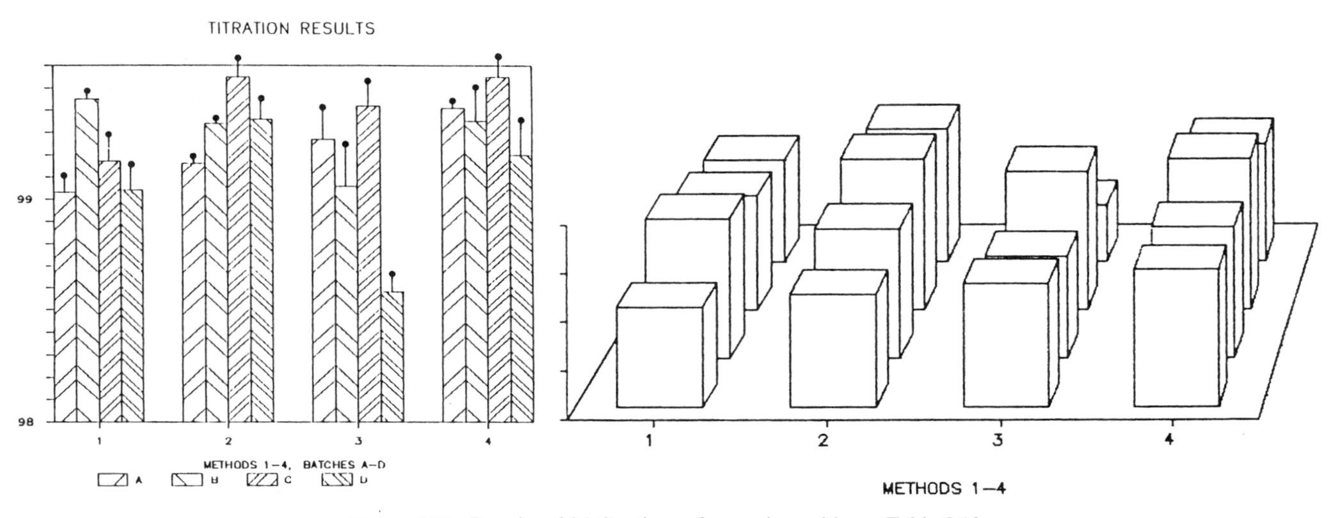

Figure 2.22. Results of 96 titrations of an amino acid; see Table 2.13.

The same information is given in a pseudo-three-dimensional representation in Fig. 2.22 (right). In both figures the ordinate spans the range 98–100%. Especially the latter figure ought to make one think before calculations are started:

(1) It is easy to prove that there are differences between two given means using the t test (case b_1 or c).
(2) A one-way (simple) ANOVA with six replicates can be conducted by either regarding each titration technique or each batch as a group, and looking for differences between groups.
(3) A two-way ANOVA (not discussed here) would combine the two approaches under (2).

Finding differences in one aspect of the problem; the second is to integrate problem-specific chemical know-how with statistics:

(a) The chloride titration seems to give the highest values, and the direct titration the lowest. Back-titration to the first equivalence point appears to be the most precise technique.
(b) The four batches do not appear to differ by much if averages over methods or batches are compared.
(c) Batches A, B, and C give similar minimal values.
(d) The average standard deviation is $\pm 9.5/100/100 \approx \pm 0.1\%$, certainly close to the instrumentation's limits.

By far the most parsimonious, but nonstatistical, explanation for the observed pattern is that the titrations differ in selectivity, especially as regards basic and acidic impurities. Because of this, the only conclusion that can be drawn is that the true values probably lie near the lowest value for each batch, and everything in excess of this is due to interference from impurities. A more selective method should be applied, e.g., polarimetry or ion chromatography. Parsimony is a scientific principle: Make as few assumptions as possible to explain a observation; it is in the realm of wishful thinking and fringe science that combinations of improbable and implausible factors are routinely taken for granted.

The lessons to be learned from this example are clear:

- Most statistical tests, given a certain confidence level, only provide clues as to whether one or more elements are different from the rest.

- Statistical tests incorporate mathematical models against which reality, perhaps unintentionally or unwillingly, is compared, for example:

  additive difference: t-test, ANOVA
  linear relationship: linear regression, factorial test.

- More appropriate mathematical models must be specifically incorporated into a test, or the data must be transformed so as to make it testable by standard procedures.
- A decision reached on statistical grounds alone is never as good as one supported by (chemical) experience and/or common sense.
- Never compare apples and oranges if this distinction is evident or plausible.
- Refrain from assembling incomplete models and uncertain coefficients into a spectacular theoretical framework without thoroughly testing the premises. Definitive answers thus produced all too easily take on a life of their own as they are wafted through top floor corridors.

2.4.2. Full Factorial Experiments

In planning experiments and analyzing the results, the experimenter is confronted with a series of decisions:

classical experiment $\Leftarrow ? \Rightarrow$ statistically guided experiment

full analysis after all experiments are finished $\Leftarrow ? \Rightarrow$ on-the-run analysis

functional relationships $\Leftarrow ? \Rightarrow$ correlations

orthogonal factors, simple mathematical system $\Leftarrow ? \Rightarrow$ correlated factors, complex mathematics

What strategy is one to follow? In the classical experiment, one factor is varied at a time, usually over several levels, and a functional relationship between experimental response and factor level, e.g., concentration, is established. The data analysis is carried out after the experiment(s). If several factors are at work, this approach is successful only if they are more or less independent, i.e., do not strongly interact. If the aim is to optimize some conditions in order to maximize a target variable, e.g., yield, it might even be that the classical approach would fail, except if—and this is an expensive proposition—the number of experiments is sharply increased as in the brute force approach. Figure 2.23 explains the

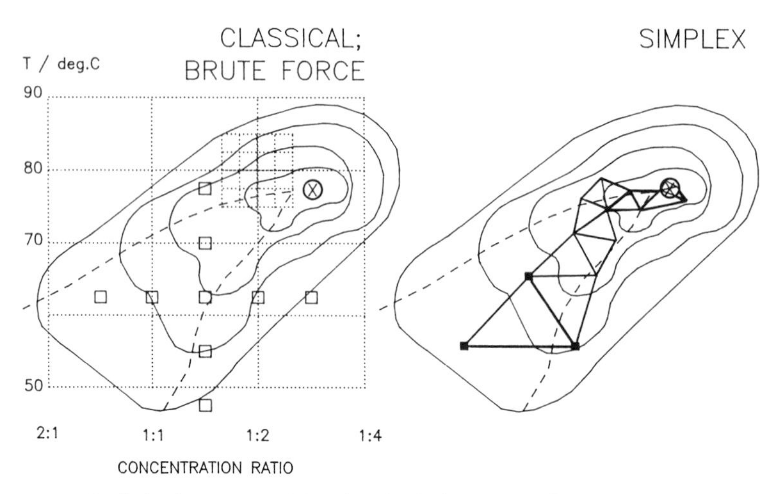

Figure 2.23. Optimization approaches. The classical approach fixes all factors except one, which is systematically varied (rows of points in left panel); the real optimum (x) might never be found this way. The brute-force approach would prescribe experiments at all grid points (dotted lines), and then further ones on a finer grid centered on $80/1:2$, for example. A problem with the simplex-guided experiment (right panel) is that it does not take advantage of the natural factor levels, e.g., molar ratios of $1:0.5$, $1:1$, $1:2$, but would prescribe seemingly arbitrary factor combinations, even such ones that would chemically make no sense, but the optimum is rapidly approached. If the system can be modeled, simulation[110] might help. The dashed lines indicate ridges on the complex response surface. The two figures are schematic.

problem for the two-factor case. As little as three factors can confront the investigator with an intractable situation if he chooses to proceed classically. The way out is to use the factorial approach, which can just be visualized for three factors. An example from process optimization work will illustrate the concept. Assume that temperature, the excess concentration of a reagent, and the pH have been identified as probable factors that influence yield: A starting value (40°C, 1.0% concentration, pH 6) and an increment (10, 1, 1) is decided upon for each; this gives a total of eight combinations that define the first block of experiments; see Figure 2.24. Experiment 1 starts off with 40°C, 1% concentration, and pH 6; that is, all increments are zero. Experiments 2–8 have one, two, or three increments different from zero. The process yields are given in Table 2.14, second row. These eight combinations, after proper scaling, define a cube in 3-space.

Program FACTOR in Section 5.2.6 lists the exact procedure, which is only sketched here. Table 2.14, for each factor and interaction, lists the

Table 2.14. Data and Results for the Discussed Example; for Calculations, See Program FACTOR.

[If all specified effects are statistically significant, the estimates equal the measurements, and the deviations (residuals) are all zero. Here, a standard deviation of ± 0.1 was assumed, which means that the only the factors **a**, **b**, **c**, and **abc** are significant; there are no effects due to interactions **ab** and **bc**.]

Factor	Progr. {5.2.6}	1	a	b	c	ab	ac	bc	abc
Measurement	$Y(\,)$	51	62	54	46	68	59	52	62
Obs. effect	$E(\,)$	0	12.0	4.5	−4.0	0.0	−0.5	0.0	−1.5
Spec. effect	$F(\,)$	±0.0	+1.2	+4.5	−4.0	+0.0	−0.1	+0.0	−0.6
Model	$M(\,)$	—	+1.2	+4.5	−4.0	—	—	—	−0.6
Estimate	y	51.25	61.75	54.25	45.75	67.75	59.25	51.75	62.25
Deviation	$Y(\,) - y$	−0.25	+0.25	−0.25	+0.25	+0.25	−0.25	+0.25	−0.25

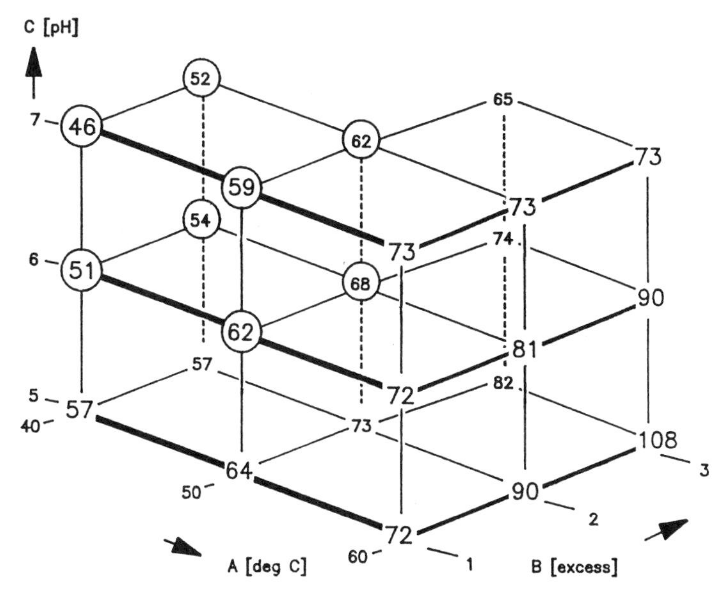

Figure 2.24. Factorial space. Numbers in circles denote process yields actually measured (initial data set); all other numbers are extrapolated process yields used for planning further experiments (assumption: repeatability $s_y = \pm 0.1$; all values rounded to the nearest integer); the estimated yield of 108% shows that the simple linear model is insufficient.

observed effect and the specific effect, the latter being a slope or slopes that make(s) the connection between any factor(s) and the corresponding effect. A t test is conducted on the specific effects, i.e., $t = E/(\Delta x \cdot s_E)$, where E is the observed effect, s_E its (estimated) standard deviation, and Δx the change in the factor(s) that produced the effect. If t is larger than the critical $t(p, f)$, the specific effect is taken to be significant; otherwise it is not included in the model; a model effect for a combined factor, e.g., **ac**, indicates an interaction between factors **a** and **c**. Specific effects found to be nonsignificant are given as a dash (—). In Figure 2.24 (page 131), the yields measured for a reaction under different conditions of pH, temperature, and excess of a reagent are circled. The extrapolation to the adjacent cubes is straightforward, the geometrical center of the original cube serving as a starting point; in this fashion the probable direction of yield increase is easily established. Note that the limitations of the linear three-factor model immediately become apparent: A yield cannot exceed 100%.

Example: For $T = 55$°C, 1.7% excess, and pH 7.4 the estimated yield is found as: $A = 55 - (40 + 50)/2 = 10$; $B = 1.7 - (1 + 2)/2 = 0.2$; $C = 7.4 - (6 + 7)/2 = 0.9$; A, B, and C are the coordinates relative to the center of the cube defined by the first eight experiments:

$$y = (51 + 62 + 54 + 46 + 68 + 59 + 52 + 62)/8 + 1.2 \cdot (A) + 4.5 \cdot (B)$$

$$- 4.0 \cdot (C) + 0.0 \cdot (A) \cdot (B) + 0.0 \cdot (A) \cdot (C) + 0.0 \cdot (B) \cdot (C)$$

$$- 0.6 \cdot (A) \cdot (B) \cdot (C);$$

$$y = 56.75 + 12 + 0.9 - 3.6 - 1.08 = 64.97.$$

The course of action is now to do either of two things:

(a) Continue with measurements so that there are always eight points as above. This choice would have the advantage of requiring only minimal mathematics, but one is limited by having to choose cubically arranged lattice points.

(b) Program FACTOR on the disc uses a brute force technique to estimate the changes necessary in each of the three factors in order to move closer to the local maximum. Since the precise direction is sensitive to the noise superimposed on the original eight measurements, these suggested changes should only be taken as indications to plan the next 2–4 experiments, whose coordinates do not necessarily lie on the previously established lattice. All available results are entered into a multilinear regression program, e.g., Ref. 109 (programs of this type are available in many program packages, even on hand-held computers, so that further elaboration is not necessary here). In essence, the underlying model is an extension of the linear regression model explained in Chapter 2, in that more than one independent variable x_i exists:

$$Y = a_0 + a_1 \cdot x_1 + a_2 \cdot x_2 + a_3 \cdot x_3 + \cdots. \qquad (2.35)$$

Because the above factor experiment suggests **a**, **b**, **c**, and **abc** as independent variables, cf. bottom row in Table 2.14, the data table would take on the form:

$$\begin{array}{lllll} |a_1 & b_1 & c_1 & a_1b_1c_1|, & |y_1|, \\ |a_2 & b_2 & c_2 & a_2b_2c_2|, & |y_2|, \\ |\cdot & \cdot & \cdot & \cdot\ |, & |\cdot|, \\ |\cdot & \cdot & \cdot & \cdot\ |, & |\cdot|, \\ |a_n & b_n & c_n & a_nb_nc_n|, & |y_n|. \end{array}$$

Some experimenting might be necessary if it turns out that quadratic terms such as b^2 improve the fit between the model and the data. However, the relevant point is that such a model is only a means to refine and speed up the process of finding optimal conditions. For this purpose it is counterproductive to try for a perfect fit; it might even be advantageous to keep the model simple and throw out all the best 5–10 experiments.

The factorial experiment sketched out above is used in two settings:

(1) If very little is known about a system, the three factors are varied over large intervals; this maximizes the chances that large effects will be found with a minimum of experiments, and that an optimal combination of factors is rapidly approached (example: new analytical method to be created, no boundary conditions to hinder investigator).

(2) If much is known about a system, such as an existing production process, and an optimization of some parameter (yield, purity, etc.) is under investigation, it is clearly impossible to endanger the marketability of tons of product. The strategy is here to change the independent process parameters (in random order) in small steps within the operating tolerances (cf. Fig. 1.8). It might be wise to conduct repeat experiments so as to minimize the overall error. In

Table 2.15. Number of Experiments as a Function of the Number of Factors and Levels.

$n = L^f$		with L = Number of Levels per Factor f = Number of Factors, Eq. (2.36)			
		$L = 2$	3	4	5
Number	$f = 3$	$n = 8$	27	64	125
of	4	16	81	256	625
factors	5	32	243	1024	3125

this way a process can evolve in small steps, nearly always in the direction of an improvement. The prerequisites are a close cooperation between process and analytical chemists, a motivated staff willing to exactly follow instructions, and, preferably, highly automated hardware. The old-style and undocumented "let's turn off the heat now so we can go to lunch" mentality introduces unnecessary, and even destructive, uncertainty.

The choice of new vertices should always take into consideration the following aspects:

- A new point should be outside the lattice space already covered.
- The expected change in the dependent (= target) variable (effect) should be sufficient to distinguish the new point from the old ones.
- The variation of a factor must be physically and chemically reasonable.
- Extrapolations should never be made too far beyond terra cognita.

The optimization technique embodied in program FACTOR could easily be expanded from three purported factors to four or more, or to three, four, or even five levels per factor. Mathematically, this is no problem, but the number of experiments rises sharply. The most economical approach is then to pick out three factors, study these, and if one of them should prove to be a poor choice, replace it by another. The necessary number of experiments for the first round of optimization is as shown in Table 2.15; see also Fig. 2.24.

CHAPTER

3

ANCILLARY TECHNIQUES

3.0. INTRODUCTION

Chapters 1 and 2 introduced the basic statistical tools; ready application of these tools requires a computer. If the computer were employed only to run statistics packages, a fine instrument would be poorly utilized. In this chapter, a number of techniques are explained that tap the benefits of fast data handling and simulation in connection with getting to know one's data: Visual presentations in various formats elucidate character and limits of many a problem or a problem-solving approach much better than tables. Also, options concerning the way a model is fit to data are presented, and some advice on writing or using programs is given.

3.1. OPTIMIZATION TECHNIQUES

Once a mathematical model has been chosen, there is the option of either fixing certain parameters (see Section 4.10) or fixing certain points, e.g., constraining the calibration line to go through the origin.[48–51,56,101]

When one tries to fit a mathematical function to a set of data one has to choose a method or algorithm for achieving this end, and a weighing model with which to judge the goodness of fit.

As for the method, four classes can be distinguished:

- graphical fit
- closed algebraic solution
- iterative technique
- brute force approach

Each of these four classes has its particular advantages: The *graphical* fit technique makes use of a flexible ruler (or a steady hand): Through a graphical representation of the data a curve is drawn that gives the presumed trend. While simple to apply, the technique is very much subject to wishful thinking. Furthermore, the found curve is descriptive in nature, and can only be used for rough interpolations or estimation of confidence intervals. A computerized form of the flexible ruler is the *spline* function[111, 112] that exists in one- and two-dimensional forms. The gist is that several successive points (ordered according to abscissa value) are approximated by a quadratic or a cubic polynomial; the leftmost point is dropped, a further one is added on the right side, and the fitting process is repeated. These local polynomials are subject to the condition that at the point where two of them meet, the slope must be identical. Thus only a very small element out of each constrained polynomial is used, which gives the overall impression of a smooth curve. The programs offer the option of adding "tension," which is akin to stiffening a flexible ruler. The spline functions can be used wherever a relatively large number of measurements is available and only a phenomenological, as opposed to a theoretical, description is needed. Splines can, by their very nature, be used in conditions where no single mathematical model will fit all of the data, but, on the other hand, the fact that every set of coefficients is only locally valid means that comparisons can only be visually effected.

The *algebraic* solution is the classical fitting technique, as exemplified by the linear regression (Chapter 2). The advantage lies in the clear formulation of the numerical algorithm to be used and in the uniqueness of the solution.

The *iterative* method encompasses those numerical techniques that are characterized by an algorithm for finding a better solution (set of model coefficients), starting from the present estimate[113]: Examples are the regula falsi, Newton's interpolation formula, and the simplex techniques.[93, 105] The first two are well known and are amenable to simple calculations in the case of a single independent variable. Very often though, several independent variables have to be taken account of; while two variables (for example, see Section 4.2) barely remain manageable, further ones tax imagination. A simplex-optimization program (available at most computing centers, as PC software, and also in the Curve ROM module of the HP-71 hand-held computer) works in a $(k + 1)$-*dimensional* space, k dimensions being given by the coefficients $(a_1, a_2, \ldots, a_k)$, and the last one by χ^2. The global minimum in the $\chi^2 = f(a_1, a_2, \ldots, a_k)$ function is sought by repeatedly determining the "direction of steepest descent" (maximum change in χ^2 for any change in the coefficients a_i), and taking a "step" to establish a new vertex. A numerical example is found in Table 1.17.

Here a problem enters that is already recognizable with Newton's formula: If pure guesses instead of graphics-inspired ones are used for the initial estimate, the extrapolation algorithm might lead the unsuspecting user away from the true solution, instead of toward it. The higher the number of dimensions, the larger the danger that a local instead of the global minimum in the goodness-of-fit measure is found. Under the key word "ruggedness" (Section 3.4) a technique is shown that helps reduce this uncertainty. Simplex methods and their ilk are dependent on powerful computing devices and, because of their automatic mode of operation, demand thoughtful use and critical interpretation of the results.

The *brute force* method depends on a systematic variation of all involved coefficients over a reasonable parameter space. The combination yielding the lowest goodness-of-fit measure is picked as the center for a further round with a finer raster of coefficient variation. This sequence of events is repeated until further refinement will only infinitesimally improve the goodness-of-fit measure. This approach can be very time consuming and produce reams of paper, but if carefully implemented, the global minimum will not be missed; cf. Figs. 2.23 and 4.3.

The algebraic/iterative and the brute force methods are numerical respectively computational techniques that operate on the chosen mathematical model. Raw residuals r are weighed to reflect the relative reliabilities of the measurements.

The weighing model with which the goodness of fit or figure of merit ($GOF = \Sigma(u_i)$) is arrived at can take any of a number of forms:

$Y = f(x/a \cdot \cdot)$,	Y designates the model, with the independent variable(s) x, and parameters $a \ldots$
y_i	measured value
$r_i = y_i - Y(x/a \cdot \cdot)$	residual
$w_i = w(x)$	weighing function, e.g., reciprocal variance; cf. Section 2.2.10
$u_i = w_i \cdot g(r_i)$	statistically and algebraically weighed residual, for example:
$u_i = w_i \cdot \mathrm{ABS}(r_i)$	linear
$u_i = w_i \cdot (r_i)^2$	quadratic, absolute (least squares)
$u_i = w_i \cdot (r_i/Y(x/a \cdots))^2$	quadratic, relative (least squares)
$u_i = w_i \cdot (r_i)^2/Y(x/a \cdots)$	intermediate (χ^2).

These continuous functions can be further modified to restrict the individual contributions u_i to a certain range, for instance r_i is minimally equal to the expected experimental error, and all residuals larger than a given number r_{max} are set equal to r_{max}. The transformed residuals are then weighed and summed over all points to obtain the goodness of fit, GOF.

A good practice is to use a weighing model that bears some inner connection to the problem and results in GOF figures that can be physically interpreted. The residual standard deviation s_{res}, which has the same dimension as has the reproducibility s_y, might be used instead of χ^2.

The chosen weighing model should also be applied to a number of repeat measurements of a typical sample. The resulting GOF figure is used as a benchmark against which those figures of merit resulting from parameter fitting operations can be compared; see Table 1.17. The most common situation is that one compares the residual standard deviation against the known standard deviation of determination (reproducibility or repeatability).

Once a fitted model is refined to the point where the corresponding figure of merit is smaller than the benchmark (Table 1.17), introducing further parameters or higher dimensions is (usually) a waste of time and only nourishes the illusion of having "enhanced" precision.

If several candidate models are tested for fit to a given data set, it need not necessarily be that all weighing models $w(x) \cdot g(r_i)$ listed as examples above would indicate the same model as being the best (in the statistical sense), nor that any given model is the best over the whole data range.

It is evident that any discussion of the results rests on three premises: constraints, the fitted model, and the weighing model. Constraints can be boon or bane, depending on what one intends to use the regression for.[49] The employed algorithm should be of secondary importance, as long as one does not get trapped in a local minimum or by artifacts. While differences among fitted models can be (partially) judged by using the goodness-of-fit measure (graphical comparisons are very useful, especially of the residuals), the weighing model must be justified by the type of the data, by theory, or by convention.

3.2. EXPLORATORY DATA ANALYSIS

Natural sciences have the aura of certainty and exactitude. Despite this, every scientist has experienced the situation of being befuddled:

- Analytical projects must often be initialized when little or nothing is known about the system to be investigated. Thus data are generated under conditions believed to be appropriate, and after some numbers have accumulated, a review is undertaken.

- Well-known processes sometimes produce peculiar results; data covering the past 20 or so batches are collated for inspection. Frequently, one has no idea what hypotheses are to be tested.

Both situations lend themselves to exploration: The available data are assembled in a data file along the lines "one row per batch, one column per variable," or an equivalent organization. Programs are then used that search for correlations (use program CORREL) or allow one to choose any two (see program VIEW_XY, or three, program XYZ) variables for display in a two- (or pseudo three-) dimensional format. Combinations of variables are tried according to intuition or any available rule of thumb. Perceived trends can be modeled; that is, a mathematical function is fitted. These models can then be substituted for or subtracted from the real data. What ideally remains are residuals commensurable in size with the known analytical error and without appreciable trend.

Exploratory data analysis is a form of a one-man brainstorming session; the results must be accordingly filtered and viable concepts can then be turned into testable hypotheses. The filtering step is very important, because spurious and/or erroneous data turn up in nearly every data set not acquired under optimally controlled conditions. Global figures of merit, such as the correlation coefficient (see Section 2.1), if not supported by visual trend analysis, common sense, and plausibility checks, may foster wrong conclusions. An example is worked in Section 4.11, "Exploring a Data Jungle," and in Section 4.22.

The EDA technique cannot be explained in more detail because each situation needs to be individually appraised. Even experienced explorers now and then jump to apparently novel conclusions, only to discover themselves the victims of some trivial or spurious correlation.

3.3. ERROR PROPAGATION AND NUMERICAL ARTIFACTS

The flow sheet shown in the introduction and that used in connection with a simulation (Section 1.4) provide insights into the pervasiveness of errors: At the source, errors are experienced as an inherent feature of every measurement process. The standard deviation is commonly substituted for a more detailed description of the error distribution (see also Section 1.2), as this suffices in most cases. Systematic errors due to interference or faulty interpretation cannot be detected by statistical methods alone;

control experiments are necessary. One or more such primary results must usually be inserted into a more or less complex system of equations to obtain the final result (for examples, see Refs. 17, 72–75, 84, 85, and 113). The question that imposes itself at this point is: "How reliable is the final result?"

Two different mechanisms of action must be discussed:

(a) The act of manipulating numbers on calculators of finite accuracy leads to numerical artifacts (see Table 1.2).[9,10]

(b) The measurement uncertainty is transformed into a corresponding uncertainty of the final result due to algebraic distortions and weighing factors, even if the calculator's accuracy is irrelevant.

An example for mechanism (a) is given in Section 1.1.2: Essentially, numerical artifacts are due to truncation, as when a computational operation results in a number the first digits of which are irrelevant, and the remaining ones were corrupted by numerical overflow or truncation. The following general rules can be set up for simple operations:

(a1) Ten nearly identical numbers of s significant digits each, when added, require at least $s + 1$ significant digits for the result. Note that when the numbers are different by orders of magnitude, more significant digits are needed.

(a2) Two numbers of s significant digits each, when multiplied, require $2 \cdot s$ digits for the result.

(a3) The division of two numbers of s significant digits each can yield results that require nearly any number of digits.

(a4) Differences of nearly identical sums are particularly prone to truncation; see Eqs. (1.3b) and (2.7)–(2.9).

(a5) Even the best calculators or software packages work with only so many digits: For 12-digit arithmetic, cf. Table 1.1, adding the squares of five-digit numbers already causes problems.

A few **examples** will illustrate this (significant digits in parentheses):

$$1.2345678 + 9.8765432 = 11.1111111 \qquad (8), (8), (9)$$

$$1.23457 \cdot 8.56789 = 10.5776599573 \qquad (6), (6), (12)$$

$$3.333333333333333 \cdot 3 = 10 \qquad (\infty), (1), (2)$$

$$0.0333/0.777777 = 0.42814322529.. \qquad (3), (6), (\infty)$$

$$1024/256 = 4 \qquad (4), (3), (1)$$

Classical error propagation (b) must not be overlooked: If the final result R is arrived at by way of an algebraic function

$$R = f(x_1, x_2, \ldots, x_k, A, B, ..),$$

with $x_1 \cdots x_k$ factors and $A \cdots$ parameters, the function f must be fully differentiated with respect to every x:

$$d_i = \left\{ \frac{\partial f}{\partial x_i} \right\}, \qquad i = 1 \ldots k, \tag{3.1}$$

and typical errors e_i must be estimated for all factors.[13,31] The linear terms of the full expansion of $\partial f / \partial x$ are then used to estimate the error in R:

$$e_R^2 = \Sigma(d_i \cdot e_i)^2. \tag{3.2}$$

If, as is usual, standard deviations are inserted for e_i, e_R has a similar interpretation. Examples are provided in Refs. 17, 57, 70–74, and 113–116 and in Section 4.17.

3.4. RUGGEDNESS AND SUITABILITY OF A METHOD

In Section 2.2.11 the validation of an analytical method is addressed. One point raised there concerns the question whether a method is robust or not, or, rephrased, "can this method be trusted to work in my own hands or in the hands of my colleagues?". This is no idle question: An involved protocol is not likely to be followed for very long, especially if it calls for unusual, tricky, or complicated operations, unless these can be automated. Unrecognized systematic errors,[11] if present, can frustrate the investigator or even lead to totally false results being taken for true.[11,18,71] The ruggedness of an analytical method thus depends on the technology employed, and, just as important, on the reliability of the people entrusted to apply it.

Statistical methods, in contrast, are rugged when algorithms are chosen that on repetition of the experiment do not get derailed by the random analytical error inherent in every measurement,[86,117] that is, when similar coefficients are found for the mathematical model, and equivalent conclusions are drawn. Obviously, the choice of the fitted model plays a pivotal role. If a model is to be fitted by means of an iterative algorithm, the initial guess for the coefficients should not be too critical. An extensive introduction into robust statistical methods is given in Ref. 118.

In the following, an example from Chapter 4 will be used to demonstrate the concept of statistical ruggedness, by applying the chosen fitting model to data purposely "corrupted" by the Monte Carlo technique. The data are normalized TLC peak heights from densitometer scans; see Section 4.2:

- An exponential and several polynomial models were applied, with the χ^2 measure of fit serving as arbiter.
- An exponential function was fitted using different starting points on the parameter space (A, B); see Fig. 4.3.

For the purpose of making the concept of ruggedness clear, a polynomial of order 2 suffices. The fitted parabola was subtracted from the data and the residuals were used to estimate a residual standard deviation: $s_{res} = \pm 6.5$. Then, a Monte Carlo model was set up according to the equation

$$Y = 4 + 25 \cdot x - 1.12 \cdot x^2 \pm \mathrm{ND}(0, 6.5). \tag{3.3}$$

A y_i value was then simulated for every x_i value in Table 4.3. This new, synthetic data set thus had statistical properties identical (n, S_{xx}), or very similar $(s_{xy}, s_{yy}, s_{res})$ to those of the measured set, the difference residing in the stochastic variations of the y_i' around the estimate Y; see program SIMILAR. The regression coefficients were calculated as they had been for the measured data, and the resulting parabola was plotted. This sequence of steps was repeated several times. This resulted in a family of parabolas [each with its own set of coefficients similar to those in Eq. (3.3)] that very graphically illustrates the ruggedness of the procedure: Any individual parabola that does not markedly deviate from the confidence band thus established must be regarded as belonging to the same population. By extension, the simulated data underlying this nondeviant curve also belongs to the same population as the initial set. The perceived confidence band also illustrates the limits of precision to which interpolations can be carried, and the concentration range over which the model can safely be employed. The coefficients so obtained must be investigated for dispersion and significance: The standard deviation helps to define the number of

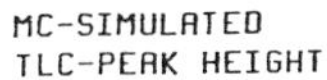

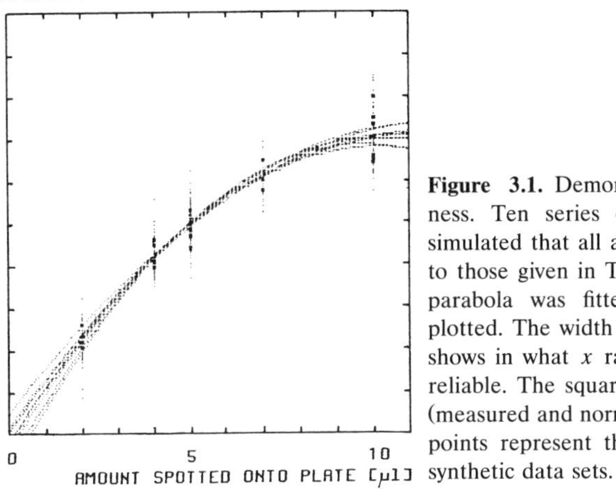

Figure 3.1. Demonstration of ruggedness. Ten series of data points were simulated that all are statistically similar to those given in Table 4.3. A quadratic parabola was fitted to each set and plotted. The width of the resulting band shows in what x range the regression is reliable. The squares depict the original (measured and normalized) data, and the points represent the statistically similar synthetic data sets.

significant digits, and a t test carried out for H_0: $\mu = 0$ reveals whether the coefficient is necessary at all, see Fig. 3.1.

3.5. SMOOTHING AND FILTERING DATA

All techniques introduced so far rely on recognizable signals; what happens if the signal more or less disappears in the baseline noise?

If the system is *static*, repeating the measurement and averaging according to Eq. (1.1) will eventually provide a signal-to-noise ratio high enough to discriminate signal from background.

If the system is *dynamic*, however, this will only work if the transient signals are captured and accumulated by a computer, as in FT-NMR. A transient signal that can be acquired only once can be smoothed if a sufficiently large number of data points is available.[66] It must be realized that the procedures that follow are cosmetic in nature and only serve to enhance the presentability of the data. Distortion of signal shape is inevitable.[20] An example will illustrate this: In Fig. 3.2 three Gaussian signals of different widths (full width at half maximum, FWHM) are superimposed on a sloping background. With Monte Carlo simulation noise is added and individual "measurements" are generated. How is one to extract the signals? There are various digital filtering techniques[94, 119–124]

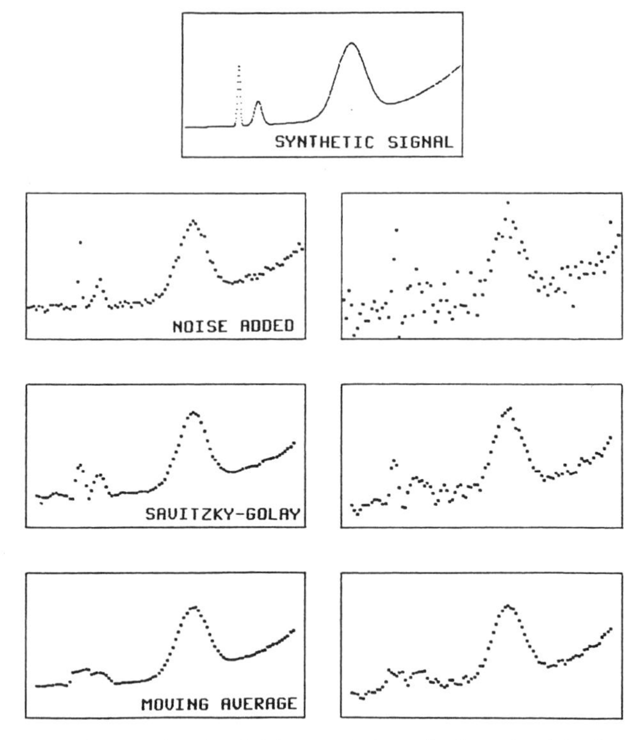

Figure 3.2. Smoothing a noisy signal. The synthetic, noise-free signal is given at the top. After the addition of noise by means of the Monte Carlo technique, the panels in the second row are obtained (little noise, left, five times as much noise, right). A 7-point Savitzky-Golay filter of order 2 (third row) and a 7-point moving average (bottom row) filter are compared.

such as boxcar averaging, moving average, Savitzky-Golay filtering, Fourier transformation,[125,126] correlation analysis,[127] and others.[128–130] The first three of these are related, differing only in the number of calculations and the weighing scheme. The filter width must be matched to the peak widths encountered.

Boxcar averaging is the simplest approach. The name can be traced to an image that comes to mind: Imagine a freight train (locomotive and many boxcars) running across the page. All measurements that fall within the x range spanned by one freight car (one "bin") are averaged and depicted by a single resulting point in the middle of this range.

The *moving average* can be explained as above. After having obtained the first set of averages, the "train" is moved by one point and the operation is repeated. This simply provides more points to the picture. Wider bins result in more averaging. There is a trade-off between an increase of the signal-to-noise ratio and distortion of the signal shape. Note that narrow signals that are defined by only a few measurements become distorted or even disappear.

Savitzky-Golay filtering[131–134] operates by the same mechanism, the difference being that instead of just averaging the points in a bin, a weighing function is introduced that prefers the central relative to the peripheral points. There are Savitzky-Golay filters that just smooth the signal, and there are others that at the same time calculate the slope[135–137] or the second derivative.[135] Depending on the coefficient sets used, "tension" can be introduced that suppresses high-frequency components. The ones that are presented here have a width of 7 points, see Fig. 3.2, Ref. 20, and use the coefficients -2, 1, 6, 7, 6, 1, and -2. Program SMOOTH incorporates a SG filter (restricted to smoothing polynomials of order 2 and 3, filter width $3 \cdots 11$) that provides filtered output over the whole data range; BC and MA filters are also available.

An extension to two- resp. multidimensional filters is presented in Refs. 102, 138, and 139. Matched filters[102] are the most elegant solution, provided one knows what form the signal has. The choice between different filter types is discussed in Ref. 102. The CUMSUM technique (Section 1.8.5, program SMOOTH) can also be regarded as a special type of filtering function, one that detects changes in the average y.

3.6. MONTE CARLO TECHNIQUE

In Section 1.4, Simulation of a Series of Measurements, the Monte Carlo technique was introduced in general terms as an important numerical tool for studying the relationship of variables in complex systems of equations. Here, the algorithm will be presented in detail, and a more complex example will be worked.

The general idea behind the MC technique is to use the computer to roll dice, that is, to generate random variations in the variables, which are then inserted into the appropriate equations to arrive at some result. The calculation is repeated over and over, the computer "rolling the dice" anew every time. In the end, the individual results average to that obtained if average variables had been inserted into the equations, and the distribution approximates what would have been found if the propagation-of-errors procedure had been applied. The difference between classical propagation of errors and MCT is that in the former the equation has to be in closed form [i.e., result $= f(x, y, z \cdots)$, the result variable appearing only once, to the left of the equals sign], and the equation has to be differentiable. Also, propagation-of-errors assumes that there is a characteristic error ΔX, symmetrical about $\bar{x}$, as in mean/standard deviation, that is propagated. Often one of these conditions is violated, either because one does not want to introduce simplifying assumptions in order to obtain closed solutions (e.g., n is assumed to be much larger than 1 to allow the transition from $n - 1$ to n), or because differentiation is impossible (step functions, iterative algorithms, etc.).

MCT allows one to choose any conceivable error distribution for the variables, and to transform these into a result by any set of equations or algorithms, such as recursive (e.g., root-finding according to Newton) or matrix inversion (e.g. , solving a set of simultaneous equations) procedures. Characteristic error distributions are obtained from experience or the literature, e.g., Ref. 31.

In practice, a normal distribution is assumed for the individual variable. If other distribution functions are required, the algorithm $z = f(\mathrm{CP})$ in Section 5.1.1, respectively, the function FNZ(R) in Section 5.2.10 have to be appropriately changed.

The starting point is the (pseudo-) randomization function supplied with most computers; it generates a rectangular distribution of events; that is, if called many times, every value between 0 and 1 has an equal probability of being hit. For our purposes, many a mathematician's restraint regarding randomization algorithms (the sequence of numbers is not perfectly random because of serial correlation, and repeats itself after a very large number of cycles,[140] $c \approx 10^9$ for a PC) is irrelevant, as long as two conditions are met:

- fewer than that very large number of cycles are used for a simulation (no repeats), and
- the randomization function is initialized with a different seed every time the program is run.

The rectangular distribution generated by the randomize function needs to be transformed into the appropriate distribution for each variable, generally a Normal Distribution.

Since the transformation function is symmetrical in the case of the Normal Distribution, only the lower left portion needs to be defined; R values larger than 0.5 are reflected into the 0–0.5 range, and the sign of the result z is accordingly changed. Note that the $z = f(\mathrm{CP})$ function described in Sections 1.2.1 and 5.1.1 is used. The algorithm is (cf. program

MCSIM in Section 5.2.10)

$$(1) \qquad R = \text{random number}, 0 \le R \le 1,$$

$$(2) \qquad R\begin{cases} \le 0.5 \Rightarrow \text{CP} = R, & \Rightarrow \text{sign}\, w = -1, \\ > 0.5 \Rightarrow \text{CP} = 1 - R, & \Rightarrow \text{sign}\, w = +1, \end{cases}$$

$$(3) \qquad \text{Calculate } u = \log_{10}(\text{CP}),$$

$$(4) \qquad \text{Calculate } v = f(u),$$

$$(5) \qquad \text{Calculate } z = w \cdot v,$$

$$(6) \qquad \text{Calculate ND variable} = \bar{x} + z \cdot s_x,$$

where $\bar{x}$ and s_x are parameters of the MC subroutine. $f(u)$ is the second polynomial given in Section 5.1.1. See Figure 3.3 and Table 3.1.

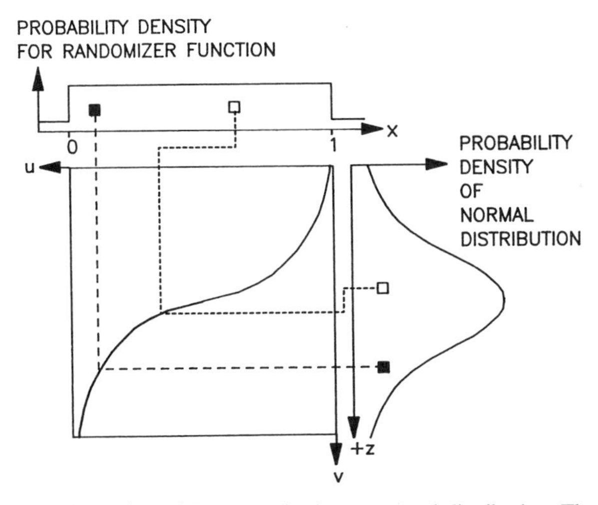

Figure 3.3. The transformation of a rectangular into a normal distribution. The rectangle at the upper left shows the probability density (idealized observed frequency of events) for a random generator. The curve at the lower left is the transposed cumulative probability CP versus deviation z function introduced in Section 1.2.1. At the lower right, a Normal Distribution probability density PD is shown. The dotted line marked with an open square indicates the transformation for a random number larger than 0.5; the dashed line starting from the filled square is for a random number less than or equal to 0.5. The upper right quadrant is not used with symmetrical distribution functions.

Table 3.1. Four Examples of Monte Carlo Calculations for $\bar{x} = 1.23$ and $s_x = 0.73$. (The last line gives the position of the first uncertain digit and an indication of the probable error, e.g., the first result is between 1.1316 and 1.1324.)

Four examples for $\bar{x} = 1.23$, $s_x = 0.073$:

R	0.0903	0.7398	0.995	0.999
sign w	−1	+1	+1	+1
CP	0.0903	0.2602	0.005	0.001
u	−1.0443	−0.5847	−2.3010	−3.0000
v	+1.3443	+0.6375	+2.5761	+3.0944
z	−1.3443	+0.6375	+2.5761	+3.0944
$\bar{x} + s_x \cdot z$	1.132	1.277	1.418	1.456
Uncertainty	±0.0004	±0.0004	±0.000	±0.007

Simple examples for the technique are provided in Figs. 1.9, 1.10, and 1.17 and in program SIMGAUSS. Additional operations are introduced in Figs. 1.2, 1.3, and 1.18, namely, the search for the extreme values for a series. A series of interesting applications, along with basic facts about MTC, is to be found in Ref. 140.

3.7. COMPUTER SIMULATION

"What if?" This question epitomizes both serious scientific thinking and little boys' dreams. The three requirements are (a) unfettered fantasy, (b) mental discipline, and (c) a thinking machine. The trick to get around the apparent contradiction between points (a) and (b) is to alternately think orderly and freely. The unimpeded phase is relaxing: Think up all those "impossible" situations that were never dealt with in school because there are no neat and simple answers: That is precisely why there is a problem out there waiting to be solved. Then the bridled phase: Systematically work through all those thought-up combinations. This can be enormously tiring, can tax memory and common sense, and demands precise recording. For this reason a computer is a valuable assistant (c).

What are the mechanics of simulation[30, 110, 141]? In Fig. 3.4 the classical sequence experiment/raw data/evaluation is shown. A persistent but uninspired investigator might experiment for a long time without ever approaching an understanding of the system under scrutiny, despite his attempts to squeeze information from the numbers. A clever explorer will with a few experiments gather some fundamental facts, e.g., noise and signal levels, limits of operation of the instrumentation and the chemistry (boiling points, etc.), and combine this knowledge with some applicable mathematical descriptions, like chemical equilibria, stoichiometry, etc.

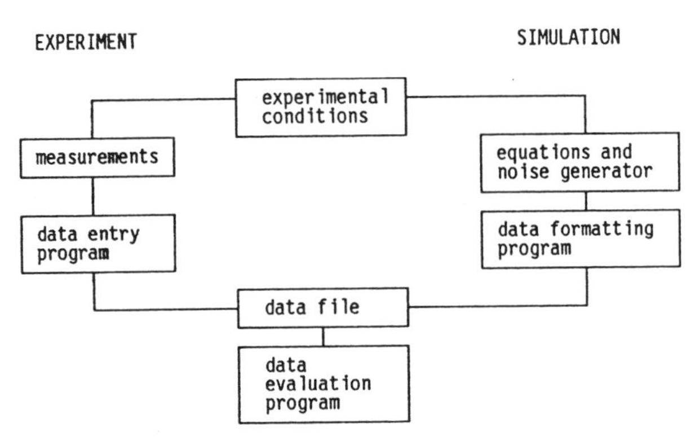

Figure 3.4. Schematic depiction of the relation between experiment and simulation. The first step is to define the experimental conditions (concentrations, molecular species, etc.), which then form the basis either for the experiment or for simulation. Real data are manually or automatically transferred from the instruments to the data file for further processing. Simulated values are formatted to appear indistinguishable from genuine data.

This constitutes a rough model; some parameters might still have to be estimated. The model is now tested for fit against the available experimental data, and refinements are applied where necessary. Depending on the objective of the simulation procedure, various strategies can be followed: What is common to all of them is that the experiment is replaced by the mathematical model. Variation of the model parameters and/or assumptions allows "measurements" to be simulated (calculated) outside the experimentally tested parameter domain. The computer is used to sort through the numbers to either find conditions that should permit pivotal experiments[110] to be conducted, or to determine critical points. In either case, experiments must be done to verify the model. The circle is now closed, so that one can continue with further refinements of the model.

What are the advantages to be gained from simulation? Some are given in Section 1.4 and others are listed below:

- Refined understanding of the process under scrutiny
- Optimization of the experimental conditions, so that reliable data can be acquired, or that data acquisition is more economical
- The robustness of assumptions and procedures can be tested
- If a well-developed theory is available, the limits of the experimental system (chemistry, physics, instrumentation) can be estimated
- Very complex systems can be handled
- Deterministic and random aspects can be separately studied

A typical problem for simulation is the investigation of a reaction mechanism involving coupled equilibria, such as a pH-dependent dissociation in the presence of metal ions, complexing agents, and reagents. Some of the interactions might be well known, and for others bounds on association constants and other parameters can at least be roughly estimated. A model is then set up that includes the characteristics of UV absorption and other observables for each species one is interested in. Thus equipped, one begins to play with the experimentally accessible concentrations to find under what conditions one of the poorly known values can be isolated for better characterization. In the end, various combinations of component concentrations can be proposed that with high probability will permit successful experiments to be conducted, in this way avoiding the situation where extensive laboratory effort yields nothing but a laconic "no conclusion can be drawn."

Here again, no precise instructions can be given because each situation will demand a tailored approach. Note: Numerical simulation in many ways resembles the "what-if" scenario technique now made possible by spread-sheet programs.

3.8. PROGRAMS

Computer programs have become a fixture of our lives. In the following a few comments will be given on using and writing programs.

Using Program Packages

Many fine program packages for the statistical analysis of data and untold numbers of single-task programs are commercially available for a variety of computers ranging from mainframes to programmable pocket calculators. The potential user first faces the job of finding and settling for a hardware/software combination that should also offer high flexibility and fit the budget. No small task! Much commercial software is offered for an array of more or less equivalent machines, e.g., PCs of different makes and model lines; while most of the listed functions (data input, data editing, statistical tests, graphical and tabular output, etc.) might work as described, there is nearly always room for surprises. Hardware improvements and operating system updates carry the risk of unwanted interactions with specific program parts, for instance, because of altered memory allocations, and software revisions are often released so hastily that some low-priority functions are incompletely coded.

The user is urged to

- Buy a hardware configuration identical to that for which the software is known to satisfactorily run on, or
- Employ data sets for which verified intermediate and final results are available to check out alternative hardware/software combinations.

All software should be treated with suspicion until completely exonerated for reasons that reside in human imperfection:

- Software can be poorly designed: Data format specifications are perhaps incompatible, intermediate results are inaccessible, all data are lost if an input error is committed, or results and data are not transferable to other programs. The division into tasks (modules, menu positions) might reflect not human but computer logic, and the sequence of entries the user is forced to follow might include unnecessary or redundant steps.
- Software designed for general or specific use might be useless in the intended application: Sociologists, mathematicians, clinical researchers, and physicists, just to name a few user categories, all gather and analyze data, to be sure, but each one has particular data structures and hypotheses. Tests must be selected to fit the application. A renowned program package is likely to bolster the authority of the user, providing he knows how to properly use it; overreliance on such packages, without the user having a solid grounding in statistics, will in the end cast a shadow of doubt on his mastery of the subject.
- The software's author, as is often the case, is unfamiliar with the particulars of analytical chemistry or unaware of what happens with the numbers after the instrument has issued a nice report. Is it surprising then, that the correlation coefficient between signal and concentration is standard fare, but error propagation and the estimation of confidence limits of an unknown's concentration is singularly lacking? Most chromatography software might be fine for manipulating chromatograms acquired in a short time span, but the comparison of stacks of chromatograms is accomplished only if there are no nonproportional shifts in retention times, a phenomenon often observed when HPLC columns age.
- Instrument suppliers are well acquainted with the design, construction, promotion, and sale of their products. The analytical problem-solving capabilities thereof are more often than not demonstrated on textbook-variety problems that are only remotely indicative of a ma-

chine/software combination's usefulness. If software is tailored toward showroom effectiveness, the later user suffers.

- Software is rarely completely free of coding errors: While manifest errors are eliminated during the debugging stage, the remaining ones crop up only when specific data patterns or sequences of instructions are encountered.

There is a two-stage procedure to help judge the veracity of a program's output:

- First, validated data are input, and the results are compared on a digit-by-digit basis with results from a reference program. If possible, the same is done with intermediate results; see example on Table 1.1.
- Second, the above data base is systematically modified to force the program to its limits. For example, the number of measurements is multiplied by entering each datum several times instead of only once, etc. That this step is necessary is shown by the authors' occasional experience that commercial programs as simple as a sorting routine would only work with the supplied demonstration data file.
- The next step is obviously to write one's own programs, to make sure that the arithmetic and the graphics exactly correspond to the personal needs.

Writing Programs

Writing programs is partly an art, but the organizational aspect should not be underestimated, namely,

- Intraprogram organization: modularity, choice between function and subroutine options, documentation, etc.
- Interprogram organization: interface protocols for subroutines, data formats, systematic use of variable names, etc. This point becomes painfully obvious when a program library inexorably grows to the point where even the programmer gets confounded by the multitude of extant assumptions, conventions, and assignments. Modifying programs to conform to a common convention around which they were never designed is much harder than to set down a convention beforehand. This includes setting aside some variable names for special purposes such as ordinary, array, index, or scratch-pad variables, and creating a common data format for the vast majority of the programs (some odd formats might be inevitable). A number of ancillary pro-

grams are necessary to interconvert data file formats, to create test data, etc.

Programs should be developed to be of the single-task variety. This keeps them short and easy to document. If necessary, small program elements can be merged into a larger one, or can be called up in the correct order from a master program. This approach makes it much easier to validate a package (GLP concern) than if a large, custom-designed program had been written.

The programs provided on the appended disc were written with the following in mind:

- Transparent and modular structure; all subroutines are in the same sequence as in the menu; there is a full menu and an abbreviated one; each subroutine starts off with a "####REM----function" statement (and is subdivided by further "####REM.....function" statements, where appropriate); comments are appended to individual lines; a given module may be used in various programs; the same variable names are used throughout: raw data: $R(I, J)$, row index: $I = 1 \cdots N$, column index: $J = 1 \cdots M$, abscissa: column K, ordinate: column L, boundaries of graph window in user units: X0, X9, Y0, Y9, or XMIN, XMAX, YMIN, YMAX, x range analyzed: X1–X2, statistical sums SX, SY, SXX, SXY, SYY, variance: V, etc.
- The source code is not compiled so the reader can examine it
- Results are presented in graphical form whenever possible; high-resolution screen displays are backed up by printable intermediate-resolution screens.
- A rounding function is employed that makes the most of the available column width when printing results (automatic switch to scientific format, when appropriate), but the number of significant digits displayed is limited to what common sense dictates. All calculations are performed using the double-precision mode.
- All algorithms used follow the precise scheme given in the text, with as many of the variables as possible having the same name. All math was checked at least five times, with the full program on two different computers, with a "kernel" program on a third computer, and twice manually using calculators.

CHAPTER

4

COMPLEX EXAMPLES

4.0. INTRODUCTION

Life is a complicated affair; otherwise five-year-old toddlers would have voting rights and none of us would have had to attend school for so long. In a similar vein, the correct application of statistics has to be learned gradually through experience and guidance. Usually a combination of several simple tests is required to decide a case. For this reason, a series of more complex examples was assembled. The presentation closely follows the script as the authors experienced it; these are real-life examples straight from the authors' fields of expertise with only a few of the more nasty or confusing twists left out.

4.1. TO WEIGH OR NOT TO WEIGH

Situation and Design. A photometric assay for an aromatic compound prescribes the following steps:

(1) Accurately weigh about 50 mg of compound.
(2) Dissolve and dilute to 100.0 ml (master solution MS).
(3) Dilute 2.00 ml of MS to 100.0 ml (sample solution SS).
(4) Measure the absorbance A of SS.

The operations are repeated for samples 2 and 3; the final result is the average of the three individual results per batch.

The original method uses the classical "weigh powder, dilute volumetrically" scheme. In order to rationalize the procedure and at the same time guarantee delivery of results of optimal quality, an extensive review was undertaken. Essentially, all analyses in a fixed period of time were done according to a modified protocol that incorporated a weighing step after every volumetric operation. The rationale was simple: Weighing on an analytical balance is inherently much more accurate and precise than any operation involving pipettes and graduated flasks, but is also more labor-

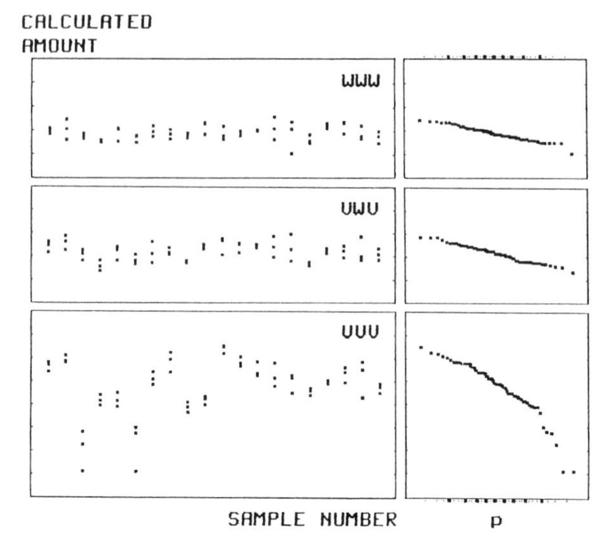

Figure 4.1. Assay results calculated according to three schemes W/W/W (top), V/W/V (middle), and V/V/V (bottom). The raw values (left-hand panel) are plotted chronologically; in the right-hand panels, probability charts—rotated by 90 degrees, with CP = 0 ··· 1 going from right to left—are shown. The V/V/V scheme obviously does not produce normally distributed results, whereas the other two schemes do.

intensive. Twenty batches were thus investigated and the evaluation of the absorbance measurements (data files VVV.001, VWV.001, and WWW.001) was carried out according to three schemes: For steps (2) and (3) above, namely, to either do the dilution volumetrically, or to do it gravimetrically, or use a combination thereof:

W/W/W, **V/W/V**, and **V/V/V** for the sequence S/A/D,

where **W**: **W**eight to five significant digits, in grams
V: nominal **V**olume of pipetted aliquot or full flask
S: solvent used for MS, nominal 100 ml, approx. 100 g
A: aliquot, nominal 2 ml, approx. 2 g
D: diluted SS, nominal 100 ml, approx. 100 g

The results are presented in Fig. 4.1.

Questions

(1) Which scheme is best; which is sufficient?
(2) What part of the total variance is due to the analytical methodology, and how much is due to variations in the product?

(3) Can an assay specification of "not less than 99.0% pure" be upheld?

(4) How high must an in-house limit be set that at least 95% of all analyzed batches are within the above specification?

Results. Figures of merit are found in Table 4.1.

The variances were calculated according to Tables 1.8 and 1.9, respectively, Eqs. (1.3). Use program MULTI to obtain:

Test	VVV	VWV	WWW
Bartlett	H_0	H_0	H_1
ANOVA	H_1	H_1	—
MRT (No. of groups)	10	6	—

The reason that H_1 was found for WWW is that the intrinsic standard deviation is very small but two values fall out of line: 0.015 and 0.71; the quotient of the $s_{x,\text{ before}}$ and the $s_{x,\text{ after}}$ is about 46!

Evidently, replacing volumetric work by weighing improves the within-group variance from 0.192 (**VVV**) to 0.0797 (**WWW**, factor of 2.4) (see Table 4.1) and the standard deviation from ± 0.44 to ± 0.28 (± 0.28 is a good estimate of the analytical error); much of the effect can be achieved by weighing only the aliquot (**VWV**), at a considerable saving in time. The picture is much more favorable still for the between-groups variance: The improvement factor can be as large as 34. A look at Fig. 4.1 shows why this is so: The short-term reproducibility of the fixed volume dispenser (pipetor) used for transferring 2.0 ml of solution MS cannot be much inferior to

Table 4.1. Decomposition of Variance by ANOVA; the Column Marked VVV · f Shows by Way of Example how Variances Are Added.

	WWW	VWV	VVV		VVV · f	f
Average	99.92	100.19	99.81		—	
Variance						
Within groups*	0.0797	0.0884	0.192	→	7.68	40
					+	
Between groups**	0.111	0.235	3.82	→	72.58	19
Total	0.0899	0.136	1.36	←	80.26	59

*Analytical repeatability.

**Made up of analytical artifacts and/or production reproducibility; *f*: Degrees of freedom; the arrows indicate in which direction the calculation proceeds.

that of the corresponding weighing step **W** because the within-group residuals are similar for **WWW** and **VVV**. However, some undetermined factor produces a negative bias, particularly in samples 3–6, 9, 10; this points to either improper handling of the pipetor or clogging. The reduction of the within-group variance from **VVV** to **VWV** is to a major part due to the elimination of the large residuals associated with samples 3 and 6. The extreme standard deviations in **WWW** are 0.015 and 0.71, $F > 2100$, which means the ANOVA and the MRT tests cannot be carried out.

The total variance (corresponding standard deviations ± 0.3, ± 0.37, resp. ± 1.17) is also improved by a factor of 15, which means the specifications could be tightened accordingly.

The process mean with the **VVV** scheme was 99.81; that is essentially indistinguishable from the nominal 100%. The proposed SL is "more than 99.0%"; the corresponding z value (see Section 1.2.1) is $(99.81 - 99.00)/\sqrt{0.192} = 1.85$, which means that about 3.2% [use $\text{CP} = f(z)$ in Section 5.1.1] of all measurements on good batches are expected to fall below the specification limit (*false negative*). The mean of three replicates has a z value of $1.85 \cdot \sqrt{3} = 3.20$ [use Eq. (1.5c)], giving an expected rejection rate of 0.13%. The corresponding z values and rejection rates for the **WWW** scheme are only minimally better; however, the reliability of the **WWW** and **VWV** schemes is vastly improved, because the occurrence of systematic errors is drastically reduced. The same calculation could have been carried out using the t instead of the Normal Distribution; because the number of degrees of freedom is relatively large ($f = 19$), virtually the same results would be obtained.

False negative responses of 0.13–3.2% are an acceptable price. What are the chances of *false positives* slipping through? Four alternative hypotheses are proposed for the **VVV** scheme (compare μ to the SL = 99.0!):

$$
\begin{aligned}
&H_1\colon \mu = 98.8, && z = (98.8 - 99.0)/\sqrt{0.192} = 0.46, && p = 0.323,\\
&H_2\colon \mu = 98.5, && z = (98.5 - 99.0)/\sqrt{0.192} = 1.14, && p = 0.127,\\
&H_3\colon \mu = 98.0, && z = (98.0 - 99.0)/\sqrt{0.192} = 2.28, && p = 0.0113,\\
&H_4\colon \mu = 97.7, && z = (97.7 - 99.0)/\sqrt{0.192} = 2.97, && p = 0.0015.
\end{aligned}
$$

Interpretation. While good batches of the quality produced ($\bar{x} = 99.81\%$ purity) have a probability of being rejected (false negative) less than 5% of the time, even if no replicates are performed, false positives are a problem: An effective purity of $\mu = 98.5\%$ will be taxed "acceptable" in

12.7% of all cases because the found $\bar{x}$ is 99% or more. Incidentally, plotting $100 \cdot (1 - p)$ versus μ creates the so-called power curve.

If the much better and still reasonably workable **VWV** scheme ($V = 0.0884$) is chosen, the probabilities for false positives (cf. above) drop to 0.25, 0.046, 0.0005, and 0.0005; thus $\mu = 98.5\%$ will be found "acceptable" in less than 5% of all trials. The **WWW** scheme would bring only a small improvement at vastly higher cost: For the above case 0.046 would reduce to 0.038.

Conclusion. The **VWV** scheme will have to be implemented to make precision commensurable with the demanding limits proposed by the marketing department. If the production department can keep up to its promise of typically 99.8% purity, all is well. For the case of lower purities, quality assurance will have to write a SOP that contains an action limit for $\bar{x}$ at AL $\approx$ 99.5%, and an error recovery procedure that stipulates retesting the batch in question so as to sufficiently increase n and decrease V_x. The production department was exonerated by the discovery that the high between-groups variances were analytical artifacts that could be eliminated by the introduction of a weighing step. The problem remains, however, that there is no margin for error, negligence, or lack of training.

4.2. NONLINEAR FITTING

Initial Situation. A product contains three active components that up to a certain point in time were identified using TLC. Quantitation was done by means of extraction/photometry. Trials to circumvent the time-consuming extraction steps by quantitative TLC (diffuse reflection mode) had been started but were discontinued due to reproducibility problems. The following options were deemed worthy of consideration:

(1) HPLC (ideal because separation, identification, and quantitation could be accomplished in one step).
(2) Diode-array UV spectrophotometer with powerful software incorporated. (Although the spectra overlapped in part, quantitation could be effected in the first derivative mode. The extraction/separation step could be circumvented.)
(3) Conventional UV/VIS spectrophotometer (manual measurements from strip-chart recorder traces and calculations on the basis of fitted polynomials; the extraction/separation step would remain).
(4) Quantitative TLC (an internal standard would be provided; the extraction/separation step could be dropped).

Table 4.2. Peak Heights Taken from Densitometric Scans for Various Amounts of a Given Sample Applied onto Seven TLC Plates Taken from the Same Package.

Peak Height (mm)	TLC Plate 1	2	3	4	5	6	7
2 μl/spot				42.5	31.5	39.5	
2				44.5	33.2	39.5	
4	62	67	74			75	
4	61	68	69			75	
5	63	80	85.5	92	67	74.8	93.5
5	71	80.5	83	99	75.5	75.5	95
7	87						109
7	81						114
10				124	95	110.8	
10				153	97	112.8	

Options (1) and (2) had to be eliminated from further consideration because the small company where this happened was producing insufficient cash flow. Despite the attractiveness of photometric quantitation, option (3) did not hold up to its promise; careful revalidations revealed a disturbing aspect: The extractions were not fully quantitative; the efficiency varied by several percent depending on the temperature and (probably) the origin of a thickening agent. Thus one had the choice of a multistep extraction, at further cost in time, or a large uncertainty that obviated the advantage of the photometric percision. Option (4) was still beset by "irreproducibility": Table 4.2 contains reflection-mode densitometer peak-height measurements for one of the three components from seven TLC plates; cf. Fig. 4.2.

Data Normalization. Since the general trend for every set of numbers is obviously similar, a simple noralization was applied: The average over all 14 height measurements at 5 μl/spot was set equal to 100%, yielding the reduced height values given in bold **numbers** in Table 4.3. This data normalization reduces the plate-to-plate variability from ± 11.6 to ± 6.4 residual standard deviation and permits a common calibration curve to be used; see Fig. 4.2.

Curve Fitting. It is immediately apparent that a straight line would be a poor approximation, but a polynomial of order 2 or 3 would be easy to fit; this was done in the example in Section 3.4. From a theoretical perspective, a disadvantage of polynomials, especially the quadratic one, is their

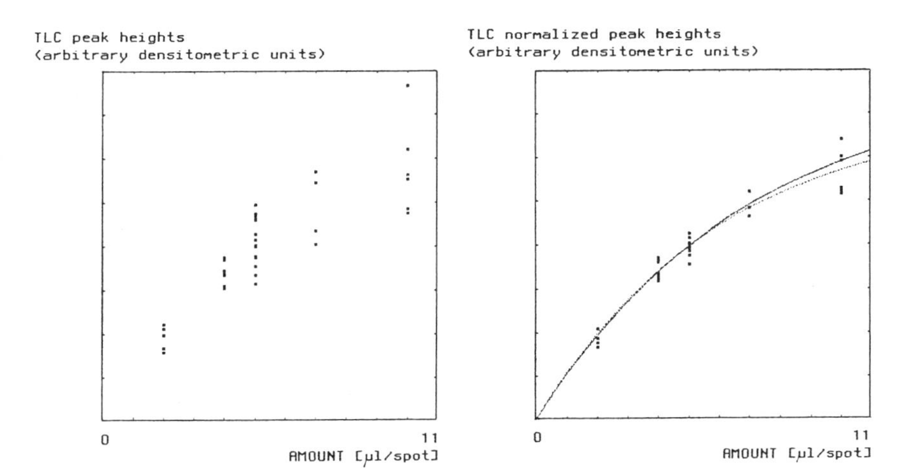

Figure 4.2. Raw (left) and normalized (right) peak heights versus the amount spotted on the plates. The average of all 5-μl spots per plate was set to 100% (right); the improvement due to the normalization is evident. The solid curve is for $A = 176.65$ and $B = -0.16$. The dotted curve is for the best quadratic parabola; cf. Fig. 3.1 and Eq. (3.3).

Table 4.3. Data from Table 4.2 Normalized to Average of 5-µl / Spots (Bold) on a Plate-by-Plate Basis. A Quadratic Regression was Applied to these Data, Yielding Eq. (3.3).

Peak Height (% of avg. 5-μl Spot)	TLC Plate 1	2	3	4	5	6	7	
2 μl/spot				44.5	44.2	52.6		
2				46.6	46.6	52.6		
4	92.5	83.5	87.8			99.8		
4	89.6	84.7	81.9			99.8		
5	**94.0**	**99.7**	**101.5**	**96.3**	**94.0**	**99.5**	**99.2**	**avg.**
5	**106.0**	**100.3**	**98.5**	**103.7**	**106.0**	**100.5**	**100.8**	**= 100%**
7	129.9						115.6	
7	120.9						121.0	
10				129.8	133.3	147.4		
10				160.2	136.1	150.1		

nonasymptotic behavior: While one can easily impose the restriction $Y(0) = 0$ for the left branch, the strong curvature near the focus normally falls into a region that physically does not call for one, and at larger concentrations the peak height would decrease. Also, while each x_i can unambiguously be assigned a single Y_i, the reverse is not necessarily true.[102] An equation more suited to the problem at hand is $Y = A \cdot (1 - \exp(B \cdot X))$ in accord with the observation that calibration curves for TLC measurements are highly nonlinear and tend to become nearly horizontal for larger amounts of sample spotted onto the plate. The disadvantage here is that there is no direct method for obtaining the coefficients a and b.

Three paths can be advanced: (1) expansion, e.g., Taylor series; (2) trial and error, e.g., generating curves on the plotter; and (3) simplex optimization algorithm; see Section 3.1.

(1) An expansion brings no advantage in this case because terms at least quartic in coefficient b would have to be carried along to obtain the required precision.
(2) If a graphics screen or a plotter is available a fairly good job can be done by plotting the residuals $r_i = y_i - Y$ versus x_i, and varying the estimated coefficients a and b until suitably small residuals are found over the whole x range; this is accomplished using program TESTFIT.
(3) If iterative optimization software is available, a result of sufficient accuracy is found in short order. This is demonstrated in the following.

In Fig. 4.3 the starting points (initial estimates) are denoted by squares in the A, B plane; the evolution of the successive approximations is sketched by dotted lines that converge on the optimum, albeit in a roundabout way. The procedure is robust: No local minima are apparent; the minimum, marked by the circle, is rapidly approached from many different starting points within the given A, B plane; note that a B value larger than about 0.1, an A value smaller than about -20, and combinations in the lower right corner pose problems for the algorithm or lead to far-off relative minima. Despite the fact that a unique solution was found [Eq. (4.1)], good practice is followed by graphing the reduced data together with the resulting curve (Fig. 4.2); the residuals are plotted separately. A weighing scheme could have been implemented to counter heteroscedacity, but this, in the light of the very variable data, would amount to statistical overkill.

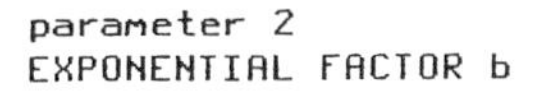

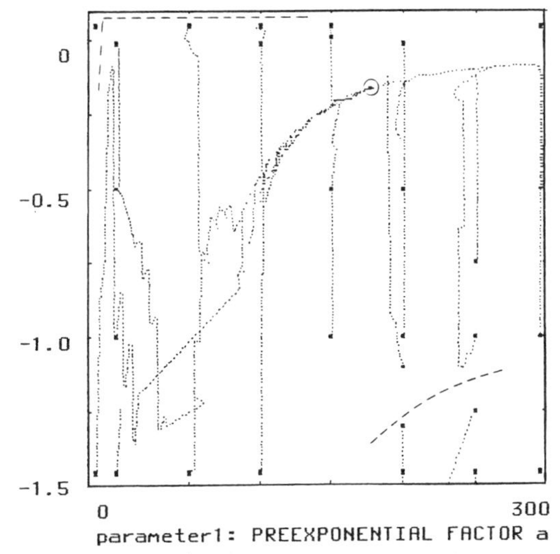

Figure 4.3. Optimization of parameters. The exponential equation given below was fitted to the normalized data from Table 4.3. Each square corresponds to a starting point in the A, B plane for the simplex algorithm employed. The dotted traces show how the system converges on the optimum (circle). The gradient "north" of the "east-west" valley is exceedingly steep. The dashed lines indicate ridges, outside of which other minima are found. A brute force approach (Section 3.1) would entail calculating the sum of (squared) residuals between measurement and model for every intersection of a grid superimposed on the parameter space, and finding the minimum.

Equation fitted to data in Table 4.3:

$$Y = A \cdot (1 - e^{B \cdot X}), \tag{4.1}$$

with $A = 176.65$ and $B = -0.16$.

In this specific case the predictive power of the polynomial P^2 [see Eq. (3.3)] and the exponential function are about equal in the x interval of interest. The peak height corresponding to an unknown sample amount would be divided by that of the standard (5 μl/spot) run on the same plate. If possible, each sample and the standard would be spotted several times on the same plate, e.g., in alternating sequence, and the average of the sample, respectively, the standard peak heights would be used for improved precision. The resulting quotient is entered on the ordinate and the unknown amount is read off the abscissa after reflection. Despite the normalization the results are not exceptional in terms of reproducibility;

thus it is perfectly acceptable to use a graphical approximation, by constructing upper and lower limits to the calibration points with the help of a flexible ruler, to find confidence limits for X.

As long as the health authorities accept 90–110% specification limits on the drug assay, the normalization method presented above will barely suffice for batch release purposes. Since there is a general trend toward tightening the specification limits to 95–105% (this has to do with the availability of improved instrumentation and a world-wide acceptance of GMP standards), a move toward options 1 (HPLC) and 2 (DA-UV) above is inevitable.

For an example of curve fitting involving classical propagation of errors in a potentiometric titration setting, see Ref. 113.

4.3. UV-ASSAY COST STRUCTURE

Problem. The drug substance in a pharmaceutical product is to be assayed in an economical fashion. The following constraints are imposed:

- UV-absorbance measurement at a given wavelength, either on a photometer, which is known to be sufficiently selective, or on a HPLC.
- The extraction procedure and the solvent.
- The number of samples to be processed for every batch produced: Six samples of 13 tablets each are taken at prescribed times after starting the tablet press (10 tablets are ground and well mixed, two average aliquots are taken, and each is extracted); the additional 3 tablets are used for content uniformity testing; this gives a total of $6 \cdot (2 + 3) = 30$ determinations that have to be performed.

Available Lab Experience

- The relative standard deviation of the determination was found to be ±0.5% (photometer) resp. ±0.7% (HPLC[142]) for samples and references.
- The relative content varies by nearly ±1% due to inhomogeneities in the tablet material and machine tolerances.
- Duplicate samples for the photometer involve duplicate dilutions because this is the critical step for precision; for the HPLC the same solution is injected a second time because the injection/detection/integration step is more error prone than the dilution.

Requirements

- The mean content must be 98–102% for 9 out of 10 mixed samples.
- The individual contents must be 95–105% for 9 out of 10 tablets and none outside 90–110%.
- The linearity of the method must be demonstrated.

Options. The analyst elects to first study photometry and place the three reference concentrations symmetrically about the nominal value (= 100%). The initial test procedure consists of using references at 80, 100, and 120% at the beginning of the series, and then a 100% reference after every fifth determination of an unknown sample.

Known Cost Factors for the UV Method

- A technician needs an average of 10 minutes to prepare every photometer sample for measurement.
- The necessary reference samples are produced by preparing a stock solution of the pure drug substance and diluting it to the required concentrations. On the average, 12 minutes are needed per reference.
- The instrument-hour costs 20.-.
- A lab technician costs 50.-/hour (salary, bonus, etc.). The price levels include the necessary infrastructure.
- The instrument is occupied for 5 minutes per determination, including changing the solutions, rinsing the cuvettes, etc.
- The technician is obviously also occupied for 5 minutes, and needs another 3 minutes per sample for the associated paperwork.

First Results. The confidence interval from linear regression for $\mathrm{CI}(X)$ was found to be too large; in effect, the plant would have to produce with a 0% tolerance for drug substance content; a tolerance of about $\pm 1\%$ is considered necessary.

Refinements. Since the confidence interval must be small enough to allow for some variation during production without endangering the specifications, the additivity of variances is invoked: Since 90% of the results must be within $\pm 2\%$ of nominal, this can be considered to be a confidence interval

$$\tfrac{1}{2}\mathrm{CI} = t \text{ factor} \cdot \sqrt{(V_{\text{prod}} + V_{\text{analyt}})} \leq 2\%.$$

Trial calculations are done for $V_{\text{analyt}} = (0.5)^2$ and $V_{\text{prod}} = (0.9)^2 \cdots (1.1)^2$;

the required t factors for $p = 0.1$ turn out to be $1.94 \cdots 1.66$, which is equivalent to demanding $n = 7 \cdots 120$ calibration samples. Evidently, the case is critical and needs to be underpinned by experiments. Twenty or 30 calibration points might well be necessary if the calibration scheme is not carefully designed.

Solution. 20–30 calibration points are too many, if only for reasons of expended time. The analyst thus searches for a combination of perhaps $n = 8$ calibration points and $m = 2$ replications of the individual samples. This would provide the benefit of a check on every sample measurement without too much additional cost. An inspection of the various contributions in Eq. (2.17) toward the CI(X) in Table 2.9 reveals the following for $n = 8$ and $m = 2$:

- About 70.9% contribution due to the $1/m$ term.
- About 17.7% contribution due to the $1/n$ term.
- About 11.3% contribution due to the $1/S_{xx}$ term ($S_{xx} = 1250$, $\Delta x = 10$).

Various calibration schemes similar to those given in Section 2.2.8 were simulated. The major differences were (1) the assumption of an additional 100% calibration sample after every fifth determination (including replications) to detect instrument drift, and (2) the cost structure outlined in Table 4.4, which is summarized in Eq. (4.2) below. The results are

Table 4.4. Cost Components.
(The numbers give the necessary time in hours. Amortization of the equipment and infrastructure is assumed to be included in the hourly rates.)

Item	Photometric		HPLC	
	Machine	Operator	Machine	Operator
Rate per hour	20.-/h	50.-/h	30.-/h	50.-/h
Prepare calibration solutions	—	$k/5$	—	$k/4.6$
Measure calibration Points	$n/12$	$n/12$	$n/4$	$n/60$
Prepare samples	—	$30 \cdot m/6$	—	$30/5.5$
Measure samples	$30 \cdot m/12$	$30 \cdot m/12$	$30 \cdot m/4$	$30/60$
Evaluate results	—	$30 \cdot m/20$	$30 \cdot m/120$	$1/12$

Where n is the number of calibration points, m is the number of replicates per sample, and k is the number of calibration solutions prepared.

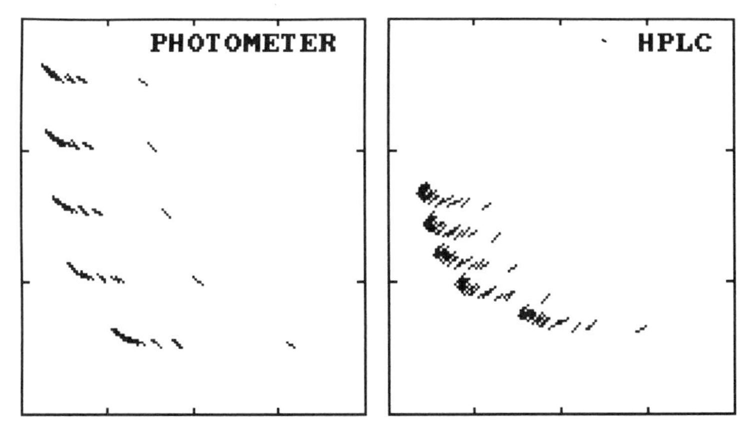

Figure 4.4. Estimated total analytical cost for one batch of tablets versus the attained confidence interval CI(X). 336 (UV) resp. 640 (HPLC) parameter combinations were investigated (some points overlap on the plot; ordinate: cost; abscissa: precision).

depicted graphically in Fig. 4.4, where the total cost per batch is plotted against the estimated confidence interval CI(X). This allows a compromise involving acceptable costs and error levels to be found.

Interpolations at 110% of nominal were simulated; if the interpolations are envisaged closer to the center of mass, the third term will diminish in importance, and the sum of the first two terms will approach 100%.

Besides the photometric method, a HPLC method could also be used (cf. first constraint under "Problem" above). The HPLC has a higher relative standard deviation for two reasons: The transient nature of the signal results in a short integration time, and the short path length makes for a very small absorption. The available HPLC is equipped with an autosampler, so that a different calibration scheme can be implemented: Instead of a complete calibration curve comprising n points at the beginning of the series, two symmetrical calibration samples, e.g., 80/120% are run at the beginning, and after 5–10 determinations a double calibration is again performed, either at the same two or at two different concentrations. In this way, if no instrumental drift or change of slope can be detected during the run, all calibration results can be pooled. Overall linearity is thus established. Also, S_{xx} increases with every calibration measurement. In the photometer case above, the additional 100%-of-nominal measurement after every 5–10 determinations does not contribute toward S_{xx}. The number of replicate sample injections, m, was taken to be 1, 2, 3, 4, or 5. Preparing HPLC samples takes longer because vials have to be filled and capped.

Known Cost Factors for the HPLC Method

- The technician needs 11 minutes to prepare every sample solution and 13 minutes to prepare a calibration solution. A vial is filled in 1 minute.
- The instrument costs 30.- per hour.
- The instrument is occupied for 15.5 minutes per determination, including 30 seconds for printing the report.
- The technician needs 1 minute per determination to load the sample into the autosampler, type in sample information, start the machine, etc., and 20 minutes for preparing the eluent per batch.
- The technician needs only 5 minutes per batch for evaluation, including graphics and tables because the computerized instrument and the integrator do the rest.

The HPLC instrument is more expensive and thus costs more to run; because it is used 12–16 hours per day, versus only 2–5 hours/day for the photometer, the hourly rates are not proportional to the initial investment and servicing needs. The cost structure given in Table 4.4 can now be derived.

The total costs associated with the 30 determinations are calculated to be

$$5.83 \cdot n + 500.0 \cdot m + 10.0 \cdot k \qquad \text{(photometer)}, \tag{4.2}$$

$$8.33 \cdot n + 232.5 \cdot m + 10.87 \cdot k + 301.89 \qquad \text{(HPLC)}. \tag{4.3}$$

The number of man-hours expended is equal to the sum over all items in the columns marked 50.-/hour.

A few key results are given in Fig. 4.4. It is interesting to see that the two curves for the photometer and the HPLC nearly coincide for the above assumptions, HPLC being a bit more expensive at inferior precisions. Note the structure that is evident especially in the photometer data: This is primarily due to the number m of repeat determinations that are run on one sample ($m = 1$ at bottom, $m = 5$ at top). One way for reducing the photometer costs per sample would be to add on an autosampler with an aspirating cuvette, which would allow overnight runs and would reduce the time the expensive technician would have to spend in front of the instrument (a time interval of less than about 5 minutes between manual operations cannot be put to good use for other tasks). In order to reduce the confidence interval of the result from $\pm 1\%$ to $\pm 0.6\%$ relative, HPLC costs would about double on a per batch basis.

This simulation shows that the expected analytical error of $\pm 0.7\%$ (HPLC) will only be obtained at great cost. Experiments will be necessary to find out whether this assumption was overly optimistic or not. A trade-off between tolerance toward production variability and analytical error might be possible and should be investigated to find the most economical solution.

4.4. PROCESS VALIDATION

Situation. During the introduction of a new tablet manufacturing process, the operation of a conditioner had to be validated; the function of this conditioner is to bring the loaded tablets to a certain moisture content for further processing.

Question

Does the conditioner work in a position-independent mode; that is, all tablets in one filling have the same water content no matter into which corner they were put, or are there zones where tablets are dried to a larger or lesser extent than the average?

Experiment. Ten different positions within the conditioner representing typical and extreme locations relative to the air inlet/exhaust openings were selected for analysis. Eight tablets were picked per position; their water content was accurately determined on a tablet-to-tablet basis using the Karl Fischer technique. Table 4.5 gives an overview over all results.

Data Evaluation. The Bartlett test (Section 1.7.3; cf. program MULTI using data file MOISTURE.001) was first applied to determine whether the within-group variances were homogeneous, with the following intermediate results:

$$A = 0.1719, \qquad E = 3.50,$$

$$B = -424.16, \qquad F = 1.052,$$

$$C = 1.4286, \qquad G = 3.32,$$

$$D = 70.$$

For $f = k - 1 = 9$ degrees of freedom the χ^2 value comes to 3.50

Table 4.5. Water Content per Tablet in Percent, and Some Intermediate Results.

	Position										
Tablet	1	2	3	4	5	6	7	8	9	10	
1	1.149	1.082	1.847	1.096	1.109	1.181	1.191	1.175	1.152	1.169	
2	1.020	1.075	1.761	1.135	1.084	1.220	1.019	1.245	1.073	1.126	
3	1.106	1.112	1.774	1.034	1.189	1.320	1.198	1.103	1.083	1.050	
4	1.073	1.060	1.666	1.173	1.170	1.228	1.161	1.235	1.076	1.095	
5	1.108	1.068	1.779	1.104	1.252	1.239	1.183	1.101	1.021	1.041	
6	1.133	1.003	1.780	1.041	1.160	1.276	1.109	1.162	1.051	1.065	
7	1.108	1.104	1.816	1.141	1.148	1.232	1.158	1.171	1.078	1.127	
8	1.026	1.049	1.778	1.134	1.146	1.297	1.140	1.078	1.030	1.023	
	1.090	1.069	1.775	1.107	1.157	1.249	1.145	1.159	1.071	1.087	average
	0.047	0.034	0.052	0.049	0.051	0.045	0.059	0.062	0.040	0.051	SD
	8	8	8	8	8	8	8	8	8	8	n
	8.723	8.553	14.20	8.858	9.258	9.993	9.159	9.270	8.564	8.696	Σx
	0.0155	0.0081	0.0190	0.0168	0.0181	0.0144	0.0240	0.0266	0.0114	0.0180	$\Sigma(x - \bar{x})^2$

(uncorrected) resp. 3.32 (corrected); this is far below the critical χ^2 of 16.9 for $p = 0.05$. Thus the within-group variances are indistinguishable.

Because of the observed homoscedacity, a simple ANOVA test can be applied to determine whether the means all belong to the same population. If there were any indication of differences among the averages, this would mean that the conditioner worked in a position-sensitive mode and would have to be mechanically modified. See Table 4.6.

Interpretation. Because the calculated F value ($0.36108/0.00246 = 148$) is much larger than the tabulated critical one (2.04, $p = 0.05$), the group means cannot derive from the same population.

The multiple range test was used to decide which means could be grouped together. First, the means were sorted and all possible differences were printed, as in Table 4.7. Next, each difference was transformed into a q value according to Eq. (1.25). With $70 - 9 = 61$ degrees of freedom, the critical q value for the longest diagonal (adjacent means) would be 2.83,

Table 4.6. Variance Components.

$S_1 = 0.1719$	$f_1 = 70$	$V_1 = 0.00246$	Variance within groups
$S_2 = 3.2497$	$f_2 = 9$	$V_2 = 0.36108$	Variance between groups
$S_T = 3.4216$	$f_T = 79$	$V_T = 0.043311$	Total variance

Table 4.7. Table of Differences between Means.
(Rounding errors may occur, e.g., 1.070500 − 1.069125 = 0.001375 is given as 1.071 − 1.069 = 0.001 in the top left cell.)

Ordered Means ▼ ▶	1.071	1.087	1.090	1.107	1.145	1.157	1.159	1.249	1.775
1.069	0.001	0.018	0.021	0.038	0.076	0.088	0.090	0.180	0.706
1.071		0.017	0.020	0.037	0.074	0.087	0.088	0.179	0.705
1.087			0.003	0.020	0.058	0.070	0.072	0.162	0.688
1.090				0.017	0.055	0.067	0.068	0.159	0.685
1.107					0.038	0.050	0.052	0.142	0.668
1.145						0.012	0.014	0.104	0.630
1.157		differences x_{ji} ···					0.002	0.092	0.618
1.159								0.090	0.616
1.249									0.526

Table 4.8. Reduced q Values for the Differences between Means.
(The critical reduced q values pertinent for each diagonal (↖) are given in the right-hand column. The number of decimal places was reduced here so as to highlight the essentials).

Ordered Means ▼ ▶	1.071	1.087	1.090	1.107	1.145	1.157	1.159	1.249	1.775	Reduced Critical q^*
1.069	0.03	0.36	0.43	0.77	1.53	1.78	1.81	3.64	14.3	
1.071		0.33	0.40	0.74	1.51	1.76	1.79	3.62	14.3	↖
1.087			0.07	0.41	1.17	1.42	1.45	3.28	13.9	↖ 1.18
1.090				0.34	1.10	1.35	1.38	3.21	13.9	↖ 1.17
1.107					0.76	1.01	1.04	2.87	13.5	↖ 1.16
1.145						0.25	0.28	2.11	12.8	↖ 1.15
1.157		Reduced q_{ij}					0.03	1.86	12.5	↖ 1.13
1.159								1.83	12.5	↖ 1.11
1.249									10.6	↖ 1.09
										↖ 1.05
										1.00

*See Table 1.7 for $f = 61$. For the diagonal 0.43 − 0.74 − 1.17 − 1.35 − 1.04 − 2.11 − 12.5 the tabulated reduced critical q_c is 1.09; program MULTI uses a uniform q_c = 1.1 for all diagonals because the authors are not aware of an algorithm for calculating q_c with sufficient precision. Combinations of means to the left of the vertical and below the horizontal lines can be grouped together.

Table 4.9. Assignment of Means to Groups.

Group										
1	1.069	1.071	1.087	1.090	1.107					
2					1.107	1.145	1.157	1.159		
3									1.249	
4										1.775

that for the top right corner (eight interposed means) 3.33; see Table 1.6. For this evaluation separate tables would have to be used for different error probabilities p. The conversion to reduced q values eliminated this inconvenience with only a small risk (see Section 1.5.4); this was accomplished by dividing all q values by $t(61, 0.05) \cdot \sqrt{2} = 1.9996 \cdot \sqrt{2} = 2.828$; cf. Tables 1.7 and 4.7, yielding reduced critical q values in the range $1.00 \cdots 1.15$. See Table 4.8.

Which means could be grouped together? A cursory glance at the means would have suggested (values rounded) $1.07 \cdots 1.16$ for one group, 1.25 resp. 1.78 as second, respectively, third one.

An inspection of the bottom diagonal $0.03 \cdots 10.6$ shows that the two largest means differ very much ($10.6 \gg 1.83 > 1.1$), whereas each of the other means could be grouped with at least its nearest neighbor(s). The line drawn into the table distinguishes significant from nonsignificant reduced q values: The mean 1.249 is larger than 1.159, and the means 1.145, 1.157, and 1.159 are indistinguishable among themselves, and could be grouped with 1.107, but not with 1.087 (e.g., $1.35 > 1.1$). The values 1.069–1.107 form a homogeneous group, with 1.107 being the connecting element to the higher values. Thus one can symbolically depict Table 4.9.

On the evidence given here, the tablet conditioner seemed to work in order, but the geometrical positions associated with the means $(1.249, 1.775)$ differ from those with the means $(1.069 \cdots 1.107)$; indeed, five of these positions were near the entry port of the controlled-humidity airstream, and the other two were situated in corners somewhat protected from the air currents. The 1.159 group marks the boundary of the acceptable region.

Tablet samples were pulled according to the same protocol at different times into the conditioning cycle; because the same pattern of results emerged again and again, enough information had been gained to permit mechanical and operational modifications to be specified that eliminated the observed inequalities to such a degree that a more uniform product could be guaranteed.

4.5. REGULATIONS AND REALITIES

Situation. Low limits on impurities are a requirement that increasingly is being imposed on pharmaceuticals and high-quality chemicals. A specification that still commonly applies in such cases is the "2% total, 0.5% individual" rule, which is interpreted to mean that any one single impurity must remain below 0.5%, while the sum of all impurities shall not exceed 2.0% w/w. Conversely, the purity of the investigated compound must come to 98.0% or higher. In practice, this numerical reconciliation will not work out perfectly due to error propagation in the summation of the impurities, and the nonzero confidence interval on the $>98\%$ purity figure; thus it is considered inappropriate to measure impurity by way of the (100 minus %- purity) difference.

The Case. In 1986 there were semiofficial indications that a major registration authority might tighten the above requirement on pharmaceutical-grade chemicals by a factor of 4 to "0.5 weight-% total and an individual upper limit on every impurity found in an 'accepted sample'." The plant manager, after hearing this, asks the analyst to report whether a certain product characterized by a large number of small impurities would comply.

Review of Data. The analyst decides to review the critical HPLC and GC impurity methods by retrieving from his files the last 15 or so sets of chromatograms. On the average, 14.5 (range: 11–22) HPLC impurity peaks ranging from 0.004 to 0.358 area-% are found per chromatogram. Some components are better characterized by GC, so these results are used. The fact that most of these impurities are either insufficiently characterized because of the small amounts found, or no clean reference samples exist, means that area-% (integrated absorbance) instead of weight-% (integrated concentration) are used for the evaluation; this is accepted by the authorities. There is a numerical problem, though: Because the sum of all detected signal areas is equated with 100%, the effect of closure[143] must be taken into account; that is to say, since both the result *and* the components of the addition are known, some algebraic operations are inadmissible. Also, the numbers bear out a large variability for any one impurity from batch to batch; a typical impurity might yield $0.1 \pm 0.02\%$ ($n = 16$). The purity given by the main peak averages to $98.7 \pm 0.14\%$. From these first results the operations manager draws the

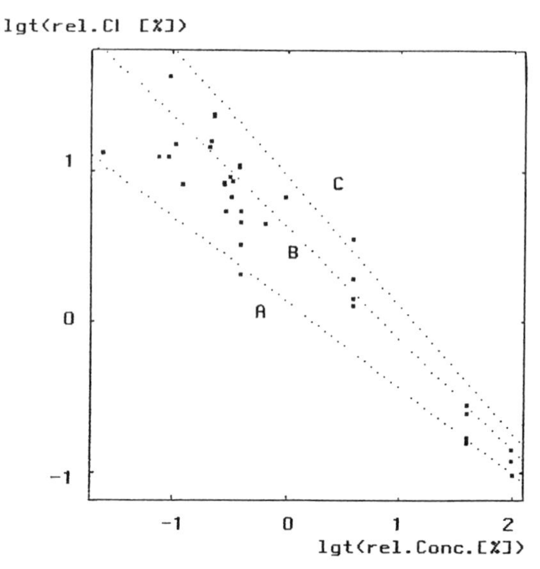

Figure 4.5. Trend of one-half of the relative confidence interval, $\pm 100 \cdot t \cdot s/c$, versus relative concentration c in log-log depiction. The regression lines are given by A: $y = 0.426 - 0.573 \cdot x$; B: $y = 0.920 - 0.740 \cdot x$; C: $y = 1.246 - 0.853 \cdot x$ (best, average, worst case).

conclusion that an additional crystallization step should push the two critical impurities below the 0.1% limit, and the sum of impurities to just below 0.5%.

Experience. The analyst is skeptical, and for good reason: Because of the low average levels, a higher relative standard deviation must be expected for the small peaks than for the large ones. This would force the analyst to do multiple purity determinations to keep confidence intervals within reasonable limits, and thus incur higher analytical overhead, and/or force the plant to aim for average impurity levels far below the present ones, also at a cost. The analyst then assembles Fig. 4.5: The abscissa is the logarithm of the impurity concentration in percent, while the ordinate gives the logarithm of half the relative confidence interval on the impurity, i.e., $\frac{1}{2}$(rel. CI) = $\pm \mathrm{lgt}(100 \cdot t \cdot s_x/c_x)$.

The trend $\mathrm{lgt}(0.5 \cdot \mathrm{CI}/c)$ vs. $\mathrm{lgt}(c)$ appears reasonably linear (compare this with Ref. 144; some points are from the method validation phase where various impurities were purposely increased in level). A linear regression line (B) is used to represent the average trend, while lines (A) and (C) give lower and upper estimates. Furthermore, it is assumed that the target level for any given impurity is given by a simple model:

$$\begin{array}{ccccc} \text{target level} & = & \text{allowed impurity level} & - & t_{(\alpha=0.1,f)} \cdot s_x \\ \text{TL} & & \text{AIL} & & \tfrac{1}{2} \cdot \text{CI}. \end{array} \quad (4.4)$$

This amounts to stating "the analytical results obtained from HPLC purity determinations on one batch are not expected to exceed the individual limit more than once in 20 batches." Since a one-sided test is carried out here, the $t(\alpha = 0.1, f)$, for the two-sided case corresponds to the $t(\alpha/2 = 0.05, f)$ value needed. Compare Fig. 2.12: The "target level" is related to the AIL as is the lower end of the production tolerance range to the lower specification limit for the "probably inside" case.

Example: Assume that a certain impurity had been accepted by the authorities at the AIL = 0.3% level; what level must one now aim for to assure that one stays within the imposed limit? Equation (4.4) is rewritten with both TL and CI/2 in percent of AIL, $x = \log(\text{AIL})$, $y = \log((100 \cdot t \cdot s_y)/\text{AIL})$, and the right-hand expression being replaced by the linear estimate from Fig. 4.5:

$$t = t(f = n - 1, p = 0.1), \qquad \text{Student's } t,$$

$$y[\%] = 100 \cdot \text{TL}/\text{AIL} = 100 - 100 \cdot t \cdot s_x/\text{AIL}, \qquad \text{linear estimate},$$

$$y[\%] = 100 - 10^{\{a + b \cdot \text{lgt}(\text{AIL})\}}, \qquad \text{substitution},$$

$$y[\%] = 100 - 10^{\{0.92 - 0.74 \cdot \text{lgt}(0.3)\}} = 79.7\%.$$

Table 4.10. Target Concentrations for Impurities, under Assumption of the Middle Line B in Fig. 4.6 (B: $a = 0.92$, $b = -0.74$, $m = 1$).
(If the LOQ of the method were 0.03%, the target concentration in the last line (0.012) would be inaccessible to measurement!)

Accepted Impurity Level (%)	Target Concentration (%)	Target Concentration (% of AIL)
10.0	9.85	98.5
5.0	4.87	97.5
3.0	2.89	96.3
2.0	1.90	95.0
1.0	0.917	91.7
0.5	0.431	86.1
0.3	0.239	79.7
0.2	0.145	72.6
0.1	0.054	54.3
0.05	0.012	23.7

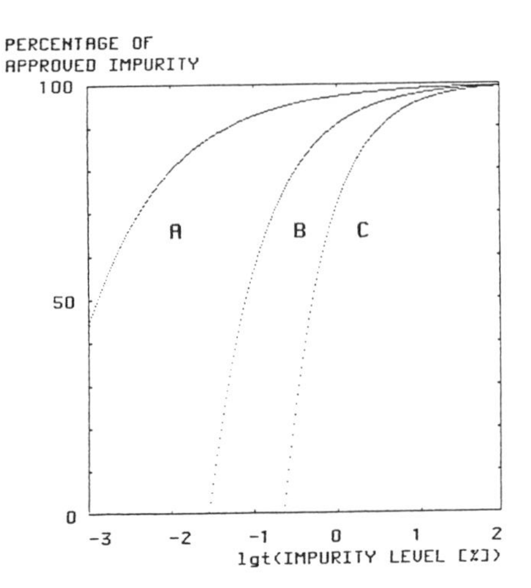

Figure 4.6. Consequences for the case that the proposed regulation is enforced: The target for each impurity is shown in percent of the level found in the official reference sample that was accepted by the authorities. Whereas the largest impurity today (0.5%) calls for a target level of (86% $\cdot$0.5) = 0.43%, the proposed upper limit on individual impurities for c = 0.1% calls for a target level of 0.05%, which is close to the LOD of the HPLC method.

For repeat measurements (m = 2, etc.), one has to subtract the logarithm of the square root of m from the sum ($a + b \cdot \text{lgt}(c)$). For m = 2 resp. m = 3, the result 79.7% would change to 85.7% resp. 88.3%. See Table 4.10.

Consequences. While this may still appear reasonable, lower accepted impurity limits AIL quickly demand either very high m or then target levels TL below the LOQ, as is demonstrated in Fig. 4.6.

Actually, it would be reasonable for the authorities to replace by 0.1% the individual limit concept for all impurities lower than about 0.1% in the accepted sample, provided that toxicity is not an issue, because otherwise undue effort would have to be directed at the smallest impurities. Various modifications, such as less stringent confidence limits, optimistic estimates [line (A) in Fig. 4.5], etc. somewhat alleviate the situation the plant manager is in, but do not change the basic facts. The effect of such well-intentioned regulations might be counterproductive: Industry could either be forced to withdraw products from the market despite their scientific merits because compliance is impossible, or then might dishonestly propose analytical methods that sweep all but a scapegoat impurity below the carpet.

4.6. DIFFUSING VAPORS

Situation. Two different strengths of plastic foils are in evaluation for the packaging of a moisture-sensitive product. Information concerning the

Table 4.11. Water Vapor Absorbed by Product.

Day	Weight Gained in mg Thin Foil	Thick Foil	Ratio
0	0	0	
2	2.0	1.4	0.70
5	4.1	3.4	0.83
8	6.5	5.4	0.83
12	10.2	8.1	0.79
			0.79 ± 0.06

diffusion of water vapor through such foils is only sparsely available for realistic conditions. To remedy this lack of knowledge, samples of the product are sealed into pouches of either foil type and are subjected to the following tests:

- Normal storage (results expected after some months to years).
- Elevated temperature and humidity (accelerated test, approximate results expected after two weeks).

Experimental. The pouches are individually weighed with a resolution of 0.1 mg every few days and the average weight gain is recorded versus time. Because the initial value is a measurement (as opposed to a theoretical assumption), there are $n = 5$ coordinates. Subtraction of the initial weight corresponds to a translation in weight-space, with no loss of information. Empty pouches serve as controls (these absorbed such small quantities of water that the water content of the plastic foil could safely be ignored). See Table 4.11.

Analysis. From the graph of the stress test results, linear regression is seen to be appropriate (this in effect means that at these small quantities of water vapor the product is far from thermodynamic saturation). The slopes, in mg of water per day per pouch, are compared:

	Regression with $a \neq 0$	Regression with $a = 0$
b_{thin}:	0.831 ± 0.028	0.871 ± 0.088
b_{thick}:	0.673 ± 0.003	0.688 ± 0.012

Results. The uncertainties associated with the slopes are very different and $n_1 = n_2$, so that the pooled variance is roughly estimated as $(V_1 + V_2)/2$, see case (c) in Table 1.5; this gives a pooled standard deviation of 0.020: A simple t test is performed to determine whether the slopes can be distinguished: $(0.831 - 0.673)/0.020 = 7.9$ is definitely larger than the critical t value for $p = 0.05$ and $f = 3$ (3.182), so that the difference must be regarded as significant. This is what would have been expected, so a one-sided test must be used to estimate the probability of error: $p < 0.001$. Because of this, $p = 0.002$ in a conventional two-sided t table would correspond to the entry we are looking for. Many t tables do not provide the $p = 0.002$ column, however, so that interpolation becomes necessary: $\log(t)$ vs. $\log(p)$ is reasonably linear over the $p = 0.1 \cdots p = 0.001$ interval (check by graphing the data, see program LR and data file INTERPOL.001). At $f = 4$: $t = 2.132$ ($p = 0.1$), 2.776 ($p = 0.05$), 3.747 ($p = 0.02$), 4.604 ($p = 0.01$), resp. 8.610 ($p = 0.001$); interpolation at $x = \log(p = 0.002) \approx -2.7$ yields $\log(t) = 0.8583$, resp. $t \approx 7.22$ (there is a small uncertainty due to the slight curvature); the tabulated value is $t = 7.17$, which shows that the interpolation is acceptably accurate. Because the experimental t value is 7.9, the probability of having made an erroneous decision is $p \leq 0.001$.

The ratio is $0.673/0.831 = 0.81$, which means the thicker foil will admit about 80% of what the thinner one would. The potentially extended shelf life will have to be balanced against higher costs and alternative designs. A very similar result is found if the regression lines are forced through the origin ($a = 0$): The ratio of the slopes is 0.79, but due to the high standard deviation on the first one (± 0.088), the ratios cannot be distinguished.

Comment. The linear regression technique here serves to average the two trends before comparing them. Under these ideal conditions it would have been just as correct to calculate the ratio of weight gains at every point in time; see Table 4.11, last column, and then averaging over all ratios. However, if for one foil a nonlinear effect had been observed, the regression technique, employing only the linear portion of those observations, would make better use of the data.

4.7. STABILITY À LA CARTE

Pharmaceutical products routinely undergo stability testing; the data base thus accumulated serves to substantiate claims as to shelf life and storage conditions. Every year, sufficient material of a production batch is retained to permit regular retesting. In the following case of Fig. 4.7 that occurred many years ago in an obscure and old-fashioned company, stability data pertaining to two batches of the same product manufactured 27 months apart were compared. The visual impression gained from the graph is that

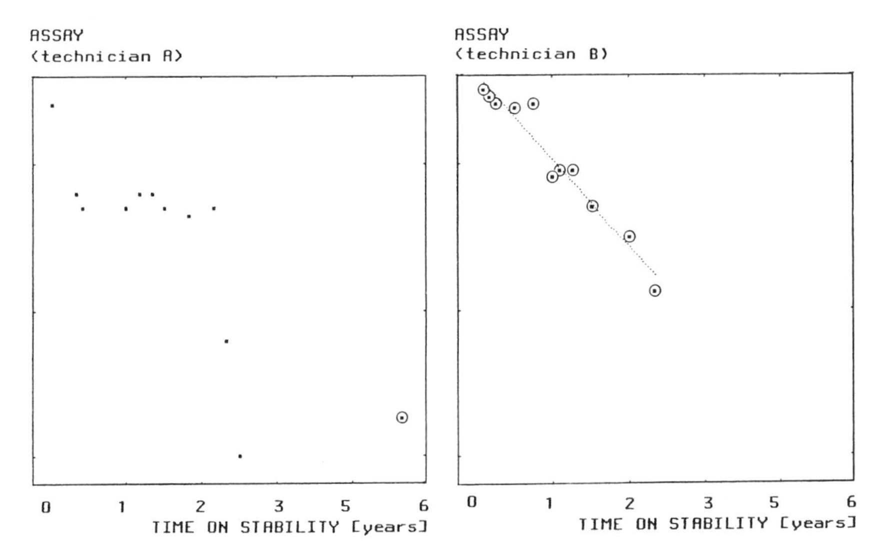

Figure 4.7. Product deterioration according to technicians A (left) and B (right) using the same analytical method. Technician A's results are worthless when it comes to judging the product's stability and setting a limit on shelf life.

something was out of control. Two lab technicians and the hitherto uncontested production manager were confronted, with the result that three culprits were named:

(1) "Technician B fabricated data to make himself look more competent."
(2) "Technician A works sloppily."
(3) "The QC manager ought to keep out of this, everything has worked so well for many years."

Being 26 months into the program for the later batch (53 months for the earlier one), the method was checked and found to be moderately tricky, but workable. Technician B (circles) did the final tests on both batches. When the same slope is used for both sets of data, technician A's residual standard deviation (combined error of the method and that of the operator) is ± 0.88, versus ± 0.11 for technician B. This, together with other information, convinced the new head of quality control that technician A had for too long gone unsupervised, and that statistics and graphics as a means of control were overdue. The production manager who, in violation of the GMP spirit, had simultaneously also served as head of QC for many years, had the habit of glancing over tabulated data, if he inspected at all. The evident conflict of interest and the absence of an appropriate depiction of data inhibited him from finding irregularities, even when a moderately critical eye would have spotted them.

4.8. SECRET SHAMPOO SWITCH

The quality control unit in a cosmetics company supervised the processing of the weekly batch of shampoo by determining, among other parameters, the viscosity and the dry residue. Control charts showed nothing spectacular (see Fig. 4.8, top). The CUMSUM charts were just as uneventful, except for that displaying the dry residue (Fig. 4.8, middle and bottom): The change in trend in the middle of the chart was unmistakable. Since the analytical method was very simple and well proven, no change in laboratory personnel had taken place in the period, and the calibration of the balances was done on a weekly basis, suspicions turned elsewhere. A first hypothesis, soon dropped, stated that the wrong amount of a major component had been weighed into the mixing vat; this would have required a concomitant change in viscosity, which had not been observed. Other components that do not influence the viscosity could have inadver-

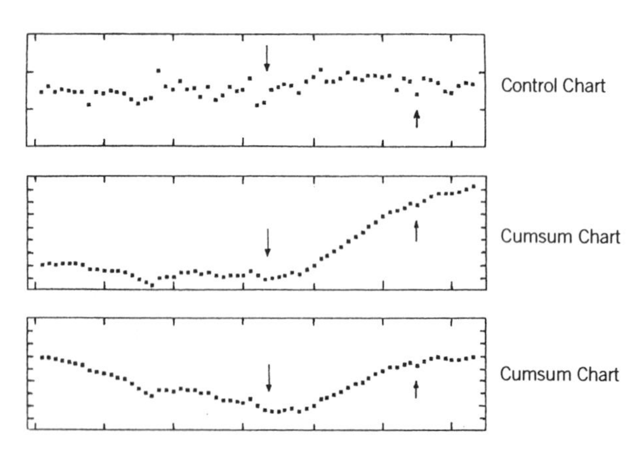

Figure 4.8. In the top panel the raw data for dry residue for 63 successive batches is shown in a standard control chart format. The fact that as of batch 34 (arrow!) a different composition was manufactured can barely be discerned. CUMSUM charts (**a** = average 1 ··· 40, middle panel; **a′** = average 1 ··· 63, bottom panel) make the change fairly obvious, but the causative event cannot be pinpointed without further information. Starting with batch 55 (second arrow!), production switched back to the old composition.

tently been left out, but this second hypothesis also proved unviable because further characteristics should have been affected. A check of the warehouse records showed that the production department, without notifying quality control, had begun to use a thickening agent from a different vendor; while the concentration had been properly adjusted to yield the correct viscosity, this resulted in a slight shift in dry residue (+0.3%), which nicely correlated with what would have been expected on the basis of the dry residues of the pure thickening agents. Twenty-one batches later, the production department canceled the experiment (the trend is again nearly parallel to that on the left of the chart). This case very nicely shows the superiority of the CUMSUM chart over simple control charts for picking up small changes in process average.

That improper weighing (first hypothesis) is not an idle thought is demonstrated by observations of how workers interact with balances. It does not matter whether digital or analog equipment is available: Repetitive work breeds casual habits, such as glancing at a (certain part of) a displayed digit, or the angle of an indicator needle, instead of carefully reading the whole. A 1 and a 7 or a 0 and an 8 do not differ by more than a short horizontal bar on some LCD displays. On a particular model of multiturn scales the difference between 37 and 43 kg resides on a 36 or a 42 display in a small window; the needle in both cases points to the 2 o'clock position because 1 kg corresponds to $\frac{1}{6}$ of a full turn. Mistaking the 4 with the 10 o'clock position is not unknown either; the angle of the needle with respect to the vertical is the same in both cases. There must be a reason for a GMP requirement that stipulates a hard-copy output of the tare and net weights!

4.9. TABLET PRESS WOES

Initial Situation. An experimental granulation technique is to be evaluated; a sample of tablets of the first trial run is sent to the analytical laboratory for the standard batch analysis prescribed for this particular product, including content uniformity (homogeneity of the drug substance on a tablet-to-tablet basis), tablet dissolution, friability (abrasion resistance), hardness, and weight. The last two tests require little time and were therefore done first.

Results

Hardness: 6.9, 6.1, 6.5, 7.6, 7.5, 8.3, 8.3, 9.4, 8.6, 10.7 kg,
Weight: 962, 970, 977, 978, 940, 986, 988, 993, 997, 1005 mg.

Conclusions. Because both parameters give very variable and inconsistent results, more tablets are requested:

Results

Hardness:	7.3, 6.2, 8.4, 8.9, 7.3, 10.4, 9.2, 8.1, 9.3, 7.5 kg,
Weight:	972, 941, 964, 1008, 1001, 988, 956, 982, 937, 971 mg.

Conclusion. Something is wrong with either the product or the analytical procedure; measuring hardness and weighing are such simple procedures, however, that it is hard to place the blame on the very reliable laboratory technician.

Strategy. Since the tablet press is still in operation, an experiment is devised to test the following question: Could the granulate form soft clumps, thus changing the flow properties, so that a few tablets in a row would be heavier or lighter than average (the amount of granulate is known to influence the final hardness)? A suitable plastic tube is connected to the exit chute of the tablet press: The tablets accumulate at the lower end of the tube strictly in the sequence they come off the press.

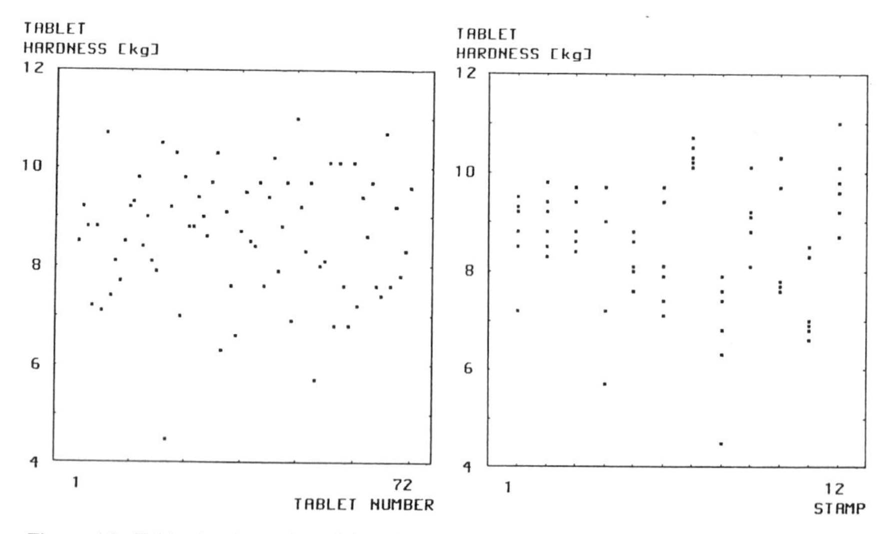

Figure 4.9. Tablet hardness found for the sequence of 72 tablets (left). There is a certain amount of regularity apparent in the hardness data; the data are therefore grouped by stamp (right). Because of the limited hardness resolution (0.1 kg), more than one tablet can yield the same value.

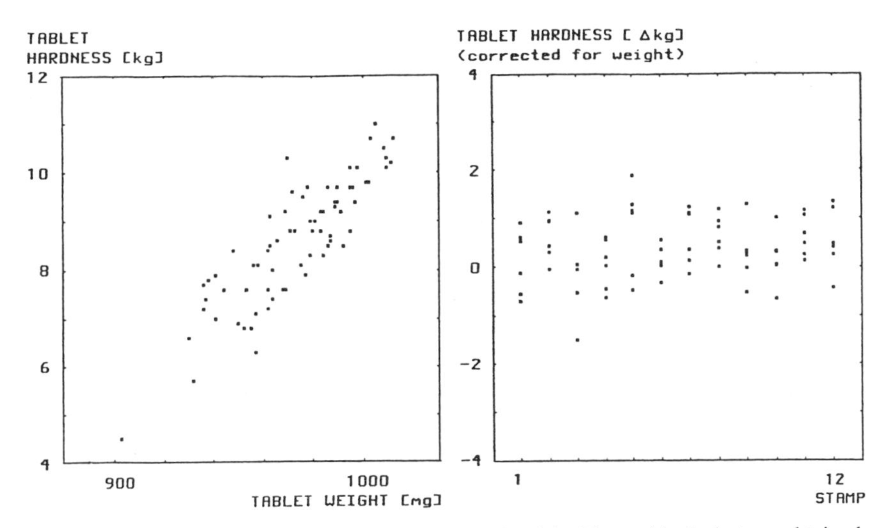

Figure 4.10. Correlation between tablet hardness and weight. The residuals that are obtained after correction for the tablet weight should be compared to Fig. 4.9 (right).

Results. The hardness and the weight are measured (for six tablets from each of the 12 stamps), i.e., for $6 \cdot 12 = 72$ tablets and are graphed (Fig. 4.9).

Conclusion. Evidently 1 out of 12 stamps is not behaving according to expectations. The data for stamp 7 appears suspicious. A hardness vs. weight graph is plotted (Fig. 4.10).

As is known from experience, higher weight yields higher hardness values because of the constant tablet volume the granulation is compressed to. The weight variation is much larger than it should be, and thus the above question of clump formation is partially answered. A linear regression is calculated, yielding

$$H\ [\text{kg}] = -32.6 + 0.0424 \cdot W\ [\text{mg}], \qquad \text{see Fig. 4.10 (left)}.$$

Data Reduction and Interpretation. The average hardness associated with each stamp (Fig. 4.9, right) is subtracted from the corresponding hardness data. Using the above linear regression, the weight-dependent part of the hardness is also subtracted, yielding the residual standard deviations $s_{res} = \pm 0.69$ (kg), a somewhat high, but reasonable dispersion in view of the still preliminary nature of the experiment. Thus it is improbable that the granulation is fully at fault.

The stamp associated with the extreme hardness values (number 7, Fig. 4.9, right) is the next suspect: It is identified and inspected on disassembly of the tablet press. Due to mechanical wear, the movement of the stamp assembly is such that an above-average amount of granulate drops into cavity number 7, and is thus compressed to the limiting hardness supported by the granulate. The tablets from stamp 7 "contaminated" what would otherwise been a fairly acceptable product. Because of the small IPC sample size ($n = 10$), the chances of spotting aberrant tablets are not that good, unless curiosity is aroused through hints or suspicions.

Conclusion. The problem areas are tentatively identified; the galenical development/formulations department is asked to improve the flow properties of the granulate and thus decrease the weight dispersion. The maintenance department will now have to find a proposal for countering the excessive wear on one stamp. Note: On more modern, instrumented, and feedback controlled tablet presses, the described deviations would have become more readily apparent, and mechanical damage could have been avoided.

4.10. SOUNDING OUT SOLUBILITY

Situation. A poorly soluble drug substance is to be formulated as an injectable solution. A composition of 2% w/v is envisaged, while the solubility in water at room temperature is known to be around 3% w/v. This difference is insufficient to guarantee stability under the wide range of temperatures encountered in pharmaceutical logistics (chilling during winter transport). A solubility-enhancing agent is to be added; what is the optimal composition? Two physico-chemical effects have to be taken into account:

- At low concentrations of the enhancer a certain molar ratio enhancer/drug might yield the best results.
- At the limiting concentration the enhancer will bind all available water in hydration shells, leaving none to participate in the solution of the drug substance.

Mathematical Modeling. A function $v = g(u)$ (Fig. 4.11, right) is found in the literature that roughly describes the data $y = f(x)$ (Fig. 4.11, left); since the parameter spaces x and y do not coincide with those of u and v, transformations must be introduced.

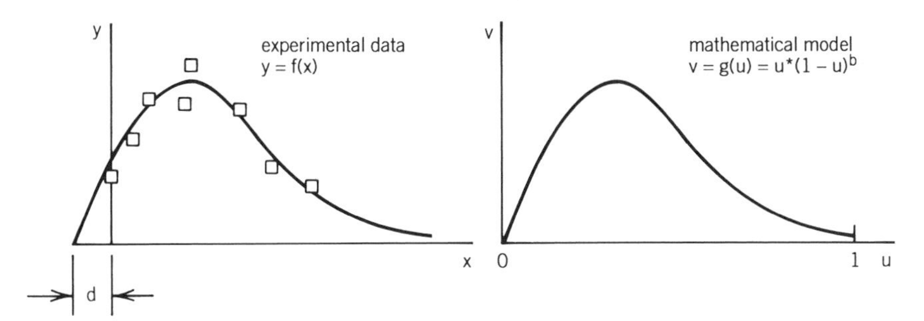

Figure 4.11. The solubility function expected for the combination of the above-mentioned physico-chemical effects (left), and a similar mathematical function (right).

Since u must be between zero and one, experimental data x from Fig. 4.11 would have to be compressed and shifted along the abscissa:

$$u = c \cdot (x + d) \qquad v = u \cdot (1 - u)^b \qquad y = a \cdot u \cdot (1 - u)^b.$$

Three parameters thus need to be estimated, namely, the scalar factor a, the compression factor c, and the shift d. Parameter b was dropped for two reasons: (1) the effect of this exponent is to be explored, so it must remain fixed during a parameter-fitting calculation, and (2) the parameter estimation decreases in efficiency for every additional parameter. Therefore the model takes on the form

$$y(\ ) = f(x(\ ), a, c, d/b),$$

where $y(\ )$ and $x(\ )$ contain the experimental data, a, c, and d are to be optimized for every b, and b is to be varied in the interval $1 \cdots 6$. This function was appropriately defined in a subroutine which was called from the optimization program. Commercial program packages that either propose phenomenological models or fit a given model to data are easily available; such equations, along with the found coefficients, can be entered into program TESTFIT. The authors emphasize that it is strictly forbidden to associate the found coefficients with physico-chemical factors *unless* there is a theoretical basis for a particular model. The optimization was carried out for the exponents $b = 1$ to 6. The resulting six curves are practically equivalent in the x interval $0 \cdots 7$.

Contrary to what is suggested in Section 2.3.2, not the simplest model of those that well represented the data was chosen, but one with a higher exponent. See Fig. 4.12. The reason becomes apparent when the curves

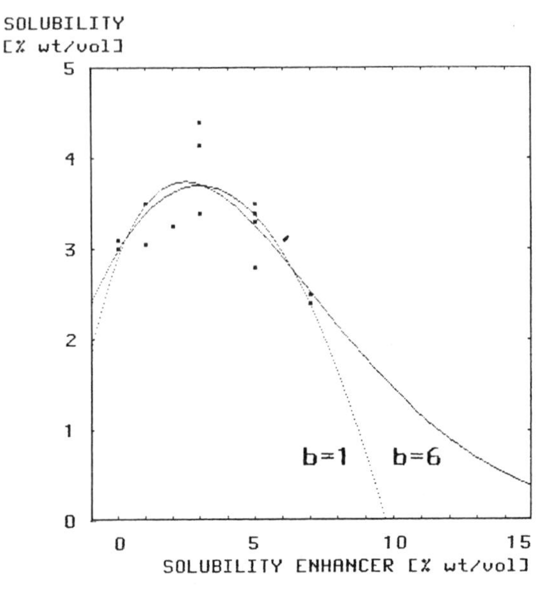

Figure 4.12. Solubility data and fitted models for parameters $a = 14.8$, $b = 1$, $c = 0.0741$, $d = 3.77$, resp., $a = 66.1$, $b = 6$, $c = 0.0310$, and $d = 2.14$.

are compared at $x > 8$: For the model with $b = 1$ a sharp drop in drug solubility at $x = 10$ is predicted, whereas the $b = 6$ model suggests a more gradual reduction in solubility, more in accord with what is expected. The issue could only have been decided with an experiment in the $10 \leq x \leq 12$ region, which was not carried out. The point of interest was the prediction that maximal solubility would be attained for 2–3% of the enhancer, and not for 5%, as had originally, and somewhat naively, been extrapolated from experiments with a drug concentration of 0.5%.

4.11. EXPLORING A DATA JUNGLE

As a new product is taken from the laboratory, through the pilot, and into the production plant, a lot of information will accumulate that helps to better understand the product and the associated production process. In general, the major characteristics are soon evident and will be exploited to optimize the production. Fine-tuning the process can only be done, however, if hitherto hidden aspects are revealed.

Situation. A case in point is a purification/drying step before the sample for in-process control (IPC) is taken. Some of the analytical parameters that were determined were water content in wt.% (Karl Fischer), various residual solvents in ppm (GC), the sum of other impurities in area-% (HPLC), the content of the major compound (HPLC and titration, as determined by comparison with an external standard), and its purity as determined from area-% figures (HPLC). The severity of the drying process had varied somewhat over 43 batches, so that an interesting question was to determine the ultimate quality that was attainable (too much heat would have provoked decomposition). See Table 4.12.

Data Analysis. Because of the danger of false conclusions if only one or two parameters were evaluated, it was deemed better to correlate every parameter with all the others, and to assemble the results in a triangular matrix, so that trends would become more apparent. The program CORREL described in Section 5.2 retains the sign of the correlation coefficient (positive or negative slope) and combines this with a confidence level (probability p of obtaining such a correlation by chance alone).

Prior to running all the data through the program, a triangular matrix of expected effects was set up (Table 4.13) by asking such questions as "if the level of an impurity X increases, will the assay value Y increase or decrease?" (it is expected to decrease). Furthermore, the probability of this being a false statement is expected to be low; in other words, a significant negative event is expected (− −). The five solvents are expected to be strongly correlated because heating and/or vacuum, which would drive off one, would also drive off the others. The titration is expected to strongly correlate with the HPLC content; because titration is less specific (more susceptible to interferences), the titration results are in general a bit higher. The HPLC purity of the major component and the sum of the other impurities, together with the residual solvents, should add to 100%, and a strong negative correlation is expected.

So much for theory: The interesting thing is to ponder what went contrary to expectations, and to try to find explanations. It is here that graphical presentations are helpful to judge the magnitude of any effect found.

Interpretation (cf. Tables 4.14 and 4.15). It seems as if the area-% purity figures derived from HPLC runs (column 9) were strongly positively correlated with the presence of residual solvents (columns 1–5, not as expected) and negatively with content determinations (columns 7–8, as expected), see Tables 4.14 and 4.15. Also, the "other impurities" category (column 6) correlates with the assays. This can be explained by the fact that solvents and some other impurities that are nearly invisible to the UV

Table 4.12. Excerpt from the IPC Results.

[Nine parameters measured on each of 43 samples. Solvent A is water (%), B–E are organic solvents (ppm), organic impurities (area-%), assays (%), and purity (area-%). The last three columns pertain to the dehydrated product. Data file JUNGLE.002 provides a (synthetic) case study]

i	1 Solvent A	2 Solvent B	3 Solvent C	4 Solvent D	5 Solvent E	6 Other Impurities	7 Assay HPLC	8 Assay Titration	9 Purity HPLC
1	4.1	6.9	11.0	13	23	0.6	98.0	98.0	98.7
2	5.1	6.2	11.3	15	34	0.8	94.9	97.0	98.1
3	6.3	7.0	13.3	18	46	0.7	94.0	94.0	98.0
.	.	.	.	.	.	.	.	.	.
.	.	.	.	.	.	.	.	.	.
43	6.1	7.1	13.2	16	30	0.2	95.5	95.6	99.4
$\bar{x}$	5.94	7.39	13.33	16.1	31.26	0.40	95.52	96.09	98.95
s_x	±1.2	±0.77	±1.55	±2.7	±14.0	±0.25	±2.30	±2.61	±0.56

detector distort the summation process, so that 100 area-% cannot be truly compared to 100 weight-%. In other words, apples and oranges should not be compared; with this in mind, though, apparent discrepancies can be used to point out weaknesses of the methodology. The issue has not been fully resolved, but essentially the concept that the sum of all impurity peak areas is indicative of purity and must be revised: Most probably, the impurity peaks will have to be individually calibrated (no small feat if the exact nature of some impurities is unknown) to obtain a sum of all impurity weights. Absolute values of $p = 5\%$ and less were taken to mean "significant effect" or even "highly significant effect"; an absolute p larger than about 20% indicates that this might well be a chance result. Three correlations that were plotted are shown in Figs. 4.13–4.15 for illustrative purposes.

The fact that in HPLC only UV-active components are registered, whereas in titration all basic functional groups are detected, constitutes a difference in specificity (quality) and sensitivity (quantity) of these two methods relative to a given impurity; see Fig. 4.13. Solvent A (water) behaves differently from the other four, as can be seen from Fig. 4.14; as it

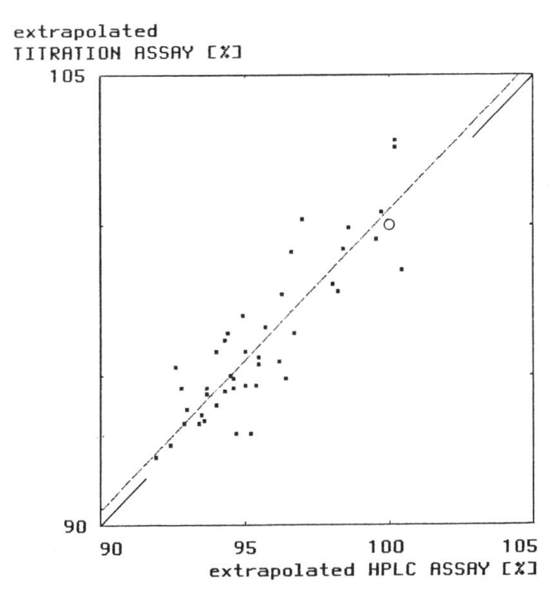

Figure 4.13. HPLC assay (column 7) and titration assay (column 8) are compared: Evidently titration yields the higher values (dashed line vs. solid line = diagonal); the reason is that one of the major impurities is of basic nature. The circle denotes the pure compound and perfect selectivity for both techniques.

Table 4.13. Expected Correlations.
[Symbolic representation with (+ +) and (− −) meaning (very) significant, (+) and (−) meaning (barely) significant.]

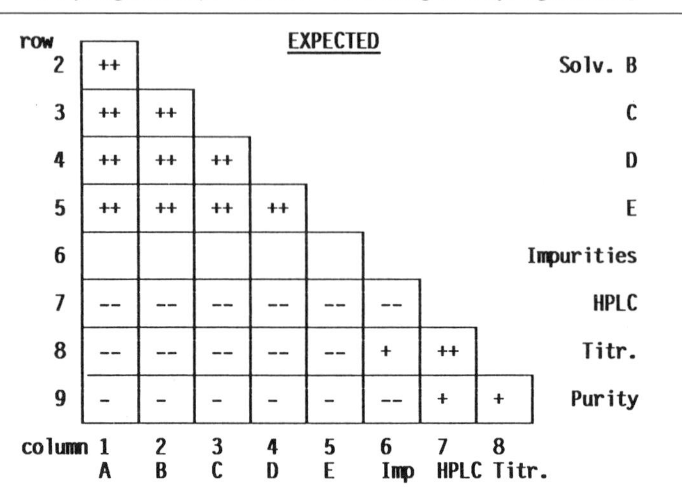

EXPECTED

row	1 A	2 B	3 C	4 D	5 E	6 Imp	7 HPLC	8 Titr.	
2	++								Solv. B
3	++	++							C
4	++	++	++						D
5	++	++	++	++					E
6									Impurities
7	--	--	--	--	--	--			HPLC
8	--	--	--	--	--	+	++		Titr.
9	-	-	-	-	-	--	+	+	Purity

Table 4.14. Results of the Numerical Analysis.
[The numbers give the probability in percent that the correlation is due to chance alone; the n in a cell indicates that a negative slope was found for the correlation line. Column and row numbers relate to the column numbers in Table 4.12.]

FOUND

row	1 A	2 B	3 C	4 D	5 E	6 Imp	7 HPLC	8 Titr.	
2	6								Solv. B
3	0	0							C
4	0	.2	0						D
5	0	35	0	0					E
6	.3n	19n	.6n	2 n	10n				Impurities
7	0 n	18n	0 n	0 n	0 n	24			HPLC
8	0 n	15n	0 n	0 n	0 n	5	0		Titr.
9	.1	10	.1	.4	.1	0 n	7 n	1 n	Purity

Table 4.15. Matrix of Raw Effects.
[Symbolic representation, as above; a low probability of chance effects ($p \leq 0.05$) means that there is a strong correlation between the parameters (+ + or − −).]

EFFECTS

row	1 A	2 B	3 C	4 D	5 E	6 Imp	7 HPLC	8 Titr.	
2	+								Solv. B
3	++	++							C
4	++	++	++						D
5	++		++	++					E
6	--	-	--	--	-				Impurities
7	--	-	--	--	--				HPLC
8	--	-	--	--	--	++	++		Titr.
9	++	+	++	++	++	--	-	--	Purity

Table 4.16. Deviation Matrix (Schematic).
[Correlations that turned out weaker (w) or stronger (s) than expected are indicated; if the slope is opposite to that expected and similar (*) or opposite and stronger (**) this is depicted by asterisks.]

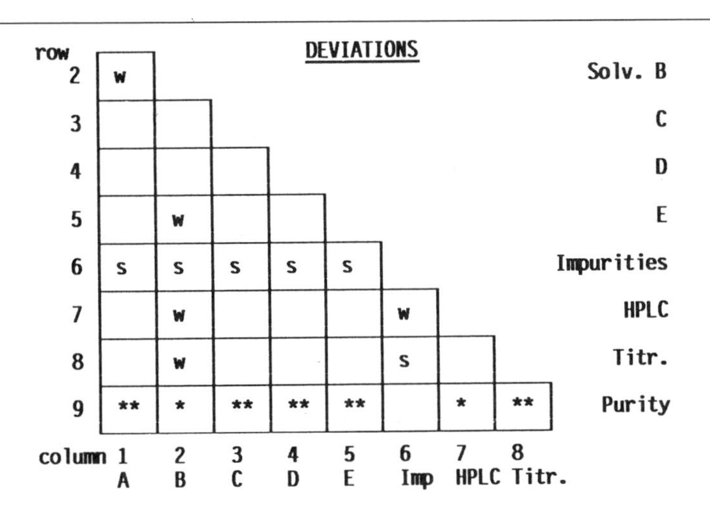

DEVIATIONS

row	1 A	2 B	3 C	4 D	5 E	6 Imp	7 HPLC	8 Titr.	
2	w								Solv. B
3									C
4									D
5		w							E
6	s	s	s	s	s				Impurities
7		w				w			HPLC
8		w				s			Titr.
9	**	*	**	**	**		*	**	Purity

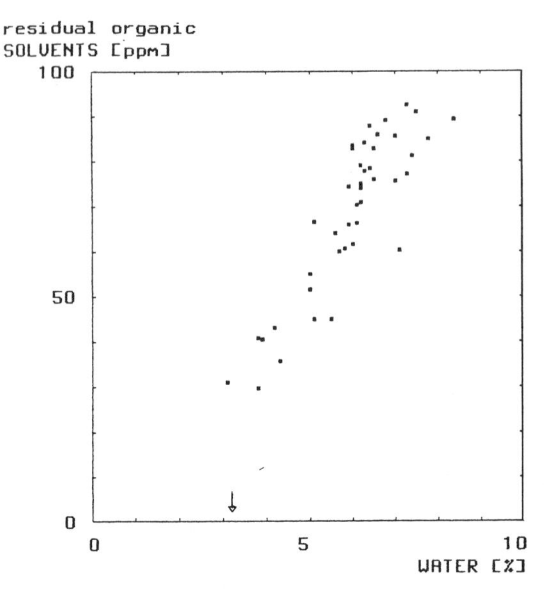

Figure 4.14. The sum of all organic solvents (columns B–E in Table 4.12) is plotted versus the residual water (column A). The drying step obviously drives off organic solvents and water to a large degree, depending on the severity of the conditions. Organic solvents can be brought down to 30 ppm, while only the excess water can be driven off, the remainder being water of crystallization (arrow: theoretically expected amount: 3.2%).

turned out, there is a crystal modification that incorporates some water, and moderate drying will drive off only the excess. This is only partly reflected in Table 4.16, column A; for this reason tabular and graphic information has to be combined. The strongly negative correlation between the "other impurities" category and the solvents C–E does not match expectations; a logical explanation could be that harsher drying increases decomposition. Solvent B, which is an alcohol behaves more like water (A) than the apolar solvents (C–E).

The technique is also used to ferret out correlations between impurities within the same HPLC chromatogram: If several reaction pathways compete for reagent, each with its own characteristic impurity profile, any change in conditions is likely to affect the relative importance of the individual pathways. Finding which impurities move in concert helps to optimize a process. Data file PROFILE.001 contains an example: 11 peak areas were determined in each chromatogram for nine production runs. Impurities 5, 6, and 8 appear to be marginally correlated to the others, if at all, while the product strongly correlates with impurities 1–4, 7, the solvent, and the reagent. Since impurities 5, 6, and 8 are far above the

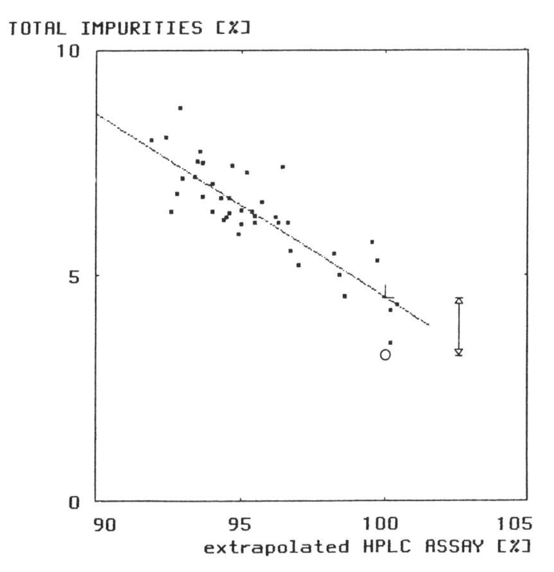

Figure 4.15. Total impurities (columns 1–6, including water of crystallization, versus the extrapolated HPLC assay of the major compound (column 7); the extrapolation takes account of the water of crystallization. The circle marks the hypothetically pure compound: 3.2% water of crystallization, but no other impurities. The arrow indicates the percentage of impurities expected (for this simple linear model) to remain in the product after all solvents and excess water have been driven off.

detection limit, analytical artifacts cannot be the reason. The correlation graph is depicted in Fig. 4.26.

These examples show that unless the interdependencies between various parameters are clearly reflected in the measurements, some interpretations may (apparently) contradict others. This should be taken as a hint that the type of analysis possible with program CORREL is of exploratory nature and should be viewed as food for thought.

4.12. SIFTING THROUGH SIEVED SAMPLES

Situation. There are two vendors for a particular bulk chemical that meet all specifications. The products are equally useful for the intended reaction as far as the chemical parameters are concerned; one physical parameter, the size distribution of the crystals, however, does not seem to match. Because the speed of dissolution might become critical under certain combinations of process variables, the chemical engineers would favor a more finely divided raw material. On the other hand, too many fines could also cause problems (dust, static charging).

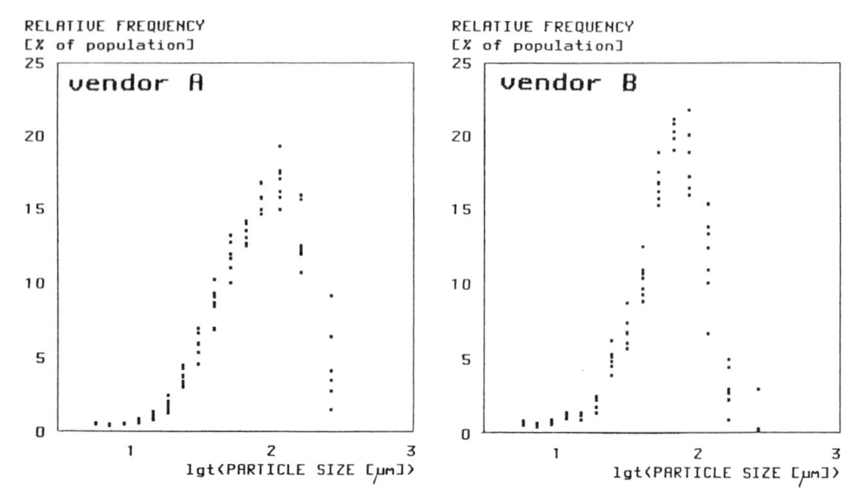

Figure 4.16. Relative weight of material per fraction for two different vendors (A: left; B: right).

Question
Are the materials supplied by the two vendors systematically different?

Course of Action. A laser-light scattering apparatus LLS is used to measure the size distributions. The results are given in weight-% per size class under the assumption that all material falls into the 15 classes between 5.8 and 564 μm. For the seven most recent deliveries from each of the two vendors samples are obtained from the retained sample storage. The 14 size distributions are measured and the average distribution for each vendor is calculated, see Fig. 4.16 and Table 4.17. As it turns out, one vendor's material contains almost no particles (0.5%) in the 261–564-μm class (bin 15); this means that the %-weight results accurately represent the situation. The other vendor's material, however, contains a sizable fraction (typically 5%, maximally 9%) in this largest size class; this implies that 1–5% "invisible" material is in the size class > 564 μm. Evidently then, the size distribution curve for this second material is accurate only on a relative basis, but not absolutely (closure plays a role here[143]); distorted data are a poor foundation for a statistical analysis. Thus, there are two ways to continue:

- Proceed as if the data were not distorted and carry in mind that any result so obtained is biased.

Table 4.17. Percentage of Material per Size Class (564-261 μm in Bin 15, 7.2-5.8 μm in Bin 1, Logarithmic Classification) for Samples 1–7 of Each of Two Vendors.

[The corresponding group averages are given in the two columns at the right. The Euclidian distance for each sample is given relative to the Group average A (*), respectively *B* (**).]

Vendor A							Vendor B								Group Average		Bin
1	2	3	4	5	6	7	1	2	3	4	5	6	7		A	B	No.
9.2	1.5	6.5	3.5	4.1	2.8	6.5	0.0	2.9	0.1	0.0	0.0	0.1	0.3		4.87	0.49	15
16.0	10.8	12.4	12.0	15.7	12.2	12.6	0.9	5.0	2.7	2.7	2.8	2.9	4.4		13.10	2.99	14
17.5	16.3	15.0	17.1	19.3	17.7	15.9	6.7	10.1	10.9	15.4	13.8	13.4	12.4		16.97	11.81	13
15.0	15.8	14.7	16.9	15.9	16.8	15.0	16.5	16.0	17.3	21.8	20.1	18.9	16.0		15.73	18.09	12
12.7	14.2	12.6	13.1	13.6	14.1	14.1	20.9	19.9	20.9	21.2	19.9	20.3	19.1		13.49	20.31	11
10.1	13.3	11.1	11.7	11.7	12.8	12.0	18.9	16.3	16.9	15.3	15.8	16.8	17.6		11.81	16.80	10
6.9	10.3	9.1	9.4	7.0	8.7	8.5	12.5	10.7	10.9	8.8	9.7	9.3	10.4		8.56	10.33	9
4.6	7.0	6.7	5.9	4.6	5.4	6.0	8.7	6.8	7.4	5.7	6.7	6.1	6.7		5.74	6.87	8
3.0	4.3	4.5	3.8	3.2	3.4	3.7	6.2	4.5	5.1	3.9	4.5	4.8	5.3		3.70	4.90	7
1.5	2.0	2.5	2.1	1.3	1.6	1.8	2.5	2.4	2.4	1.4	1.8	2.2	2.5		1.83	2.17	6
0.9	1.1	1.4	1.3	0.8	1.0	1.0	1.4	1.4	1.4	0.9	1.2	1.2	1.3		1.07	1.26	5
0.6	0.9	0.9	0.9	0.7	0.9	0.8	1.4	1.0	1.2	1.0	1.2	1.1	1.1		0.80	1.14	4
0.5	0.6	0.6	0.5	0.5	0.6	0.5	0.9	0.8	0.7	0.6	0.6	0.7	0.8		0.54	0.73	3
0.4	0.5	0.5	0.4	0.4	0.5	0.4	0.6	0.7	0.6	0.4	0.5	0.6	0.6		0.44	0.57	2
0.5	0.6	0.6	0.6	0.5	0.6	0.5	0.8	0.7	0.7	0.5	0.6	0.6	0.7		0.56	0.66	1
6.0	5.0	3.4	2.4	4.2	2.9	2.3	20.4	13.6	16.2	15.9	14.8	14.9	13.8	*			
20.3	11.7	15.5	14.2	18.1	13.7	14.9	6.9	4.2	1.7	6.0	3.2	2.2	3.0	**			

- Employ a model that mathematically describes a size distribution of this type, adjust the model parameters for best fit, and estimate the missing fraction above 564 μm; after correcting the observed frequencies, continue with a correct statistical analysis.

The second option falls out of favor for the simple reason that such a model is not available, and if it were, errors of extrapolation would be propagated into any result obtained from the statistical analysis.

It appears that one distribution is sharper than the other; a χ^2 test is applied to the group means to confirm the difference [Eq. (1.50)]: $\chi^2 = 20.6$ or 95.7 is found, depending on which distribution is chosen as reference. Since there is no theoretical distribution model to compare against, the choice of reference is arbitrary. The critical $\chi^2(p = 0.95, f = 14)$ is 23.7, which means that H_1 could have to be rejected under one perspective. The above-mentioned distortion of the data from the coarser material might have tipped the scales; this is a classical case where the human eye, used to discriminating patterns, sees something that a statistical test did not. A disturbing aspect is the fact that the individual results scatter so much so as to obscure any difference found between the means; see Fig. 4.16. The situation can be improved by regarding the cumulative (Fig. 4.17) instead of the individual frequencies, because through the summation the signal/noise ratio is improved.

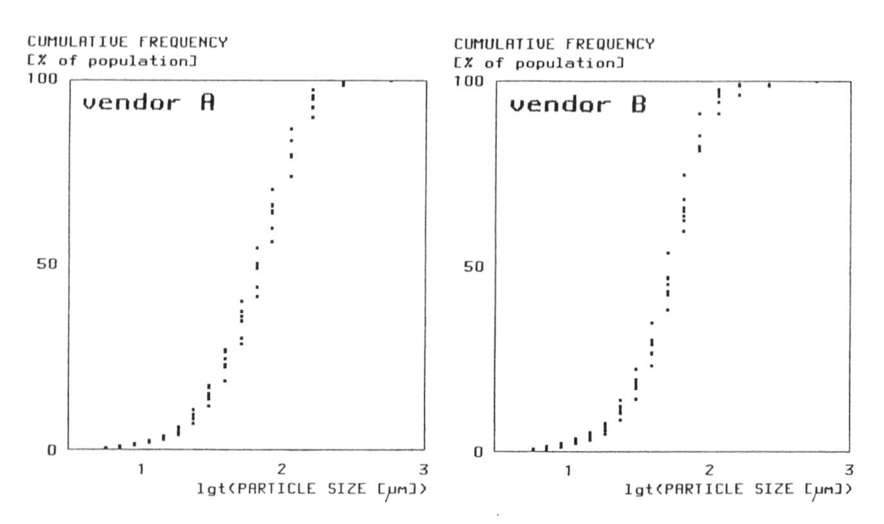

Figure 4.17. Cumulative weight per fraction for two different vendors (A: left; B: right).

Euclidian Distance. On the basis of the given evidence, the size distributions are different, but this is not fully borne out by the statistical test employed. To overcome this impasse, another technique is employed that allows each sample to be judged according to its proximity to given points. Cluster analysis (finding and comparing "distances" in 15-dimensional space) shows up a difference between the products; the disadvantage is that CA strains the imagination. In order to rectify this, the Euclidian distances separating every point from both the average of its group and that of the other group are projected into two dimensions. Four things are necessary:

(1) For each of the two groups, the average over every one of the 15 dimensions (classes, bins) is calculated (columns A and B in Table 4.17).

(2) Corresponding elements in the two vectors of averages are subtracted, and the differences are squared and added. The square root of the sum (15.21) is equal to the Euclidian distance in 15 dimensions separating the two points that represent the group averages. This distance forms the baseline in Fig. 4.18.

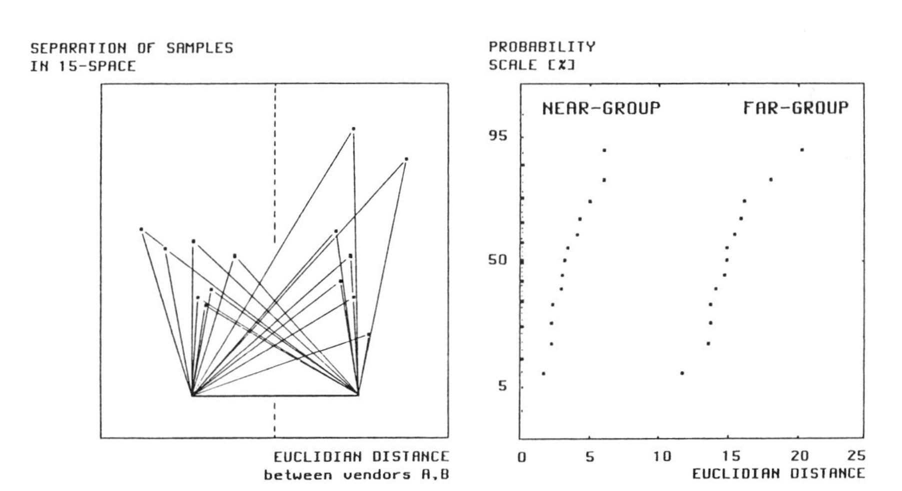

Figure 4.18. Euclidian distances of the 14 samples from their group averages; the individual points can be clearly assigned to either the left or the right group by testing against the dashed separation line. The Euclidian distances were also analyzed using the program NPP: The (absolute) distances are more or less normally distributed and yield nearly straight lines; the distinction between near- and far-group distances is evident in a horizontal separation of the two lines.

(3) Corresponding elements in the vector representing one particular sample and in the appropriate vector of averages are worked up as in (2) to find the Euclidian distance between point i and its group average (see lines marked with an asterisk and a double asterisk in Table 4.17); this forms the second side of the appropriate triangle in Fig. 4.18.

(4) In order to find the third side of the triangle, proceed as under (3) by replacing one vector of averages by the other.

Example: for sample 1, vendor A:

for point 1:	9.20, 16.00, 17.50, ..., 0.50	$A1$
vector of averages 1:	4.87, 13.10, 16.97, ..., 0.56	A
vector of averages 2:	0.49, 2.99, 11.81, ..., 0.66	B

The baseline is $b = 15.2$, with $b^2 = (4.87 - 0.49)^2 + \cdots (0.56 - 0.66)^2$,

side a is $a = 6.0$, with $a^2 = (9.20 - 4.87)^2 + \cdots (0.50 - 0.56)^2$,

side c is $c = 20.3$, with $c^2 = (9.20 - 0.49)^2 + \cdots (0.50 - 0.66)^2$.

The corresponding Cartesian coordinates are $x = -12.3$ and $y = 3.7$ if the group averages are set to $x_A = -15.21/2$ and $x_B = +15.21/2$, and $y_A = y_B = 0$.

This illustrative technique suffers from a lack of a statistically objective measure of probability. The comparison is done visually by judging the distance of the center of a group of seven points from the center line and taking into account the diameter of a group of points, or by using the right-hand panel in Fig. 4.18 or the lines marked with an asterisk or a double asterisk in Table 4.17, and looking for an overlap in near- and far-group Euclidian distances in less than, say, 1 sample out of 10 (the smallest far-group ED = 11.7, the largest near-group ED = 6.9, so there is no overlap in this particular case).

Interpretation. Using Euclidian distances, the difference between the vendor's samples shows up nicely, see data file SIEVE.001. If some samples of groups A and B had been similar, an overlapping of the two distributions would have been seen.

Instead of comparing each sample against the group averages, long-term averages or even theoretical distributions could have been employed. Program EUCLID provides additional information: the %-contribution of each bin to the total Euclidian distance. This can be used to reduce the number of dimensions included in the analysis.

In this particular example, the individual bins all carry the same dimension and are mutually coupled through the condition Σ(bin contents) = 100%. If unrelated properties were to be used in a comparison, all of the employed results must be numerically similar if each property is to contribute to the Euclidian distance to a roughly comparable amount. As an illustration, calculations involving group averages of, say, 8.3 area-%, 550 ppm, and 0.714 AU would yield wholly different interpretations than if 0.083, 0.052%, and 714 mAU had been used. For this reason, unrelated vectors are first normalized so that the overall average over groups A and B is 1.00 for each bin. An example is provided in data file EUCLID.001, where two groups of data ($n_A = 6$, $n_B = 5$) are only marginally different in each of eight dimensions, but can be almost perfectly separated visually. Normalization is achieved by using program DATA, option "Modify."

4.13. CONTROLLING CYANIDE

Situation and Criteria. A method was to be developed to determine trace amounts of cyanide (CN^-) in waste water. The nature of the task means precision is not so much of an issue as are the limits of detection and quantitation (LOD, LOQ), and flexibility and ease of use. The responsible chemist expected cyanide levels below 2 ppm.

Experimental. A photometric method was found in the literature which seemed to suit the particular circumstances. Two cyanide stock solutions were prepared, and an electromechanical dispenser was used to precisely prepare solutions of $20, 40, \ldots, 240$, respectively, $10, 30, 50, \ldots, 250$ μg $CN^-/100$ ml. 10 ml of each calibration solution were added to 90 ml of the color-forming reagent solution and the absorbance was measured using 1-cm cuvettes; see Table 4.18.

Data Analysis. The results were plotted; at first glance a linear regression of absorbance versus concentration appeared appropriate. The two dilution series individually yielded the figures of merit given in Table 4.18, bottom. The two regression lines are indistinguishable, have tightly defined slopes, and pass through the origin. The two data sets are thus merged and reanalyzed (see Table 4.19, left column). This rosy picture dissolves when the residuals (Fig. 4.19) are inspected: the residuals No. 8 (1.3 $\mu g/ml$) and 27 (2.4 $\mu g/ml$) are far from the linear regression line. The question immediately pops up, "are these points outliers?" One could now go and drop one or both points from the list, repeat the regression analysis, and determine whether (a) the residual standard deviation had become smaller, and (b) whether these points were still outside the CL(y).

Table 4.18. Raw Values and Linear Regression Results for Three Calibration Series. The Concentrations Pertain to the Stock Solutions.

No.	Conc.	Absorb.	No.	Conc.	Absorb.	No.	Conc.	Absorb.*
1	0	0.000	15	0	0.000	28	0	0.0000
2	10	0.049	16	20	0.099	29	2	0.0095
3	30	0.153	17	40	0.203	30	4	0.0196
4	50	0.258	18	60	0.310	31	6	0.0298
5	70	0.356	19	80	0.406	32	8	0.0402
6	90	0.460	20	100	0.504	33	10	0.0501
7	110	0.561	21	120	0.609			
8	130	0.671	22	140	0.708			
9	150	0.761	23	160	0.803			
10	170	0.863	24	180	0.904			
11	190	0.956	25	200	0.997			
12	210	1.053	26	220	1.102			
13	230	1.158	27	240	1.186			
14	250	1.245						

Series 1	Series 2	Series 3	Item
0.00501 ± 1.0%	0.00497 ± 1.0%	0.00504 ± 1.7%	slope b
0.00588 ± 123%	0.00611 ± 113%	−0.00033 ± 160%	intercept a
±0.0068	±0.0060	±0.00027	res. SD s_{res}
0.9997	0.9998	0.9998	coeff. determ. r^2
14	13	6	n
1.4	1.4	0.1	LOD $\mu g/100$ ml
3.0	3.0	0.2	LOQ $\mu g/100$ ml

where No.: number of measurement, Conc.: concentration in μg $CN^-/100$ ml, Absorb.: absorbance [AU], slope: slope of regression line ± one half of the relative confidence interval in percent, intercept: see slope, res. SD: residual standard deviation s_{res}, n: number of points in regression, LOD: limit of detection, LOQ: limit of quantitation, * measurements using a two-fold higher sample amount and 5-cm cuvettes; i.e., measured absorption $0 \cdots 0.501$ was divided by 10.

Table 4.19. Coefficients (Rounded) Found for Linear and Quadratic Regression over Points 1–27.

Linear	Quadratic	Item
0.005843	−0.002125	Constant term
0.004990	+0.005211	Linear term
	−0.0000009126	Quadratic term
±6.6 mAU	±4.5 mAU	res. SD

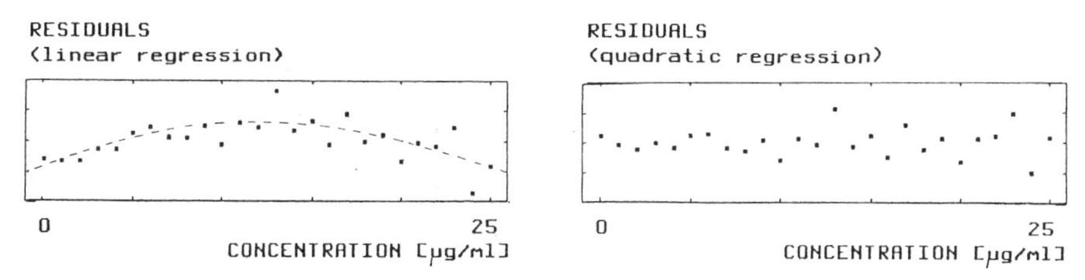

Figure 4.19. Residuals for linear (left) and quadratic (right) regressions; the ordinates are scaled ±20 mAU. Note the increase in variance towards higher concentrations (heteroscedacity). The dashed line was plotted as the difference between the quadratic and the linear regression curves. Concentration scale: 0-2.5 μg/ml, final dilution.

On the other hand, the residuals in the middle of the concentration range are positive, those at the ends negative; this, together with the fact that a photometer is being used, should draw attention to the hypothesis H_1: "curved calibration function." (Note to the nonchemist: Stray light in the photometer dominates at high absorbances, which can be responsible for lower slopes at high concentrations. The chemical workup can also produce lower yields of the chromophore at higher concentrations). The curvature is quite evident (cf. Fig. 4.19), which makes H_1 all the more probable. A quadratic regression is applied to the merged data set and the residuals are again plotted. The quadratic regression (QR) $Y = a + b \cdot x + c \cdot x^2$ is a straightforward extension of the linear regression concept[49, 102], with programs available at all computing centers, for PCs, and even for some hand-held computers; see Table 4.19, right column. There seems to be a clear reduction in the residual standard deviation, and the F test supports this notion: $F = (6.6/4.5)^2 = 2.15$, with $F(26, 26, 0.05) = 1.93$. Point No. 8 (see Fig. 4.19) is now only 0.011 AU above the parabola, which means it is barely outside the $\pm 2 \cdot s_{res}$ band; all other residuals are smaller. From the practical point of view there is little incentive for further improvement: The residual standard deviation ±4.5 mAU is now only about twice the experimental standard deviation (repeatability), which is not all that bad when one considers that two dilutions and a derivatization step are involved.

Decisions. Because quadratic regressions are more difficult to handle and the individual coefficients of a three-parameter model are less well defined than those of a two-parameter one in the case of weak curvature, any gain from using a polynomial of higher order might well be lost through error propagation. The definite course of action was to accurately calibrate a part of the given concentration interval and to either dilute samples to fit the range, or then to use thicker (5 cm) cuvettes to gain sensitivity. In case this strategy should not work, it was decided to also calibrate the 0–10-μg/100-ml region (calibration points 28–33; for results, see Table 4.18. This regression line is indistinguishable from the other two as far as the coefficients are concerned, but the LOD and LOQ are much lower). The overall operating range thus covers 2 ppb (20-ml sample amount, 5-cm cuvette) to over 200 ppm (0.1-ml sample amount, 1-cm cuvette) CN^- in the sample. In a screening run the sample is diluted according to a standard plan trimmed for speed and ease, and depending on this preliminary result, the sample is only then precisely diluted if there is impending danger of getting high cyanide levels that would require further treatment of the wastewater. A simple linear regression is used for the approximately linear portion of the calibration function. Another course of action would be to improve the chemical workup and the instrumental measurement procedures to obtain a linear calibration to higher concentrations. The economies of further method development versus occasional repetition of a dilution would have to be investigated. If a programmable sample carousel/dilutor/UV configuration were used, this repetition could be enacted automatically if an alarm limit is exceeded in the first measurement.

4.14. AMBIGUOUS AUTOMATION

A pharmaceutical specialty is produced in three dosage strengths (major component A); A and a second component B are controlled by HPLC for batch release purposes. It is decided to replace the manual injection of the sample solution by an automatic one. What is gained? Cross-validation of the methods is effected by running both methods on each of 10 samples. The mean and the standard deviation for each series of 10 measurements is calculated in Table 4.20 .

Observations

(1) It appears that automatic injection actually worsens precision (± 5.88 ▶ $\pm$ 14.2, etc.).

(2) The relative standard deviation suffers, too: ± 1.18 ▶ $\pm$ 2.87, etc.; the new variance component corresponds to $\sqrt{2.87^2 - 1.18^2} = \pm 2.6$ (2.6, 1.3, and 1.8 for component A, 0, 1.4, and 1.5 for component B). This is due to the additional spread along the diagonal because of the variability in the injected volume.

Table 4.20. Means, Individual Absolute, Relative, and Residual Standard Deviations of Components A and B for Low-, Medium-, and High-dosage Strengths of Component A; see Fig. 4.20.

The residual SD for the calibration measurements is an indicator of repeatibility (e.g., ±0.75%); the rest of the overall spread of the results (e.g., ±2.32%) is due to manufacturing variability.

	Manual Injection			Automatic Injection			
	Low	Medium	High	Low	Medium	High	
Mean:							
A:	497	750	992	493	753	1010	
B:	360	359	357	361	356	355	→ avg = **358.0**
Standard Deviation (±):							↓
A:	5.88	5.51	14.6	14.2	11.1	23.8	
B:	7.33	5.51	6.36	5.39	7.32	8.23	→ avg = **±6.79**
Relative Standard Deviation (±%):							↓
A:	1.18	0.73	1.47	2.87	1.47	2.36	
B:	2.04	1.54	1.78	1.49	2.06	2.32	avg = **±1.9%**
Residual Standard Deviation (± absolute and ±% of average value *B*):							↓
	7.03	4.22	5.66	5.56	4.46	2.68	
	1.95%	1.17%	1.59%	1.54%	1.25%	0.75%	→ avg = **±1.4%**

(3) The relative SD was a bit higher than the residual SD before the change, e.g., $\sqrt{2.04^2 - 1.95^2} = 0.6$ for component B (±0.6, ±1.0, and ±0.8; average ±0.8).

(4) After the change, the differences correspond to a larger variance component (±0, ±1.6, ±2.2; average ±1.6).

(5) The residual SD actually improves from ±1.95 ▶ ± 1.54, etc., for an average 42% decrease in the variance V_{res}.

Observations 1 and 2 are illusions due to the fact that the above numbers measure the overall spread in one dimension (vertical or horizontal), and do not take into account the correlation between the results *A* and *B* that is very much in evidence in the right side of Fig. 4.20 (automatic injection). The variability can be decomposed into three components:

- A vertical one (identical with the residual standard deviation around the regression line, to be interpreted as the uncertainty in measuring *B*).

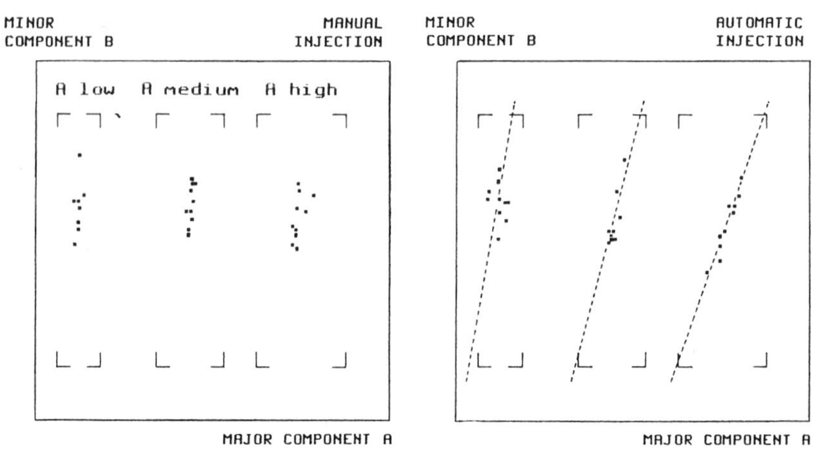

Figure 4.20. Correlation of assay values for components A and B, for three dosage levels of A, with 10 samples per group. The corner symbols indicate the $\pm 10\%$ specification limits for each component. For manual injection (left panel) only relative standard deviations of 1–2% are found, but no correlation. Automatic injection (right panel) has a lower intrinsic relative standard deviation, but the data are smeared out long the proportionality line (dashed) because no internal standard was used to correct for variability of the injected volume. The proportionality line does not go through the corners of the specification box because component B is either somewhat overdosed (2.4%), or because an interference results in too high area readings for B. The nominal values are $A_l = 500$, $A_m = 750$, $A_u = 1000$, and $B = 360$. The RSD is calculated according to Eq. (2.13).

- A horizontal one (similar to the residual standard deviation around the regression line, to be interpreted as the uncertainty in measuring A).
- One along the diagonal (to be interpreted as the uncertainty associated with the injection volume; this in effect yields a proportionality constant $k \leq 1.0$ (bubbles in the injection loop can only decrease the effectively injected amounts A', B' relative to the nominal ones A, B) so that $A' = k \cdot A$ resp $B' = k \cdot B$, with k taking on a different value for each injection. An internal standard would serve to correct k to 1.00.

From this it can be gathered that the automatic injection should eventually lead to more reproducible results (the residual standard deviation decreases by about 20%), but only if the spread along the regression line could be reduced. How repeatable could the results potentially be?

The residual standard deviation is only $\pm 0.75\%$ to $\pm 1.5\%$ relative to the average value B. The additional analytical variability is estimated as

Average (Manual + Automatic)	Range (Manual/ Automatic)	
$\sqrt{(1.9^2 - 1.4^2)} = \pm 1.3\%$	$(\pm 0.8 \cdots \pm 1.6\%)$	Excess variance
$(1.3/1.9)^2 = 47\%$	$(20 \cdots 64\%)$	% contribution to variance
$\pm 1.4 \cdot \sqrt{1 - 0.47} = \pm 1.0$	$(\pm 1.3 \cdots \pm 0.7\%)$	Potential precision of assay for component B if the internal standard helps to eliminate the variance associated with the effectively injected volume

This means that something like one-half of the B variance is due to this lack of control; it might thus be possible to achieve a repeatability of $\pm 1\%$ on both components. The fact that the potential precision ($\approx \pm 1.3\%$, with IS) for the manual injection is hardly smaller than that achieved without the benefit of an internal standard ($\approx \pm 1.8\%$) shows that skillful work was being done. The question is now why in the case of the manual injection, which shows little or no correlation between A and B, repeatabilities of no less than ± 1.5–1.8% are observed. There is an explanation: The automatic injector is more reproducible in terms of the time necessary to turn the valve; this means that the injected volume is less smeared out and yields a better integrable peak form, a notion confirmed by the actual chromatograms. The interpretation for the medium- and low-dosage forms is essentially the same. Note that at high levels of component A the repeatability (standard deviation) for A sharply rises ($\pm 11.1 \blacktriangleright \pm 23.8$), but not so for B; the reason is related to saturation effects in the HPLC column. This could be avoided by injecting two separate sample dilutions, one optimized for the reproducibility of the A peak, the other for the B peak.

The moral of the story: The satisfaction of having all points inside the specification limits should not induce inaction. An internal standard was thereupon added to the procedure, and promptly, the dispersion along the diagonal in Fig. 4.20 (right) was eliminated.

4.15. MISTRUSTED METHOD

An in-process control (IPC) of a bulk chemical was augmented by a heavy-metals test because trace quantities of a catalyst were suspected to have a deleterious effect on the following synthesis step. Because the identity of the metal was known, a simple precipitation as the sulfide was deemed to give sufficiently accurate answers in a very short time (proviso: no other heavy metal present). A test is conducted wherein a reference solution containing 20 ppm of the metal chloride is treated in parallel to the sample, and the intensities of the coloration of the suspensions are compared 3–5 minutes after mixing (the finely divided suspension later coalesces and precipitates out of solution). Concentrations much higher than 20 ppm would be accommodated by further dilution of the sample. The relative confidence interval is judged to be around $\pm 25\%$. The four batches in question were found to contain about 20, 40, 20, respectively, 90 ppm (cf. Fig. 4.21).

The production department was not amused, because lower values had been expected. Quality control was blamed for using an insensitive, unselective, and imprecise test, and thereby unnecessarily frightening top management. This outcome had been anticipated, and a better method, namely polarography, was already being set up. The same samples were run, this time in duplicate, with much the same results. A relative confidence interval of $\pm 25\%$ was assumed. Because of increased specificity, there were now less doubts as to the amounts of this particular heavy metal that were actually present. To rule out artifacts, the four samples were sent to outside laboratories to do repeat tests with different

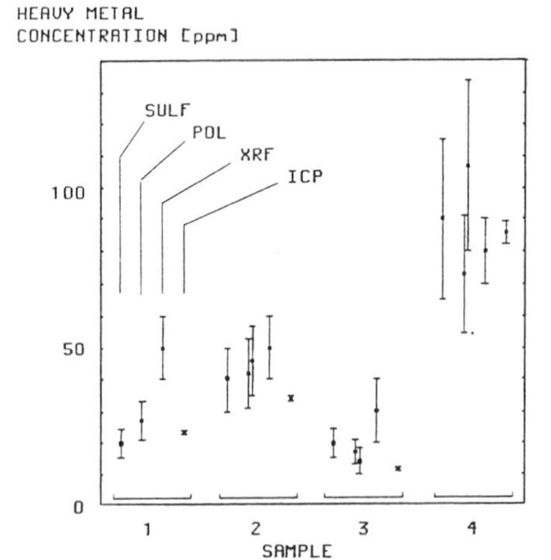

Figure 4.21. Comparison of results on four batches using four different methods. The results are grouped according to batch, and within a group, the methods are sulfide precipitation, polarography, XRF, and ICP (left to right).

methods: x-ray fluorescence (XRF[145]) and inductively coupled plasma spectrometry (ICP). The confidence limits were determined to be $\pm 10\%$ resp. $\pm 3\%$. Figure 4.21 summarizes the results. Because each method has its own specificity pattern, and is subject to intrinsic artifacts, a direct statistical comparison cannot be performed without first correcting the apparent concentrations in order to obtain presumably true concentrations. Visually, it is quite evident, though, that all methods arrive at about the same concentrations of catalyst traces: roughly 20, 40, 20, and 90 ppm. For all practical purposes, the case could be closed. QC tapped its own shoulders, secretly acknowledged its streak of luck, and vowed never again to let itself be pressured into revealing sensitive results before a double check had been run.

4.16. QUIRKS OF QUANTITATION

Situation. Suppose a (monovalent) ionic species is to be measured in an aqueous matrix containing modifiers; direct calibration with pure solutions of the ion (say, as its chloride salt) are viewed with suspicion because modifier/ion complexation and modifier/electrode interactions are a definite possibility. The analyst therefore opts for a standard addition technique using an ion-selective electrode. He intends to run a simulation to get a feeling for the numbers and interactions to expect. The following assumptions are made.

Assumptions

- The electrode shows linear behavior in the immediate vicinity of the working point on the calibration curve $\mathrm{EMF} = E0 + S \cdot \mathrm{lgt}(C)$.
- The sample has a concentration of about $C = 0.5$ mM.
- The term $E0$ remains constant over the necessary two measurements (a few minutes at most).
- The signal before digitization is sufficiently low-pass filtered so that noise is below 1 mV at the digital voltmeter (DVM).
- 50.0 ml of the sample solution will be provided. Standard additions will be carried out using a 10-mM solution of the ion. The amount to be added was, by rule of thumb, set to roughly double the concentration, for $\triangle \mathrm{EMF} \approx 20$ mV, a difference that can be accurately defined.

Note that a number of complicating factors have been left out for clarity: For instance, in the EMF equation, activities instead of concentrations should be used. Activities are related to concentrations by a multiplicative activity coefficient that itself is sensitive to the concentrations of all ions in the solution. The reference electrode necessary to close the circuit also generates a (diffusion) potential that is a complex function of activities and ion mobilities. Furthermore, the slope S of the electrode function is an experimentally determined parameter subject to error. The essential point, though, is that the DVM-clipped voltages appear in the exponent and that cheap equipment extracts a heavy price in terms of accuracy and precision (viz., quantization noise[31]).

The questions to be answered are

(1) How much 10-mM solution must be added to get reliable results? What concentration difference must be achieved to get sufficient differences in signal and burette readings?
(2) How accurately must this volume be added?
(3) How accurately must this volume be read off the burette?
(4) Must a volume correction be incorporated?
(5) Is any other part of the instrumentation critical?

A simulation program is written that varies the amount added over a small interval around the nominal 2.5 ml and

(1) Clips the simulated EMFs $E1$ and $E2$ to 1.0, 0.1, 0.01, resp., 0.001-mV resolution to emulate the digital voltmeter in the pH/ion meter.
(2) Varies the last digits of $E0$ to evade artifacts (cf. Ref.[10]).
(3) Simulates an incorrect reading of the burette.
(4) Allows for a volume correction.

First, a number of calculations are run without the above four features (*) in place (program lines 40, 110–130) to verify the rest of the program. Next, each of the features is introduced individually to capture effects, if any.

Answers. Concerning the questions posed above, the second one is easily answered by adding to or subtracting from $V2$ small volumetric errors in line 130. For the bias to remain below about 1%, the volume error must remain below 0.03 ml.

Program

```
10   V1 = 50                                   definition of parameters:
20   C1 = 5E-4                                 V1, V2, V3    : volumes
30   C2 = 0.01                                 C1, C2, C     : concentrations
40   E0 = 300 + RND                  *         E0, E1, E2    : EMFs (voltages)
50   S = 59                                    S             : slope in mV/decade
60   R = 0.1                                   R             : DVM resolution in mV
70   FOR V2 = 2.4 TO 2.55 STEP 0.01            added volume is varied
80   C = (V1*C1 + V2*C2) / (V1 + V2)           dilution factor
90   E1 = E0 + S*LGT(C1)                       calculation of reference EMF
100  E2 = E0 + S*LGT(C)                        calculation of new EMF
110  E3 = R*INT(E1 / R)              *         clipping to emulate DVM
120  E4 = R*INT(E2 / R)              *         action on E1, E2
130  V3 = V2 + 0.2                   *         simulate incorrect reading of V2
140  Q = 10↑((E4- E3) / S)                     estimate concentration
150  X = C2*V3 / ((V1 + V3)*Q- V1)
160  PRINT RESULT and PARAMETERS
170  NEXT V2
180  END
```

The first question is answered by noting that the exact volume V2 to be added is not critical as long as the DVM has "good" resolution (0.01 mV or better) and the volume is "correctly" read off (to 10 μl, or better). The volume V2 is thus set to 2–3 ml to retain sensitivity. Assume now that the instructions to the technician as far as instrumentation is concerned are ambiguous; that low-cost DVM in the corner and a plain graduated glass pipet are thought to do, and, upon repetition, some inconsistent results are obtained. Closer inspection using simulation reveals, however, that there is a systematic pattern (correlation) between volume added and estimate X (Fig. 4.22), largely because the least significant digit (LSD) of this low-resolution instrument corresponds to the 1-mV position. This is checked by varying V2 over a small interval, here from 2.35 to 2.50 ml in steps of 10 μl, a feat that is within the capabilities of a moderately priced dosimeter. The effect is due to the fact that the volume correction monotonously changes, while the clipped EMFs E3 and E4 are step functions; the interaction is commonly called a quantization effect or noise.[10] The improvement in resolution from a 0.1- to a 0.01-mV DVM is very striking in this case, which answers question 5.

A volume correction is necessary, as can be seen from a numerical experiment similar to the one above: If the increase in volume from V1 to V1 + V2 in line 80 is ignored, a bias of about 7% is produced.

Last, the calculations are repated 100 times (with variation of E0 to simulate E0 jitter within the resolution window) to obtain statistically

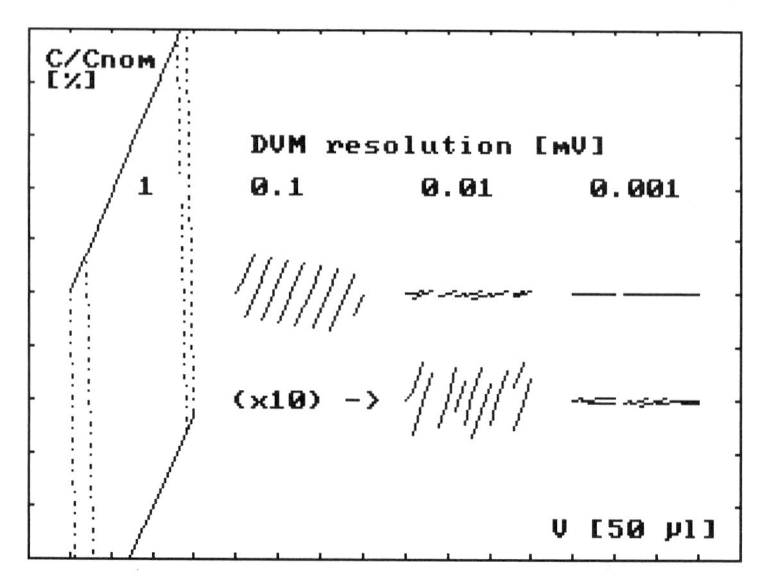

Figure 4.22. Estimated concentration of ion using the standard addition technique with an ion-selective electrode. The simulated signal traces are for DVM resolutions of 1, 0.1, 0.01, resp. 0.001 mV (left to right). For each resolution the added volume V2 is varied from 2.4 to 2.55 ml in increments of V2 = 10 μl. The burette is read off correctly. The abscissa is marked in intervals of 50 μl, the ordinate in 1% deviations from the expected 100%. The traces for 0.01- and 0.001-mV resolution are also given with an expansion factor 10. The simulation was run for five different values of E0 = 300 + RND [mV]. The dotted verticals (shown only for the 1-mV resolution case) indicate that the jump from one curve to the next occurs at unpredictable values of V2: $\Delta V = 10$ μl would in this case entail an "inexplicable" $\Delta C/C_{\text{nom}}$ of near 8%!

reliable means and standard deviations for the diverse combinations of factors (Table 4.21).

Warning. One should realize that a dishonest analyst can willingly shift the result within a range of several percent of the true value, which would certainly suffice to make a slightly out-of-specification product suddenly conform to these limits. This is accomplished with the following instrument configuration: A burette for delivering V2, a 0.1- or 1-mV-resolution pH meter, and a computer that immediately translates the actual V2, E3, E4, S, and C2 into X, and displays X. The analyst would simply have to stop the burette at the right moment to obtain highly "accurate" and "reproducible" results.

Numerical Results. Numerical results can be found in Table 4.21.

Table 4.21. Mean and Standard Deviations, in Percent of the Nominal Concentration, Found for Simulations under Various Combinations of (a) Random Variation of E0, (b) Volumetric (Reading) Error in V2, and (C) Use of a pH / ion Meter with a Resolution of 0.1 or 0.001 mV.
(For the last line the exact volume V2 added was varied in the range 2–3 ml to simulate actual working conditions, and 100 repetitions were run.)

	Numerical Results			
	E0 = 300 (mV)		E0 = 300 + RND (mV)	
Other Conditions	Res(DVM) 0.001 mV	Res(DVM) 0.1 mV	Res(DVM) 0.001 mV	Res(DVM) 0.1 mV
V3 = V2	0.00 ± 0.00	0.08 ± 0.10	0.00 ± 0.00	−0.03 ± 0.25
V3 = V2 + 0.2	7.19 ± 0.00	7.2 ± 0.12	7.18 ± 0.00	7.21 ± 0.25
V3 = V2 + 0.05	1.81 ± 0.00	1.89 ± 0.00	1.81 ± 0.03	1.80 ± 0.23
V3 = V2 + 0.01	0.37 ± 0.19	0.44 ± 0.00	0.36 ± 0.17	0.36 ± 0.25
V2 = 2 + RND V3 = V2	0.00 ± 0.09	0.08 ± 0.25	0.00 ± 0.00	−0.04 ± 0.31

Consequences

(1) A pH ion meter with a resolution of only 0.1 mV is not sufficient because the ensuing quantization noise introduces an apparent standard deviation of at least ±0.2%, and, more important in this particular case, the ensuing systematic quantization effects lead to a bias that is strongly dependent on small shifts in E0; see Fig. 4.22, left side.

(2) An electromechanical burette should be used that delivers volumes V2 with an accuracy of about 0.05 ml or better.

The above simulation can be varied within a reasonable parameter space; the critical experimental conditions should be noted, and appropriate experiments made to confirm the model.

4.17. PURSUING PROPAGATING ERRORS

The salt of a carbonic acid A is contaminated by traces of water and a second organic acid B. The content of the three components is determined as follows as in Table 4.22. The number of replicate determinations and the typical relative standard deviations are noted along with the average analytical response. Note that X is given in percent! How pure is A? The

Table 4.22. Results of an Acid Analysis.

	Component	Method	Amount	Repl	RSD
X	Water	Karl Fischer	0.85%	3	3.5%
Y	Acid B	Ion Chromatography	0.1946 mM/g	5	5.0%
Z	Sum A + B	Titration	7.522 mM/g	4	0.2%

answer is found by

(1) Subtracting B from the sum, and

(2) Correcting the difference for the water content of the sample:

$$A' = (Z - Y)/(1 - X/100)$$

$$= (7.522 - 0.1946)/(1 - 0.85/100)$$

$$= 7.3902 \text{ mM/g.}$$

(3) Error propagation is invoked by differentiating the expression for A with respect to X, Y, and Z:

$$\partial A/\partial X = (Z - Y) \cdot (-1/100)/(1 - X/100)^2,$$

$$\partial A/\partial Y = (-1)/(1 - X/100),$$

$$\partial A/\partial Z = (1)/(1 - X/100).$$

(4) Summing over the squares of the products of the partial differentials and their respective typical errors:

$$\Delta A^2 = \left((Z - Y) \cdot (1/100)/(1 - X/100)^2\right)^2 \cdot \Delta X^2$$

$$+ 1/(1 - X/100)^2 \cdot \Delta Y^2$$

$$- 1/(1 - X/100)^2 \cdot \Delta Z^2$$

(5) The typical error is here defined as a confidence limit:

$$\text{TE} = \pm\text{MEAN} \cdot \text{RSD} \cdot t\text{-factor}/\sqrt{n},$$

$$\Delta X = \pm 0.85 \cdot 0.035 \cdot 4.3027/\sqrt{3} = \pm 0.0739\%,$$

$$\Delta Y = \pm 0.1946 \cdot 0.05 \cdot 2.7764/\sqrt{5} = \pm 0.0121 \text{ mM / g},$$

$$\Delta Z = \pm 7.522 \cdot 0.002 \cdot 3.1824/\sqrt{4} = \pm 0.0239 \text{ mM / g}.$$

The confidence limits of this result are estimated to be

$$\begin{aligned}\Delta A^2 &= 0.0739^2 \cdot \left((7.522 - 0.1946) \cdot (1/100)/(1 - 0.85/100)^2\right)^2 \\ &\quad + 0.0121^2/(1 - 0.85/100)^2 \\ &\quad + 0.0239^2/(1 - 0.85/100)^2 \\ &= 0.000030 + 0.000148 + 0.000583 \\ &\quad (4\% + 19\,\% + 77\%) \\ &= (\pm 0.028)^2, \\ \Delta A &= \pm 0.028.\end{aligned}$$

The result is thus CL(A) = 7.390 ± 0.028 mM/g and should be either left as given or rounded to one significant digit in $\frac{1}{2}$CI: 7.39 ± 0.03: The percent-variance contributions are given in parentheses. Note that the analytical method with the best precision (titrimetry), because the of the particular numerical constellation, here gives rise to the highest contribution (77%).

4.18. CONTENT UNIFORMITY

Introduction. In order to assure constant tablet quality, the following requirements apply:

- Out of 20 tablets randomly pulled, one or two may deviate from the average weight by more than 5%, and none may deviate more than 10%. The average weight must be in the 95–105% range of nominal.

Note: Because with today's equipment and procedures the drug is generally very well dispersed in the granulate, especially if the drug content is high, the weight can be used as indicator. Nevertheless, the content uniformity will be determined via assays at least during the R & D phase to validate the procedure.

- The assay is conducted on 10 randomly pulled tablets. Nine out of the 10 assay results must be within 85–115% of the average, and none may be outside the 75–125% range. The average content must be within the window given below.
- The coefficient of variation must be 6% or less. This figure includes both the sampling and the analytical variance.

Situation. A table weighing 340 mg, of which 50 mg are drug, is to be produced. It is known that the c.o.v. for repeat analyses is ± 2.25 mg for the assay and ± 6.15 mg for the weight. Since the weighing operation is very accurate and has an excellent repeatability in the 300-mg range (typically $\pm \frac{1}{2}$LSD $= \pm 0.05$ mg or ± 0.005 mg, that is a resolution of 1 : 6000 or even 1 : 60000), the variability of the tablet weight must be wholly due to processing. The HPLC analysis is found to be fairly precise ($\pm 0.5\%$ or ± 1.7 mg, double determination).

Question

How much may the average content and weight deviate from the nominal values and still comply with the requirements? Two approaches will be taken:

(1) A purely statistical approach.

(2) A Monte Carlo simulation.

Statistical Approach. The minimal average weight must be

$$\bar{w} \geq \mathrm{SL}_l + t \cdot s_w = 0.95 \cdot 340 + 1.719 \cdot 6.15 = 333.6 \text{ mg},$$

which is 98.1% of the nominal weight; $t(p = 0.1, f = 19) = 1.719$. Obviously, the upper limit SL_u would be calculated analogously. The effective average must then remain in the 333.3–346.4 mg window. The minimal average drug content is similarly found to be

$$\bar{c} \geq \mathrm{SL}_l + t \cdot s_c = 0.85 \cdot 50 + 1.833 \cdot 2.25 = 46.6 \text{ mg}$$

or 93.2% of nominal. From this it can be seen that there is some leeway in terms of required composition and homogeneity: The effective average must remain in the 46.6–53.4-mg window; the sampling variance is approximately

$$\sqrt{2.25^2 - 1.7^2} = \pm 1.5 \text{ mg}$$

The true tablet composition could vary in the approximate range drug content/total weight (47 : 346) to (53 : 334) and still comply. All the same, it is imprudent to purposely to stray from the nominal values 50 : 340.

Simulation Approach. The numerical simulations were carried out using a modification of program SIMGAUSS; see data file TABLET_C.001 for content uniformity, respectively, TABLET_W.001 for weight uniformity. The mean weights and contents were varied over a range covering the nominal values. In vectors (columns) 5 and 6 of file TABLET_C.001 the results of $n = 10$ tablets subjected to individual assay are simulated for $\mu = 48$ mg and $\sigma = \pm 2.25$ mg. The observed means are 47.53 and 48.87 mg (99 resp. 102% of nominal), and the corresponding observed standard deviations are ± 1.71 and ± 1.84 mg (76, resp. 82% of nominal). All four values are within the expected confidence limits $48 \pm 2.262 \cdot 2.25/\sqrt{10} = 46.4 \cdots 49.6$ mg, respectively, $\pm 1.7 \cdots \pm 3.3$ mg (use program MSD and a $n = 10$ data file, option Disp.SD). Here, only two simulations were carried out for each μ/σ combination; this is enough to make the point, but 20+ runs would have to be carried out to obtain a representative result.

The individual values should also be examined, e.g., by using program HUBER: Vector 3 of file TABLET_C.001 is characterized by a relatively tight cluster of values, and two values (44.22, 50.08 mg) that are somewhat farther removed from the mean, namely, at $(x_i - \bar{x})/s_x = t = -2.02$ and $t = 1.62$. The probability of such deviations is assessed by using the single-sided Student's t table for $f = 9$: $p = 0.1/t_c = 1.383$, $p = 0.05/t_c = 1.833$, $p = 0.025/t_c = 2.262$. The larger of the two deviations can be accorded slight significance (Huber's $k = 4.18$). If this observation were eliminated as an outlier, the changes in the median (+0.325 mg), the mean (+0.361 mg), and the standard deviation (+0.41, $F = 1.8$) are nowhere near significance (use program TTEST). This means that the presence of this purported outlier only marginally influences the summary indicators $\bar{x}$ and s_x. The larger of the deviations is at 93.4% of nominal and at 94.5% of the observed mean; thus there is no reason to discard this batch because of nonuniformity of drug content.

The point that needs to be made is that with sample size as small as it is here ($n = 10$), the distribution can strongly vary in appearance from one sample to the next; vectors C_46 (column 1) and C_55 (column 20) of file

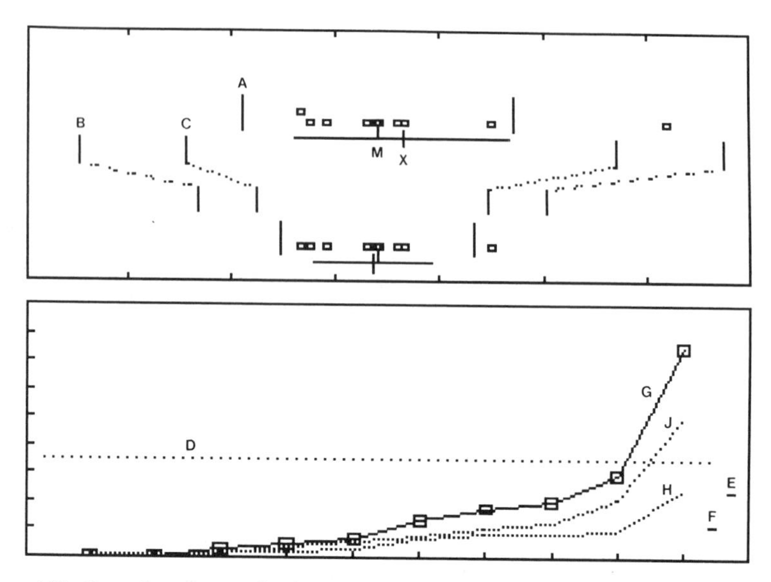

Figure 4.23. Detection of an outlier by Huber's method. Upper panel: abscissa: content [mg]; □: measurements; (A) |: Huber's cutoff at $k = 3.5$; (B, C) — · · · —: classical $\mu \pm 3\sigma$ and $\mu \pm 2\sigma$ cutoffs. The median is marked with an "up" tic (M); $\bar{x}$ is marked with an "up, down" tic (X). Lower panel: abscissa: data ordered according to $|x_i - x_m|$; (G) ———: Huber's k value; (H, J) · · · · · · · · ·: Student's t values (before, resp., after elimination of points; the elimination reduces s_x, and thus increases $z = (x_i - \bar{x})/s'_x$; (E): z value equivalent to Huber's k (D) before elimination; (F): equivalent z value after elimination.

TABLET_C.001 are the extremes, with standard deviations of ± 2.88 and ± 1.21. The corresponding Huber's k values for the largest residual in each vector are 6.56 (this looks very much like an outlier, $k_c = 3.5$) and 2.11 (far from being an outlier). The biggest k value is found for vector C_49 (column 8) at 7.49; Fig. 4.23 shows the results for this vector as they are presented by program HUBER.

In the upper half of the upper panel, the original data are plotted on the content [mg] axis. The mean is indicated by $\bar{x}$, and the median by M underneath the appropriate vertical tic marks. The $\bar{x} \pm s_x$ range is given by the horizontal. The $\pm 2 \cdot \sigma$ and $\pm 3 \cdot \sigma$ values are depicted by vertical bars below the horizontal one. The two vertical bars above the horizontal indicate Huber's $x_m \pm 3.5 \cdot \text{MAD}$ cut-off values. Since one point (□) is far outside this acceptance range (at about $+2.46 \cdot s_x$), eliminating this one yields the modified distribution pictured in the lower half of this panel. Note that the $\pm 2 \cdot \sigma$ and $\pm 3 \cdot \sigma$ values are now much closer to the mean than before because $s'_x \ll s_x$ (the appropriate cut-off values are connected by dotted lines). The $x'_m \pm 3.5 \cdot \text{MAD}$ cut-off values moved closer together also; one of the remaining $n' = 9$ points is just outside this new acceptance range. Program HUBER does not automatically repeat the elimination procedure to avoid cascades of possibly unwarranted eliminations. This example clearly shows that after the first elimination there is no justification for identifying further outliers: The long-term experience is that a standard deviation of ± 2.25 mg is typical for this production process, and at $n' = 9$ we already have $s'_x = \pm 1.1 \ll \pm 2.25$. Also, note that $\bar{x}$ and x_m moved much closer together from $n = 10$ to $n' = 9$.

The lower panel gives the points ordered according to (absolute) deviations: The abscissa is ordinal, the value with the smallest deviation $|x_i - x_m|$ being plotted at $x = 100/N\%$, the one with the highest deviation at $x = 100\%$; since the first point to be eliminated would be situated at the right-hand end, the abscissa is labelled "% points retained." The ordinate gives Huber's k factor (squares □, solid line) and the corresponding Student's t factor (dotted lines) using both s_x and s'_x (cf. option 3: Table). Huber's critical k is by default set to 3.5 (Ref. 15, dotted horizontal), but can be changed using option 2. If a $\pm 2 \cdot \sigma$ or a $\pm 3 \cdot \sigma$ cut-off rule is to be applied, the tic marks at $y = 2$ resp. $y = 3$ on the frame indicate where to draw the line. Evidently, in the example given by vector 8, nine points are closely clustered and one is far removed. The choice of Huber's k factor is not critical: One point is eliminated for $2.97 \leq k_c \leq 7.48$. If no point is eliminated ($k_c > 7.49$), we have $s'_x = s_x$, so only one dotted cut-off curve is plotted.

The analysis of data file TABLET_W.001 is analogous. Overall, using vectors 3 in the two data files TABLET_C.001 and TABLET_W.001 as an example, the following conclusions can be drawn (results rounded):

- The average content of the drug is 47.5 mg, or 2.5 mg (5.1%) below nominal.
- The standard deviation is ± 1.6 mg or c.o.v. = $\pm 3.4\%$. Note: Had measurements similar to those in vector 4 been made, the c.o.v. of $\pm 5.6\%$ would have come close to the $\pm 6\%$ limit stipulated by the USP 1985, which would have necessitated a retest.
- There are no individual values outside the 85–115% range that would signal retest and perhaps rejection.
- The average tablet weight is 334 mg, or 6 mg (1.8%) below nominal.
- The standard deviation (weight) is ± 5.6 mg.

- There are no individual values farther than 12 mg (3.6%) from the average weight.
- The (drug assay)/(tablet weight) ratio one should theoretically find is $(50.00 \pm 2.25)/(340.00 \pm 6.15) = 0.147 \pm 0.0071$. Effectively, one finds $(47.47 \pm 1.61)/(333.96 \pm 5.60) = 0.142 \pm 0.0054$ ($\pm 3.8\%$). This is obtained by error propagation as $(1.61/333.96)^2 + (5.6 \cdot 47.47/333.96^2)^2 = 0.0054^2$. Since 80% of the variance is due to the first term, analytical and sampling errors dominate. The HPLC assay contributes ± 1.7 mg, so that the sampling error comes to $\sqrt{1.61^2 - 1.7^2} \approx \pm 0$ mg; mathematically, this is an impossible situation: A variance cannot be negative! The explanation is simple: First, the HPLC precision (± 1.7 mg) is an estimate based on experience with the analytical method, and can most probably be traced to the validation report; it is to be expected that the analyst who wrote that report erred on the conservative side by rounding up all precision figures, so as not to fall into the trap of promising precision that could not be upheld in later experiments. Secondly, the ± 1.61-mg figure originated from a different measurement run (a simulation in this case) with a much lower number of observations than went into the ± 1.7-mg figure. Under such circumstances it is easily possible that the relative size of the two numbers is reversed. As an aside, the confidence intervals CI(σ) expected for $s_x = \pm 1.6/n = 10$ and, say, $s_x = \pm 1.7/n' = 50$, would be 1.1–2.9 resp. 1.4–2.1, which clearly overlap (use program MSD, data file TABLET_C.001 with $n = 10$, option dsp.SD, to first obtain $CL_u/s_x = 5.26/2.88 = 1.83$ and then $1.83 \cdot 1.61 = 2.94$, and so on). More realistically, the sampling error is estimated at $\sqrt{2.25^2 - 1.7^2} = \pm 1.5$ mg. The two quotients 0.147 and 0.142 are indistinguishable; any significant difference would have implied inhomogeneity, or, in practical terms, segregation of drug from the matrix during processing. Note: When powders of unequal size distribution, particulate shape, and/or density are well mixed, all it takes is machine vibration and the ever-present gravity to (partially) demix the components! This phenomenon is used in industrial processes to concentrate the commercially interesting component from available stock.

4.19. HOW FULL IS FULL?

Applicable Regulations. Assume that a cream is to be filled into a tube that has “20 g” printed on it. The lot size is 3000 units. The filling equipment’s repeatability is known to be $\approx \pm 0.75$ g ($\pm 3.75\%$). Two somewhat simplified regulations will be investigated that epitomize the statistical and the minimal individual fill weight approaches:

(1) (**EEC**: test $n = 50$ units)

(a) The average fill weight must not be less than 20.0 g. The effectively found averages must meet the requirement $\bar{x} \geq 20.0 - 0.379 \cdot s_x$. The factor 0.379 corresponds to $t/\sqrt{n}$ for $p = 0.005$ (one sided), $n = 50$, and $f = 49$, see Eq. (1.12a).

(b) No more than two units may be below 91% of nominal. In the case a retest is necessary (three or four units failing this criterion), no more than 6 out of the cumulated 100 units may fail.

(c) If five or more units (≥ 7 units in the retest case) fail the requirement, the product may not carry the coveted **e** logo that signals "European Quality Standard."

(2) (**CH**: Switzerland; no minimal sample number n)

(a) The average fill weight must not be less than 20.0 g. Up to 5% of the units filled may contain between 87.5 and 95% of the nominal fill weight. Units containing less than 87.5% of the nominal fill weight may not be marketed. Systematic bending of procedures to profit from these margins is forbidden.

(b) Overfilling according to given equations becomes necessary if the experimental relative standard deviation (c.o.v.) exceeds $\pm 3\%$ resp. $\pm 4.5\%$. Since a filling error of $s_x \approx \pm 0.75$ g ($\pm 3.75\%$) is associated with the equipment in question, the regulations require a minimal average fill weight of $20.00 + 1.645 \cdot s_x - 0.05 \cdot 20.00 = 20.23$ g; 20.35 g was chosen so that a margin of error remains for the line-operators when they adjust the volumetric controls.

(c) The "no marketing" proviso for seriously underfilled units forces the filler to either systematically overfill as foreseen in the regulations or to install check balances or other devices to actively control a high percentage (ideally 100%) of the containers and either discard or recycle underfilled ones.

Data file TUBEFILL.001 contains a set of 20 in-process controls (IPC) of $n = 50$ simulated weighings each. The first 10 vectors are for EEC conditions ($\mu = 20.02$ g), the others for Swiss regulations ($\mu = 20.35$ g); $\sigma = \pm 0.75$ g.

EEC CASE. The following averages were found: 20.11, 19.96, 19.82, 20.05, 19.97, 19.98, 20.04, 20.03, 20.14, and 19.94. In each of the four IPC runs, one tube was found with a fill weight below 91% of nominal. In all 10 IPC runs, the calculated average was above the 20.00-0.379 $\cdot s_x$ criterion. Thus the batch(es) conforms to regulations and can carry the **e** logo (as a matter of fact, one IPC run of $n = 50$ would have sufficed, for a sampling rate of 1.7%). If the average (target) fill weight μ were reduced to below 19.8 g, the probability of not meeting the above requirements would increase to virtual certainty.

SWISS CASE. The following averages were found: 20.32, 20.43, 20.34, 20.60, 20.35, 20.36, 20.45, 20.40, 20.30, and 20.31. The number of tubes with fill weights below the -5% limit was 4, 1, 0, 0, 4, 1, 3, 3, 3, and 3, for a total of 22, and none below the -12.5% one. Twenty-two tubes out of 500 tested correspond to 4.4%. Since the limit is 5% failures, or 2.5 per 50, fully 6 out of 10 IPC inspection runs at $n = 50$ each did not comply. At a total batch size of 3000 units, $\frac{1}{6}$ of all packages were tested. Evidently, unless the filling overage is further increased, a sampling rate of well above 10% is necessary to exclude these stochastic effects, and so the 10 inspections were combined into one test of $n = 500$.

4.20. WARRANTY OR WASTE

Introduction. Pharmaceutical preparations are exposed to all kinds of insults during their lifetime, chief among them being temperature excursions, high humidity, light, oxygen, and packaging components. Shelf lives can be as high as five years, and because the consumer must be protected from undue dangers, the health authorities have issued guidelines as to how a product must be tested. An excerpt:

- Storage at controlled room temperature (the earth is divided into four climate zones for this purpose), with sampling times at intervals of 3, 6, or 12 months.
- Storage under stress conditions, typically 26°C/60% RH, 31°C/70% RH, or 40°C, with sampling intervals of two weeks or one month.
- Storage in the same primary container as will be used on the market, e.g., blister strips for tablets or ovules, tubes for creams, or vials and ampules for injectables.
- During the development phase a series of laboratory or pilot-scale batches will be subjected to this stability program. As soon as the

process is scaled up to production-size batches, the first few, and at least one per year thereafter, will also go on stability.

Since full analyses are carried out, a lot of data are generated. Every parameter is reviewed for trends that signal product aging or outright decomposition of the active principle; this can be as cosmetic in nature as discoloration or as potentially hazardous as buildup of toxic derivatives. If the drug substance is an ester, for example, hydrolysis, particularly if moisture penetrates the primary packaging material, will decompose the compound into its acid and alcohol components. From a pharmaceutical or medical viewpoint, even if there is no toxicity issue involved, this will result in a loss of bioavailability. Even this is to be avoided because subpotency introduces therapeutic uncertainty and can go as far as lethal undertreatment.

Situation. A cream that contains two active compounds was investigated over 24 months. The assays resulted in the data given in file CREAM.001.

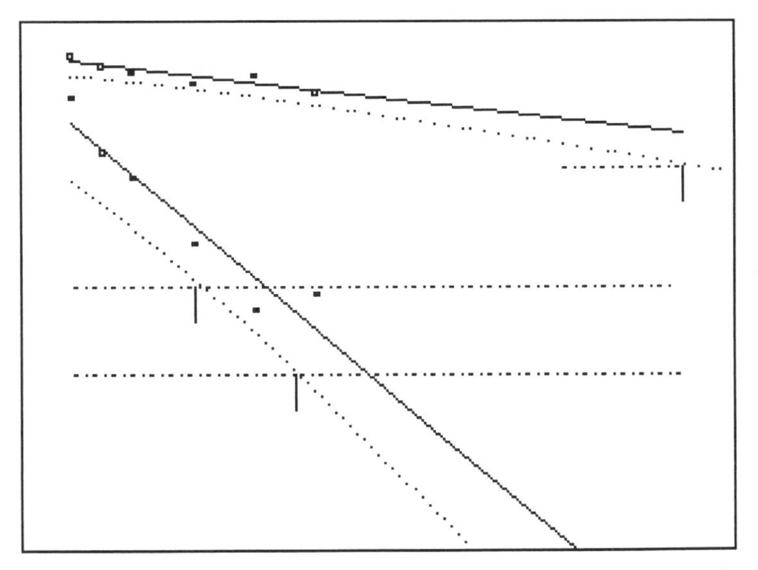

Figure 4.24. Shelf life calculation for active component 2 in a cream; see data file CREAM.001. The dotted horizontals are at the $y = 90$ resp. $y = 95\%$ levels. The data points (□) cover a 24-month stability program. The linear regression line is extrapolated until the lower 90%-confidence limit intersects the horizontals, here at 12 and 22 months. The ordinate is scaled 80–110% of nominal, and the abscissa extends from -5 to 65 months. The intercept is at 104.3%, an indication for overdosing, and the slope is ≈ -0.49 [%/month]. The data for active component 1 are given at the top, using the same scale, but vertically shifted by $+6.5\%$.

Program SHELFLIF performs a linear regression on the data and plots the (lower) 90% confidence limits for the regression line. For each time point $1 \cdots 60$ (months) it is determined whether this CL drops below levels of $y = 90\%$ resp. $y = 95\%$ of nominal. Health authorities today require adherence to the 90% standard, but it is to be expected that at least for some products the 95% standard will be introduced.

Interpretation. Active component 1, AC1, is so stable that a shelf life in excess of 60 months could be assigned. AC2, however, undergoes hydrolysis (this fact has to be independently established, i.e., by GC/MS techniques, or equivalents) and on the basis of the available data effectively limits the shelf life to 22 months; see Fig. 4.24.

The statistical interpretation is "there is a 5% chance that a further measurement at $t = 22$ months will yield a result below $y = 90\%$ of nominal." The commercial and regulatory interpretation: Since it is customary to assign shelf lives that are multiples of 6 or 12 months, given the above data, one would have problems convincing a regulatory body to grant a 24-month expiry date. Possible solutions:

(1) Ask for an 18-month expiry date (this would be a pain for the logistics and marketing departments because
   - Most companies consider it unethical to ship any goods to the wholesaler with less than 12 months of the shelf life left.
   - Quality control testing and release already take 2–4 weeks, so only 5 months remain to sell the product.
   - Reducing batch size so inventory is sold within five months increases production costs and overhead, more frequently exposes the manufacturer to out-of-stock situations, and strains the goodwill of a workforce whose production schedules are already now updated the moment they are distributed).

(2) Refer the project back to the R&D department for an improvement of the formula or the packaging (this can easily cost huge amounts of money and delay market introduction by years, especially if alternatives had not been thought of or had been prematurely deleted from the stability program).

4.21. ARRHENIUS-ABIDING AGING

Introduction. During the early stages of development of a dosage form or a drug substance one rarely has more than a notion of which temperatures the material will be processed or stored at once all has been said, done, and submitted. Also, the decomposition of a component might be subject to more than one mechanism, often with one reaction pathway dominating at lower, and another at higher temperatures. For these reasons, stress tests are conducted at a number of different temperatures that, first, are thought to bracket those that will eventually be quoted on the box ("store at 25°C or below"), and second, are as high as the dosage form will accommodate without gross failure (popping the vial's stopper because the water boiled) and still deliver results in a reasonably short time (a few weeks). Plots using program SHELFLIF would in such a case reveal increasingly negative slopes the higher the storage temperature. Program ARRHENIUS allows for all available information to be put into one file, and then calculates the slope for every temperature condition. Given that the basic assumptions underlying Arrhenius's theory hold (zero-order kinetics, i.e., straight lines in the assay-vs.-time graph, activation energy independent of temperature), the resulting slopes are used to construct the so-called Arrhenius diagram; that is, a plot of ln(slope) vs. $1/T$, where T is the storage temperature in degrees Kelvin. If a straight line results (see Fig. 4.25, left panel), interpolations (but never extrapolations!) can be carried out for any temperature of interest to estimate the slope of the corresponding assay-vs.-time plot, and then therefrom the probable shelf life. The Arrhenius analysis is valuable for three reasons:

(1) Detection of deviations from Arrhenius's theory (this would indicate complex kinetics and would lead to a more thorough investigation).
(2) Planning future experiments.
(3) Supporting evidence for the registration dossier.

Situation and Interpretation. A series of peptides was assessed for stability in aqueous solutions. The data in files ARRHEN.001/2/3 was found in a doctoral thesis.[146] Figure 4.25 shows one case where a temperature range of 50°C was covered; see also Table 4.23.

Since the Arrhenius diagram is linear and the collision parameter A is constant over the whole temperature range, the activation energy can be calculated to be $E \approx 94.8 \pm 5.6$ [KJ/mol] and it can be safely assumed that this temperature range is ruled by fairly simple kinetics.

An interpolation for 35°C on the assumption that the same sampling plan would apply as for the 80°C results yields an implausible 50 days (90% level); the reason is that both $\bar{x}$ and S_{xx} are very different for the two conditions. In this case the second option given in program ARRHEN must be used; that is, the individual time points appropriate to a test at

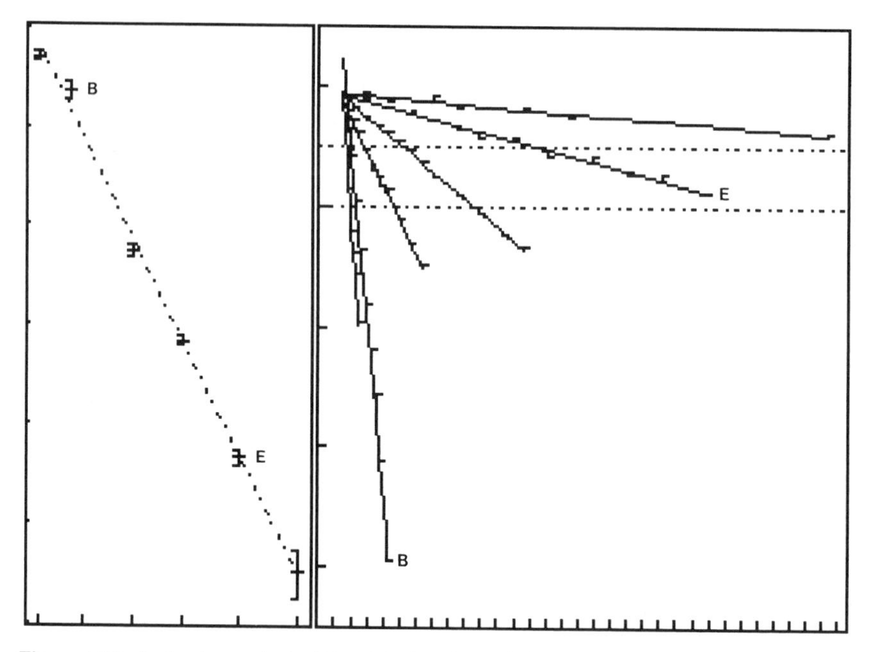

Figure 4.25. Arrhenius analysis: The right-hand panel shows the assay-vs.-time data for an aqueous solution of a peptide. The ordinate scale is 55–105% of nominal, and the abscissa scale is −14 to 312 days. The lines are for storage temperatures of 80, 73, (B), 60, 50, 40 (E), and 30°C. The left-hand panel gives the ln(-slope)-vs.-$1/T$ Arrhenius plot, with an abscissa scale of 0.00283–0.00329 and an ordinate scale of −4.76 to 0.796.

Table 4.23. Key Results of an Arrhenius Analysis; See Data File ARRHEN.002.

Item	Set_1	Set_2	Set_3	Set_4	Set_5	Set_6	DIM
Temp.	80	73	60	50	40	30	[°C]
$1000/T$	2.831	2.888	3.001	3.094	3.193	3.299	[1000/°K]
Slope	−2.104	−1.423	−0.281	−0.115	−0.035	−0.011	[%/day]
$\ln(-s)$	0.70	0.35	−1.27	−2.16	−3.34	−4.50	[−]
Shelf life 90%	4	6	28	76	240	715	[days]
$\ln(A)$	33.0	33.3	33.0	33.1	33.1	33.1	[−]

35°C are entered (0, 180, 360, 540, 720, and 900 days): The estimated shelf life is now 410 days (90% level; 170 days for 95% level).

4.22. FACTS OR ARTIFACTS?

In file JUNGLE.002 a simulated data set is presented that obeys these rules:

- All results are clipped to a specific number of digits to simulate the operation of digital readouts of analytical equipment: **p** is the precision, **d** is dimension, and $\boldsymbol{\sigma}$ is the superimposed $\mathrm{ND}(0, \sigma)$ noise in Table 4.24.
- Impurities **A**, **B**, and **C** are in the ppm range; this would be typical of residual solvents or volatile by-products of a chemical synthesis. **B** and **C** are coupled to **A**. Impurity **C** has its noise reduced by $\frac{1}{2}$ in the range above 695 ppm, an effect that can occur when a detector is switched to a lower sensitivity in anticipation of a large signal.
- Compound **D** is in the low percent range, and extremely variable, as is often the case with impurities that are not specifically targeted for control; under these circumstances, less is known about the reaction pathways that generate or scavenge these compounds than about high-impact impurities. Also, there are generally so many target variables in the optimization of a reaction that compromises are a way of life, and some lesser evils go uncontrolled.
- The **pH** is relatively well controlled, but coupled to **C**.
- **Color** is the result of an absorption measurement, commonly carried

Table 4.24. Simulation Equations Used for File JUNGLE.002.

Item	Equation for μ		$\boldsymbol{\sigma}$	**d**	**p**
A	$A = 125$		30	ppm	1
B	$B = 15.2 + (A - 125)/5$		2	ppm	0.1
C	$C = 606 + (A - 125)$	$C < 695$:	40	ppm	1
		$C \geq 695$:	20		
D	$D = 0.3$	$D \geq 0.01$	0.5	%	0.01
pH	$\mathrm{pH} = 6.3 + (C - 600)/500$		0.2	—	0.1
Color	Color = 23		7	mAU	1
HPLC	$\mathrm{HPLC} = 99 - A - B - C - D$		0.3	%	0.01
Titr.	Titr. = 98.3 + (HPLC − 99)		0.4	%	0.05

out at $\lambda \approx 410$ nm, to assess the tinge that is often found in crystallization liquors that impart an off-white to yellowish aspect to the crystalline product.

- The **HPLC** assay is fully coupled to the impurities **A–D** on the assumption that there is a direct competition between the major component and some impurity-producing reaction pathways. The basis-value 99 was introduced to simulate other concentration losses that were not accounted for by impurities **A–D**.
- The **Titr**ation result is just a bit lower than the **HPLC** result and is strongly coupled to it.

This example was set up for a number of reasons:

- Allow exploration using programs, CORREL, LR, HUBER.
- Demonstrate the effect of the size of vectors on correlation.
- Smuggle in typical artifacts.

The effect of the size of the compared vectors is shown by taking JUNGLE.002, and using program DATA, option DEL row, to cut down the number of rows to, e.g., 24, 12, or 6. The probability levels change in both directions, e.g., **A/C** from the category '■ : $p < 0.01$' to the category ':: : $p > 0.16$', that is, from highly significant to no significance whatsoever. Whenever there is a large number of measurements available, even very tenuous links turn out to be highly significant, and spurious correlations are to be expected. WARNING: Data exploration without graphical support (program CORREL, option Graph, or program VIEW_XY) and a very critical mindset can easily lead to nonsensical interpretations! Running files TABLET_C.001, TABLET_W.001, and ND_160.001 through program CORREL proves the point: These are lists of fully independent, normally distributed numbers, and correlations turn up all the same! Independent of n, the following distribution is found:

Frequency	Category	Interpretation
$\approx \frac{1}{12}$	$0.00 \leq p < 0.04$	(significant to highly significant)
$\approx \frac{1}{12}$	$0.04 \leq p < 0.08$	(marginally significant)
$\approx \frac{1}{6}$	$0.08 \leq p < 0.16$	(insignificant)
$\approx \frac{2}{3}$	$0.16 \leq p < 0.50$	(random)

Thus one must expect about 5–10% apparently significant correlations; fortunately, these false positives appear in a random arrangement, so that

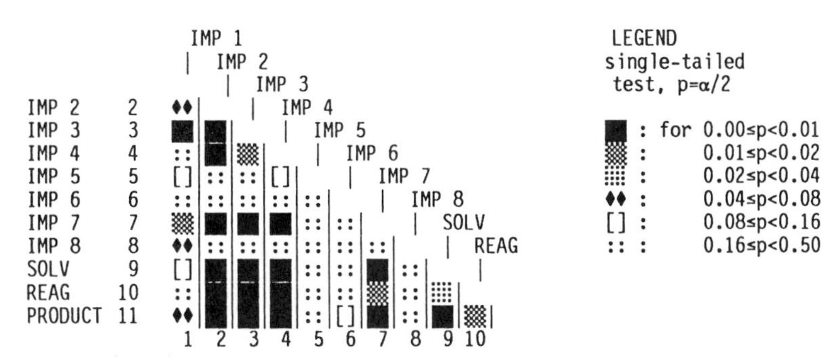

Figure 4.26. Correlation graph for file PROFILE.001. The facts that (a) 23 out of 55 combinations yield probabilities of error below $p = 0.04$ (42%; expected due to chance alone: $\approx$ 8%) and (b) that they fall into a clear pattern makes it highly probable that the peak areas [%] of the corresponding chromatograms follow a hidden set of rules. This is borne out by plotting the vectors two by two. Because a single-sided test is used, p cannot exceed 0.5.

when a really significant connection turns up, the human visual system perceives a clearly recognizable pattern; for example, for file PROFILE.001 see Fig. 4.26.

A further incentive for supplying file JUNGLE.002 is the possibility of smuggling in some numerical artifacts of the type that often crop up even in one's own fine, though just a bit hastily concocted, compilations (see JUNGLE.003):

- Rows 1–7, item **Titr**: 0.5% subtracted; typical of temporary change in the production process, calibration procedure, or analytical method, especially if values 1–7 were obtained in one production campaign or measurement run.
- Rows 9 and 10, items **A**, **B**, and **C**: Factor of 2 introduced to simulate the operation of absentmindedness in sample and hardware preparation, or computation: an extra dilution step, a sensitivity setting, or transcription errors.
- Row 14, item **B**: deleted decimal point.
- Row 38, item **Titr**: An assay value of 101.1 might be indicative of a calibration or transcriptional error if the analytical procedure does not admit values above 100.0.
- Row 45, item **pH**: A value of 4.8 in a series of values around 6.2 is highly suspicious if the calibration of the electrode involves the use of an acetic acid/sodium acetate buffer (pK $\approx$ 4.6)

When JUNGLE.002 and .003 are analyzed with program CORREL, a plausible pattern turns up: impurities **A**, **B**, and **C** and the **pH** are mutually correlated; impurity **D** is correlated with **Titr** and **HPLC**. The artifacts in JUNGLE.003 cause a reduction in the strengths of correlation in four bins (three of them over 2–4 classes), and an increase in three bins over 1 class each.

4.23. PROVING PROFICIENCY

In today's regulatory climate, nothing is taken for granted. An analytical laboratory, whether in-house or in the form of a contractor, no longer gets away with the benefit of the doubt or the self-assured "we can do that for you," but has to demonstrate its proficiency. Under the good laboratory practices (GLPs) a series of do's and don'ts have become established that are interpreted as minimal expectations. Since one never knows now which results will be declared crucial when the regulatory department collates the submission file a few years down the road, there is tremendous pressure to treat all but the most preliminary experiments as worthy of careful documentation under GLP. Besides, some manifestations of modern life—job hopping and corporate restructuring are examples—tend to impair a company's memory to the point that whatever happened more than 6 months ago is lost forever, unless it is in writing.

Requirements. As far as the bench chemist is concerned, the following nonexhaustive list of points should be incorporated into the experimental plan:

- Ensure that the actual instrument configuration conforms to what is written under 'Experimental': suppliers, models, modifications, consumables (HPLC or GC columns, gaskets, etc.), and software for the main instrument, peripherals (injectors, integrators, computers, printers, plotters, etc., and ancillary equipment (vortexer, pipettor, balances, centrifuges, filters, tubing, etc.).
- For all critical equipment log books are available that show the maintenance, validation, and calibration history and status.
- All reagents are traceable to certificates of analysis, are used within the posted shelf life, and are from known suppliers.
- Standard substances of defined purity must be available for major components and the main impurities (criteria: mole fraction, toxicity, legal requirements, etc.).

- Solvents and sample matrices must be available in sufficient amounts and appropriate purity (e.g., pooled blood plasma devoid of analytes of interest).
- An internal standard **IS** that is physico-chemically reasonably similar to the main analyte must be available.
- The analytical method must have been run using a variety of compounds to prove that there is sufficient selectivity. Materials in contact with the sample (tubing, filters, etc.) that might interfere with the analytes of interest must be proven to be innocuous.
- A series of calibration standards, **CS**, is made up that covers the concentration range from just above the limit of detection to beyond the highest concentration that must be expected (extrapolation is not accepted). The standards are made up to resemble the real samples as closely as possible (solvent, key components that modify viscosity, osmolality, etc., blood plasma).
- A series of blinded standards is made up (usually low, medium, high; the analyst and whoever evaluates the raw data should not know the concentration). Aliquots are frozen in sufficient numbers so that whenever the method is again used (later in time, different instrument or operator, other laboratory, etc.), there is a measure of control over whether the method works as intended or not. These so-called QC standards, **QCS**, must contain appropriate concentrations of all components that are to be quantified (main component, e.g., drug, and any impurities or metabolites).
- During the method validation phase, the calibration, using the **CS** solutions, is repeated each day over at least one week to establish both the within-day and the day-to-day components of the variability. To this end, at least six **CS**, evenly spread over the concentration range, must be repeatedly run (**m** $\approx$ 8–10 is usual), to yield **n** $\geq$ 50 measurements per day. If there are no problems with linearity and heteroscedacity, and if the precision is high (say, c.o.v. $\leq$ 2–5%, depending on the context), the number of repeats **m** per concentration may be reduced from the second day onwards (**m** = 2–3 is reasonable). The reasoning behind the specifics of the validation/calibration plan is written down in the method justification document for later reference.
- After every calibration, repeat **QC** samples are run (**m** = 2–3).
- At the end of the validation phase, an overall evaluation is made and various indicators are inspected, such as:
  - Variability and precision of calibration line slope
  - Variability and significance of the intercept

- Linearity, usually assessed by plotting the residuals vs. the concentration
- Back-calculated **CS** concentrations $\pm CL(X)$
- Interpolated **QCS** concentrations $\pm CL(X)$
- Residual standard deviation within-day and for pooled data
- Correlation coefficient **r** or coefficient of determination $\mathbf{r}^2$ (these indicators are so well known that some bureaucrats are unhappy if they do not have them)
- For each of the **CS** and the **QC** concentrations the overall mean and standard deviation are compared to the daily averages and SDs; from this, variance components for the within-day and day-to-day effects are estimated by subtraction of variances

- If the analytical method survives all of the above criteria (suitably modified to match the situation), it is considered to be under control.
- Changing major factors (instrument components, operators, location, etc.) means revalidation, generally along the same lines.

Setting. An established analytical method consisting of the extraction of a drug and its major metabolite from blood plasma and the subsequent HPLC quantitation was precisely described in a R&D report, and was to be transferred to three new labs across international boundaries. The originator supplied a small amount of drug standard and a number of vials containing frozen blood plasma with the two components in a fixed ratio, at concentrations termed lo, mid, and hi. The report provided for evaluations both in the untransformed (**lin**ear/**lin**ear depiction) and the doubly logarithmized (**log**arithmic/**log**arithmic) format. The three files VALID.00X supplied on the disk were selected from among the large volume of data that were returned with the three validation reports to show the variety of problems and formats that are encountered when a so-called easily transferable method is torn from its cradle and embedded in other company climates. The alternatives would have been to

(1) Run all analyses is the one and established central laboratory; the logistical nightmares that this approach convokes in the context of a global R&D effort demands strong nerves and plenty of luck (Murphy's Laws reign supreme!). Regional laboratories make sense if the expected volume of samples is large and the technology is available on all continents.
(2) Run the analyses in regional labs, but impose a military-style supply-and-control regime; this exposes the 'commander' to cultural cross-currents that can scuttle his project.

Problems

(1) In one new laboratory the instrument configuration was reproduced as faithfully as possible: The instrument was similar but of a different make.

(2) The HPLC column with the required type and grade of filling material (so-called stationary phase) was not available locally in the 2.1-mm diameter set down in the report, so the next-larger diameter (4.0 mm) was chosen, in full cognizance that this would raise the detection limit by an estimated factor of $(4.0/2.1)^2 \approx 3$–4. For the particular use, this was of no concern, however. The problem was encountered because the R & D people who developed the method had used a top-of-the-line research instrument, and had not taken the fact into account that many routine laboratories use robust but less fancy instruments, want to write them off before replacing them, and do not always have a laboratory supply house around the corner that is willing to deliver nonstandard items at reasonable prices and within a few days (some manufacturers will not, cannot, or are prohibited from delivering to all countries). As a sign of our times, some borders are very tight when it comes to smuggling x-ray-opaque metal HPLC columns across: To the customs officers, the white powder in them looks suspicious.

(3) An inadvertent, and at first sight trivial shift in conditions raised the extraction efficiency from the stated $\approx 85\%$ to nearly 100%: This together with the different model of detector (optical path!) caused the calibration line slope to become much higher. There was concern that saturation phenomena (peak broadening, shifting of retention times) would set in at lower concentrations.

Results. The raw data consisted of peak-height ratios of signal : internal standard; see data files VALID.001 (primary validation; $m = 10$ repeats at every concentration), VALID.002 (between-day variability), and VALID.003 (combination of a single-day calibration with several repeats at 35 and 350 [ng/ml] in preparation of placing QC-sample concentrations near these values). Figure 4.27 shows the results of the back-calculation for all three files, for both the lin/lin and the log/log evaluations. Figure 4.28 shows the pooled data from file VALID.002.

The data in VALID.001 show something that is characteristic of many analytical methods: The standard deviation steadily increases, particularly above 250 ng/ml, from low concentrations, where it is close to zero, to ± 0.14 at 500 ng/ml, while the c.o.v. drops, from $> 10\%$ to about 5%.

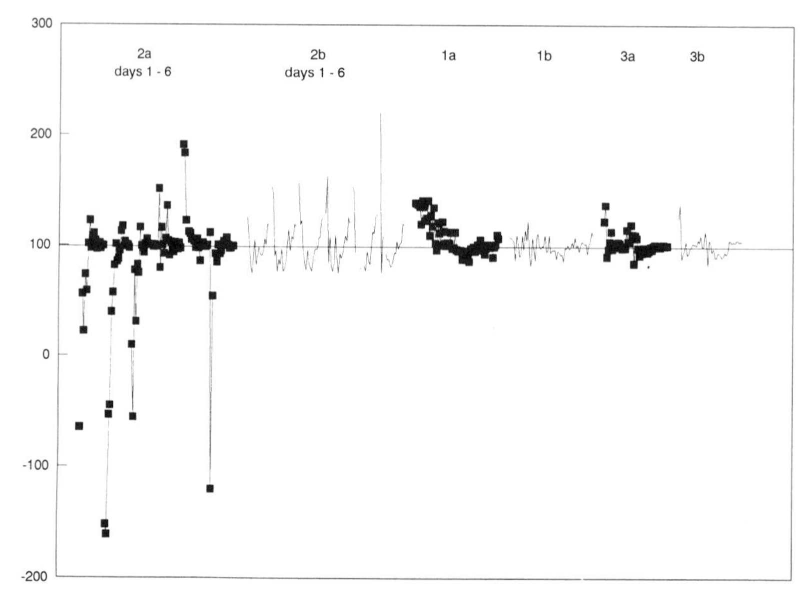

Figure 4.27. Back-calculated results for files VALID.00X. The data are presented sequentially from left to right. The ordinate is in percent of the nominal concentration. Numbers $X = 1$, 2, and 3 indicate the data file. Letters **a** and **b** indicate lin/lin, respectively, log/log evaluation. The log/log format tends to produce positive deviations at low concentrations, while the lin/lin format does the opposite, to the point of suggesting negative concentrations! The reason is that the low concentration values are tightly clustered at the left end of the lin/lin depiction, whereas the values are evenly spread in the log/log depiction, with commensurate effects on the position of $\bar{x}$, the sum S_{xx}, and the influence each coordinate has on the slope. The calibration design was optimized for the log/log format.

The residuals at $x \leq 250$ ng/ml do not appear to lie on a straight line; this notions is strongly reinforced when the data are viewed after double-logarithmic transformation (program VALIDLL; see below). Note that the back-calculated values for the two lowest concentrations are far above 100%. The same observation in two other laboratories where the same method was run confirmed that this nonlinearity is real. A cause could be assigned after the conventional three-step liquid-liquid extraction (aqueous to organic, cleanup with aqueous medium, pH change and back-extraction into water) was replaced with a single-step procedure involving selective extraction cartridges, which brought perfect linearity (the higher cost of the cartridges is more than justified by the solvent and manpower savings). Obviously the cartridges eliminated a component that interfered at low analyte concentrations. The question came up whether the method should be changed. Since this technological improvement

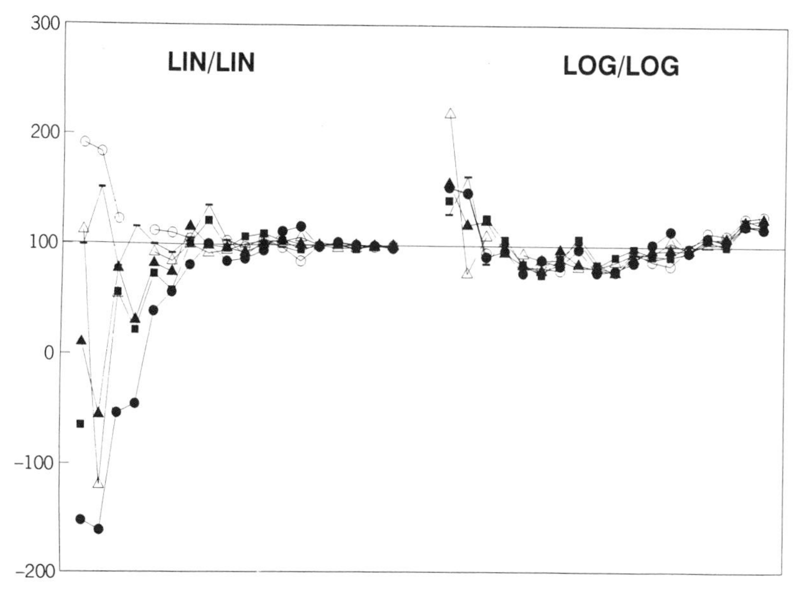

Figure 4.28. Back-calculated results for file VALID.002. The data from the left half of Fig. 4.27 are superimposed to show that the day-to-day variability most heavily influences the results at the lower concentrations. The lin/lin format is perceived to be best suited to the upper half of the concentration range, and nearly useless below 200 ng/ml. The log/log formate is fairly safe to use over a wide concentration range, but a very obvious trend suggests the possibility of improvements: (a) nonlinear regression, and (b) elimination of the lowest concentrations. Option (b) was tried, but to no avail: While the curvature disappeared, the reduction in n, $\log(x)$ range, and S_{xx} made for a larger V_{res}, and thus larger interpolation errors.

came after a lot of work had already been done, and this particular study late in the product development cycle did not require full exploitation of the available concentration range and precision, it was decided to leave things as they were and not go into the trouble of validation of the new, cross-validation of the new against the old, and registration of the new method with the health authorities in all the involved countries. Despite the obvious curvature, the coefficient of determination r^2 was larger than 0.9859 throughout.

Back calculation is achieved by equating the individual calibration signal $y(i)$ with y^*, using Eq. (2.19), and calculating $100 \cdot X(y^*)/X_{nominal}$ [%]. The estimated standard deviation on $X(y^*)$, s_X, is transformed to a coefficient of variation by calculation of either c.o.v. $= 100 \cdot s_X/X_{nominal}$ or c.o.v. $= 100 \cdot s_X/X(y^*)$; the distinction is negligible because any differ-

Table 4.25. Overview of Key Statistical Indicators Obtained from Three Different Laboratories that Validated the Transfer of the Analytical Method.

(LEGEND: NN, linear regression without data transformation; LL, idem, using logarithmically transformed axes; **, not interpretable; na, not applicable; $X15\%$, concentration x at which σ_Y/Y is 0.15; FDA, lowest calibration concentration for which $s_Y/Y < 0.15$. The two lines of bold numbers pertain to the pooled data. [X = 1, 2, or 3] identifies the appropriate data file VALID.00X.)

Slope [–]	CL(b) [%]	Intercept [–]	CL(a) [%]	Sres [–]	r^2 [–]	LOD [ng/ml]	LOQ [ng/ml]	$X15\%$ [ng/ml]	FDA [ng/ml]	Transform
0.00525	0.9	0.01790	49.5	0.0143	0.9997	1.7	3.4	15	2	NN [2]
0.00528	1.9	0.02368	77.8	0.0311	0.9987	3.5	7.0	31	1	NN
0.00521	0.6	0.01583	37.4	0.0100	0.9999	1.1	2.3	9	5	NN
0.00529	1.5	0.00690	207.8	0.0242	0.9992	2.7	5.4	25	10	NN
0.00526	1.5	0.00477	310.8	0.0228	0.9994	2.8	5.6	25	1	NN
0.00509	0.9	0.01374	59.8	0.0133	0.9998	1.6	3.2	14	5	NN
0.00523	**0.6**	**0.01400**	**40.1**	**0.0241**	**0.9991**	**1.1**	**2.1**	**25**	**2**	**NN**
0.861	4.6	**	**	0.0635	0.9930	**	**	**	**	LL [2]
0.861	5.3	**	**	0.0789	0.9901	**	**	**	**	LL
0.849	5.0	**	**	0.0740	0.9910	**	**	**	**	LL
0.874	5.3	**	**	0.0798	0.9902	**	**	**	**	LL
0.859	6.3	**	**	0.0884	0.9880	**	**	**	**	LL
0.870	6.6	**	**	0.0943	0.9859	**	**	**	**	LL
0.862	**2.0**	**	**	**0.0778**	**0.9893**	**	**	**	**	**LL**
0.00527	1.8	−0.0198	106.0	0.0616	0.9955	4.0	7.9	na	10	NN [1]
1.01	1.5	**	**	0.0347	0.9966	**	**	**	**	LL
0.00751	0.4	0.00209	295.7	0.0164	0.9998	0.8	1.6	na	1	NN [3]
0.961	1.3	**	**	0.0378	0.9983	**	**	**	**	LL

ence between $X_{nominal}$ and $X(y^*)$ that shows up here would have caused alarm above.

The options Table and Results repeat the ordinate and abscissa values and (see summary in Table 4.25) provide

(1) The absolute and the relative residuals in terms of concentration X.
(2) The back-calculated values (also in x coordinates) as absolute and relative values.
(3) Program VALID adds the symmetric c.o.v., while program VALIDLL gives the (asymmetric) low and the high values.
(4) The slope and the intercept with the appropriate relative 95% CLs, the residual standard deviation, and r^2.
(5) The LOD and the LOQ (in the case of program VALIDLL, these values are calculated as in VALID, and are given for information only, because in a double-logarithmic presentation, a LOD/LOQ cannot be calculated on the basis of an intercept).
(6) For each group of repeatedly determined signals ($m_j \geq 2$) the basic statistics are given.

Figure 4.29 gives the key indices in a graphical format.

Observations.

(a) The slopes cluster tightly around their respective means, even though they are less well defined (CL(b)) in the log/log depiction. The slope found for VALID.003 is definitely higher and very well defined ($0.007511 \pm 0.4\% > \approx 0.0052 \pm 1.3\%$). The fact that the log/log slopes for VALID.002 ($\approx 0.862 \pm 5.6\% < 1.00$) do not include $\mu = 1$ points to curvature in the nontransformed depiction.
(b) Some intercepts in the nontransformed depiction do not include $\mu = 0$; this is due to the curvature at low concentrations which is traceable to a lack of selectivity in the extraction process.
(c) As was to be expected, the coefficient of determination r^2 is not very informative. The reduction of the heteroscedacity brought about by the log/log transformation is not reflected by an improvement in r^2, which decreases from, e.g., 0.9991 to 0.9893 (S_{res} cannot be compared because of the units involved).
(d) The calibration-design-sensitive concept of calculating limits of detection yields LOD values in the range 1.1–3.5 ng/ml and LOQ

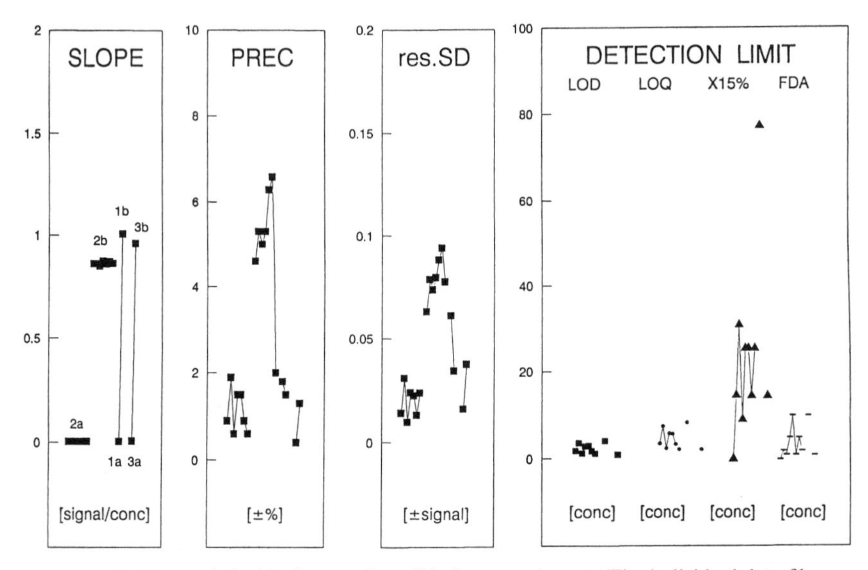

Figure 4.29. Key statistical indicators for validation experiments. The individual data files are marked in some places with the numbers 1, 2, and 3, and are in the same sequence for all groups. The lin/lin, respectively, log/log evaluation formats are indicated by the letters **a** and **b**. Limits of detection/quantitation cannot be calculated for the log/log format. The slopes are very similar for all three laboratories. The precision of the slopes is given as ±relative $\frac{1}{2}$(confidence interval) in percent; the day-to-day variability is not very large, except for day 6 in the log/log format, where the slope's error is only half as large as on the other five days. The residual standard deviation follows a similar pattern as does the precision of the slope b. The calibration-sensitive LOD conforms nicely with the evaluation as required by the FDA, except on days 4 and 5. The calibration-design-sensitive LOQ puts an upper bound on the estimates. The $X15\%$ analysis is high because it is carried out for $m = 1$; for $m = 2$, the concentration $X15\%$ in this particular case drops from 77 to 56 [ng/ml].

values in the range 2.3–7 ng/ml. The dotted curve in the option VALID (*see* point (5) below, suggests a range $X\,15\% \approx 9$–31 ng/ml, three to four times the LOQ. The LOD according to the FDA concept depends on the data set one chooses, values of 1–10 ng/ml being found (if the standard deviation for DAY_4 and $x = 10$ ng/ml, see file VALID.002, had been only 5% smaller, x_{LOD} would have been 2 instead of 10 ng/ml). The very tight definition of slope b in VALID.003 pushes the LOD to below 1 ng/ml.

(e) The analysis of the variances is inconclusive: The group SDs for the individual days are to all intents and purposes identical to the SDs

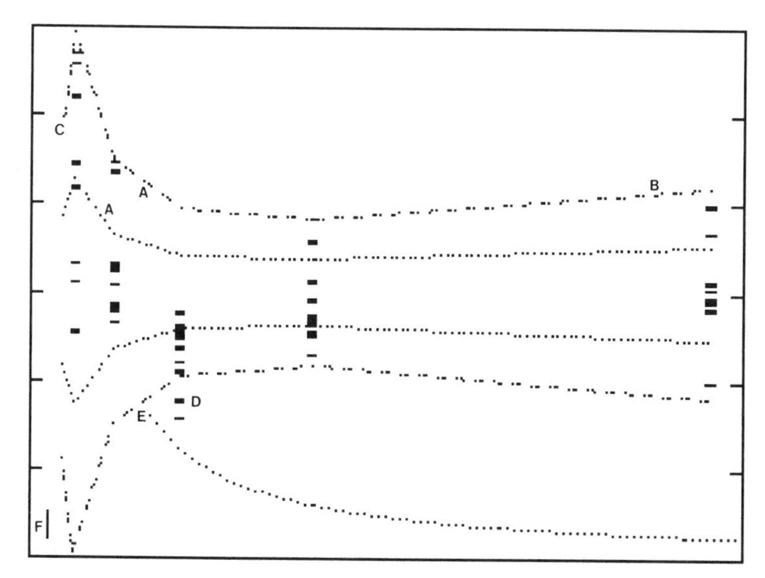

Figure 4.30. Graphical summary of validation indicators for file VALID.001. In this lin/lin depiction, the typical "trumpet" form (A) of the c.o.v./CL curves is seen; the fact that there is a spread toward the right-hand edge (B) suggests that the measurement errors grow in a slightly overproportional fashion with the concentration. The narrowing down of the "trumpet" at the lowest concentration (C) is a sign that due to the proximity to the LOD, the measurement distribution is not truly Gaussian but has the low side clipped. The data points for the lowest concentration (10 ng/ml) are off scale (119.5–141.1%), while the fourth-smallest concentration (D) yields negative biased results (effect of background interference at the lowest concentrations due to the extraction's lack of selectivity, ensuing curvature). If the same plot were made for one day's data from VALID.002 ($m = 2$ at each concentration), the dashed CL curves would be off scale because of the large Student's t at $f = 1$. The curve that starts at the lower right corner stops at the point x where the c.o.v. is 15% of the estimate Y (E). The LOQ calculated according to Fig. 2.14 appears in the lower left corner (F).

of the pooled data; the same is true for the residual SDs. If there is a between-days effect, it must be very small.

Option Valid assembles all of the important facts into one graph (see Fig. 4.30):

(1) The abscissa is the same as is used for the calibration line graph. The ordinate is $\pm 30\%$ around $Y = a + b \cdot x$.
(2) At each concentration, the relative residuals $100 \cdot (y(i) - Y(x))/Y(x)$ are plotted.
(3) At each concentration x_j the c.o.v. found for the group of m_j repeat determinations is plotted and connected (dotted lines).

(4) At each concentration x_j the relative confidence limits $\pm 100 \cdot t \cdot s_y/\bar{y}(j)$ found for the group of m_j repeat determinations is plotted and connected (dashed lines). Because file VALID.002 contains only duplicate determinations at each concentration, $n = 2$ and $d \cdot f = 1$; thus $t(f, p = 0.05) = 12.7$, the relative CL are mostly outside the $\pm 30\%$ shown. By pooling the data for all six days it can be demonstrated that this laboratory has the method under control.

(5) Beginning at the lower right corner, a dotted curve is plotted that gives the c.o.v. for the estimated signal Y and $m = 1$, namely,

$$y = 100 \cdot \sqrt{V_Y}/Y(x)\ [\%]$$

$$= 100 * \mathrm{SQR}(VRES * (1/N + 1/M + (X - XB)$$

$$* (X - XB)/SXX))/\mathrm{ABS}(A + B * X).$$

The lower edge of the graph corresponds to 0%; the value $y = 60\%$ would be reached near the top left corner, but the curve is not plotted beyond the point where it reaches $y = 15\%$; the corresponding x value is given. This result is sensitive to the calibration design, just as is the LOD/LOQ concept presented in Section 2.2.7, and is achieved through extrapolation to below the lowest calibration concentration; $x_{\min}$ can be chosen so that signals in the lower half of the size distribution are not overwhelmingly determined by noise. By running duplicates, $m = 2$, the c.o.v.-curve moves to the left and gives lower detection limits $X15\%$.

In contrast to this, the FDA requests a limit of acceptance defined by $100 \cdot s_Y/Y \leq 15\%$ for the lowest calibration standard $x \geq x_{\mathrm{LOD}}$. This, of course, is insensitive to the calibration design and depends exclusively on the precision of the measurements performed at this low concentration. GLP requires that further calibration concentrations in this region may only be added with a full six-day revalidation, even if it should turn out that the lowest x was initially chosen too high. The analyst, therefore, is torn between (a) sacrificing valuable sensitivity (this can be disastrous for a pharmacokinetic study), (b) redoing the whole validation program with an adjusted lowest concentration, or (c) adding a series of concentrations designed to cover the suspected region of the LOD from the beginning. The danger is that the signal distribution is no longer symmetric and residuals tend to become positive. Figure 4.29 shows how closely the estimates match.

CHAPTER

5

APPENDICES

5.0. INTRODUCTION

In this chapter necessary but not-so-readable lists (glossary, references, index) are provided. Furthermore, in this age of computing, algorithms for approximating statistical tables and some useful program listings have been added. Details of how to install the programs and use them are given in Section 5.3.

5.1. NUMERICAL APPROXIMATIONS TO SOME FREQUENTLY USED DISTRIBUTIONS

There are various reasons for replacing tabulated values by numerical approximations. To be useful in this context, the following criteria must be fulfilled:

- A wide range of degrees of freedom must be spanned, over which the approximated value changes appreciably.
- Relative accuracies of about 1% or better should be attained, which suffices for practical applications.
- The programmable calculator or PC that is to be used must be able to work with the number of significant digits required by the algorithm; rounding the coefficients can appreciably alter the results of an approximation.
- The use of an algorithm must conserve memory relative to a table-oriented approach. Polynomials or similar functions should be used because recursive functions tend to be slowly converging.

5.1.1. Normal Distribution

The probability density can be calculated by way of Eq. (1.7). Both a forward and an inverse function for the cumulative probability CP are needed:

- For a given deviation $z = (x - \mu)/\sigma$, CP $= f(z)$ calculates the cumulative probability of finding a deviation as large purely by chance; this corresponds to the t test for very large numbers of degrees of freedom.
- The inverse, $z = f(\text{CP})$, is particularly valuable in connection with the Monte Carlo method, because normally distributed random numbers can be generated.

Calculation of CP from z

USE. Calculate the cumulative probability CP for a given normalized deviate $z = (x_i - \bar{x})/s_x$ or $z = (x - \mu)/\sigma$.

ASSUMPTION. Normal Distribution.

REFERENCE. Reference 147, Eq. (26.2.19).

PROCEDURE. Use the algorithm given below; because of the symmetry of the function, only the first quadrant is defined. Use program CPVAL.

ACCURACY. The algorithm is extremely accurate; no deviation between calculated and tabulated (four decimals) CP being larger than 0.2% relative (**LRR**, this occurs where both z and CP are very close to zero); typically, the deviations are less than 0.00005 absolute (**TAR**), and less than 0.03% for z larger than 0.15 (**TRR**).

ASSIGNMENTS

z: normalized deviate

u: transformed deviate (in third quadrant)

y: polynomial approximation

s: sign of $z(-1, 0, +1)$

CP: 0 for $z = -\infty$, 1 for $z = +\infty$.

ALGORITHM

$$u = \mathrm{ABS}(z),$$

$$y = a + b \cdot u + c \cdot u^2 + d \cdot u^3 + \cdots g \cdot u^6,$$

$$\mathrm{CP} = 0.5 \cdot \left(1 + s \cdot (1 - y^{-16})\right)$$

with

$$\begin{aligned} a &= 1.000\,000\,000\,0, \\ b &= 0.049\,867\,347\,0, \\ c &= 0.021\,141\,006\,1, \\ d &= 0.003\,277\,626\,3, \\ e &= 0.000\,038\,003\,6, \\ f &= 0.000\,048\,890\,6, \\ g &= 0.000\,005\,383\,0. \end{aligned}$$

Example: For $z = -1.56$, CP = 0.0593798 is found (tabulated value: 0.05938),
$z = +0.80$, CP = 0.788144 is found (tabulated value: 0.78814).

Calculation of z from CP

USE. Inverse of above function; given a cumulative probability CP, the equivalent normalized deviate z is calculated.

ASSUMPTION. Normal Distribution. For the optimization of the coefficients 64 $\log_{10}(1 - \mathrm{CP})$ values (CP: four decimal places) for $z = 0.00 \cdots 3.00$ in steps of 0.05, and $z = 3.5$, 4.0, and 4.4 were used; see also the comment under Student's t, Table 5.1.

REFERENCE. None.

PROCEDURE. Use the algorithm given below; because of the symmetry of the function, only the third quadrant is defined; cumulative probability values CP lower than 0.5 are transformed to their (decadic) logarithm; the others are first subtracted from 1.00. The sign is appropriately set to -1 or $+1$. Use program ZVAL.

ACCURACY. The algorithm is fairly accurate, no calculated z value being off by more than 0.0166 up to $z = 4.4$, and with most deviations below 0.006 absolute. Monte Carlo simulations (n = 20,000 events) yield a mean of 0 and a standard deviation of ± 1.009; this is close enough for most practical purposes. Figures of merit: $\mathbf{s_{res}}$: ± 0.005; **LAR**: -0.015 for $z = 0$; **TAR**: 0.005; **LRR**: 20% at $z = 0.05$, 2% at $z = 0.15$, 0.3% at $z = 1.5$; abbreviations, see Table 5.1. Deviations for $z > 3.5$ (CP < 0.0003 or CP > 0.9997) can be larger than 0.01, but this is irrelevant due to the low probability of having to simulate such a z value; since empirical evidence points toward wider-than-ND tails, this is a step in the right direction.

ASSIGNMENTS

CP: cumulative probability in the range $0 \cdots 1$,

x: decadic logarithm of transformed CP (third quadrant),

s: sign of expression (CP − 0.5),

y: polynomial approximation,

z: normalized deviate.

ALGORITHM

$$\mathrm{CP}\begin{cases} < 0.5: x = \log_{10}(\mathrm{CP}), & s = -1, \\ \geq 0.5: x = \log_{10}(1 - \mathrm{CP}), & s = +1, \end{cases}$$

$$y = a + b \cdot x + c \cdot x^2 + d \cdot x^3 + e \cdot x^4 + f \cdot x^5 + g \cdot x^6,$$

$$z = s \cdot y,$$

with

$$a = -0.906\,906\,6,$$
$$b = -3.645\,91,$$
$$c = -2.205\,586,$$
$$d = -0.962\,350\,6,$$
$$e = -0.236\,619\,2,$$
$$f = -0.029\,213\,59,$$
$$g = -0.001\,375\,013.$$

Example: For CP = 0.05938, $z = -1.5633$ is found (tabulated value: -1.56),
= 0.78814, $z = +0.7982$ is found (tabulated value: $+0.80$).

5.1.2. Student's *t* Distributions

This empirical one-line function fits into almost any program, especially if only one significance level is needed:

Calculation of Student's *t* from df and *p*

USE. Calculate Student's t values given p and df; Student's t is used instead of the normal deviate z when the number of measurements that go into a mean $\bar{x}$ is relatively small and the assumption of μ, respectively, σ (infinite precision) has to be replaced by mean $\bar{x}$ (Normal Distribution), respectively, s_x (χ^2 distribution).

ASSUMPTIONS. Empirical polynomial approximation to t tables. A good overall fit was attempted; relative errors of less than 1% are irrelevant as far as practical consequences are concerned. The number of coefficients is a direct consequence of this approach. Polynomials were chosen in lieu of other functions in order to maximize programming flexibility.

REFERENCE. Reference 150 recently came to the authors' attention, who had independently devised the algorithm in 1974. A somewhat different equation is used for $p = 0.05$, 0.025, and 0.005 in Ref. 149.

PROCEDURE, ACCURACY. Use the algorithm given below; the figures of merit are given in Table 5.1. Use program TVAL.

ASSIGNMENTS

df: degrees of freedom (df was chosen here so as not to cause confusion, with coefficient f),

t: Student's t,

p: probability of error.

ALGORITHM

$$t = a + b/\mathrm{df} + c/\mathrm{df}^2 + \cdots j/\mathrm{df}^8,$$

Use the coefficients $a \cdots g(h, j)$ given in Table 5.1.

Table 5.1. Coefficients for Approximating t Values for Various Confidence Levels (Upper Part), and Figures of Merit (Lower Part).
(Note that the coefficients **h** and **j** are used only for $p = 0.0001$.)

p	a	b	c	d	e	f	g
0.5	0.67447220	0.24667600	0.06681826	0.01190292	0	0	0
0.2	1.281482	0.8489111	0.5543883	0.2734102	0.1197992	0	0
0.1	1.644487	1.539693	1.326133	0.9955611	0.8080881	0	0
0.05	1.959002	2.416196	2.544274	2.583332	2.598259	0.6047031	0
0.02	2.328194	3.608506	7.266717	0.6699166	11.71529	6.192514	0
0.01	2.586279	4.2351680	18.24859	−27.92441	66.51121	0	0
0.002	3.089668	8.178925	19.70815	27.40369	129.5110	−213.5485	343.9671
0.001	3.287494	9.948274	23.103690	52.92142	224.7871	−521.1793	843.7513
0.0001	3.708149	31.50907	−260.1995	2222.611	−4122.848	173.4048	7378.562
	$h = 274.0580$	$j = 665.3953$					

p	s_{res}	LAR	df	TAR	LRR %	df	TRR %
0.5	0.0003	0.0005	4	0.0003	0.072	12	0.05
0.2	0.0008	0.004	2	0.0005	0.22	2	0.03
0.1	0.0014	0.0048	3	0.001	0.20	3	0.06
0.05	0.001	0.0047	3	0.0003	0.15	3	0.02
0.02	0.003	0.02	1	0.0006	0.15	3	0.02
0.01	0.007	0.029	3	0.004	0.49	3	0.1
0.002	0.0006	0.0027	4	0.0004	0.037	4	0.01
0.001	0.002	0.011	4	0.002	0.12	4	0.03
0.0001	0.1	0.40	4	0.06	4.3	1000	1

where LAR: largest absolute residual $r_i = t_{approx} - t_{tabulated}$
TAR: typical absolute residual
LRR: largest relative residual
TRR: typical relative residual
df: number of degrees of freedom at which LAR or LRR is found

Example: For df = 5, $p = 0.05$: $t = 2.5690$ is found (tabulated value: 2.5706).

COMMENT. The precise sequence of digits in each coefficient depends on

(a) The precision of the tabulated values used (three decimal places),
(b) The degrees of freedom for which data points were taken (50 points: df = 1, 2, ..., 30, 32, 34, ..., 42, 45, 47, 50, 55, 60, 70, 80, 90, 100, 120, 500, 10,000 = ∞),
(c) The form of the optimization software used (Hewlett Packard HP71B curve fit module), and
(d) The number of coefficients chosen ($3 \cdots 8$).

Calculation of *p* from Student's *t* and *f*

COMMENT. Instead of calculating a critical t_c and comparing it to the experimental one, the experimental t is converted into an estimated error probability, which is then compared with a preset value, e.g., 0.05. The medical and social science communities prefer using the second approach. This algorithm is theoretically underpinned,[150] but uses quite a bit of memory.

Program PROBAB

USE. Given a Student's t factor and the number of measurements n, find the probability p that the result is due to chance alone.

ASSUMPTIONS. Definition of Student's t factor.

REFERENCE. Reference 150.

PROCEDURE, ACCURACY. Calculate p by either of two algorithms, depending on whether the number of degrees of freedom df is even or odd. The function is very accurate: **LAR**: -0.0003 at $p = 0.5$; **LRR**: 0.5% at $p = 0.001$. Use program PVAL.

ASSIGNMENTS

T: Student's t factor,
F: number of degrees of freedom df,
A, B, C, D: intermediate results,
P: probability p = result.

LISTING

```
     SUB PROBAB(F,T,P)

     RADIANS
     D = ATAN(T / SQR(F))
     A = COS(D)
     B = A*A
     IF INT(F / 2) # F / 2 THEN 150         if F = odd then go to 150

     A = 1                                  F: even numbered
     IF F = 2 THEN 130 ELSE C = 1
     FOR K = 2 TO F-2 STEP 2
     GOSUB 230
     NEXT K
 130 A = A*SIN(D)
     GOTO 260

 150 IF F = 1 THEN A = 0 & GOTO 210         F: odd numbered
     IF F = 3 THEN 200 ELSE C = A
     FOR K = 3 TO F-2 STEP 2
     GOSUB 230
     NEXT K
 200 A = A*SIN(D)
 210 A = 2*(D + A) / PI                     PI = 3.14...
     GOTO 260

 230 C = C*B*(K-1) / K                      subroutine
     A = A + C
     RETURN

 260 P = 1 - A                              calculate
     SUB END                                probability
```

Example: $t = 3.182$, df = 3:

$p = 0.05002$, found probability,

$p = 0.05000$, tabulated probability.

5.1.3. *F* Distributions

USE. Calculation of F from f_1 and f_2 for $p = 0.05$ and $p = 0.025$ for the F test.

used in this program, please refer to the user manuals which accompany your DOS disks. This section is followed by Section 5.3, which includes technical notes on the software for reference.

5.2.1. Hardware Requirements

In order to use the enclosed disk, you must have at minimum the following computer equipment available:

- IBM XT, AT, or PS/2 or compatible with at least a 80286 processor
- One 3.5 inch disk drive and a second floppy drive or a hard disk drive
- EGA or VGA graphics monitor and card
- A standard 80-column line or laser printer

5.2.2. Software Requirements

In addition to the hardware listed above, you must have the DOS 5.0 operating system to run the programs on this disk (see Section 5.2.3 if you have DOS version 3.3 to 4.0). You will need the following files from DOS 5.0:

- QBASIC.EXE
- GRAPHICS.COM
- ANSI.SYS
- EDIT.COM

Also, you should have an AUTOEXEC.BAT file created in the root directory of your startup disk. Option Z from the *Statistical Methods in Analytical Chemistry* (SMAC) menu will return control of your computer to the AUTOEXEC.BAT file. If you are running a DOS menu, this menu will reappear when you leave the SMAC menu.

To avoid returning control to your AUTOEXEC.BAT file, you can change option Z from the SMAC menu by using the EDIT program that is included in DOS 5.0.

5.2.3. Note to Users of DOS 3.3 to 4.0

For users of DOS 3.3 to 4.0, there is a version of these programs available for use with your computer. To run the DOS 3.3 version of this software, you must have the following programs available on your DOS disk:

- BASIC.COM or GWBASIC.COM
- ANSI.SYS
- GRAPHICS.COM

If you have these programs on your DOS disk, you can receive a free copy of the DOS 3.3 version of these programs by writing to: **Meier Version 3.3, John Wiley & Sons, Professional, Reference & Trade Division, 605 Third Avenue, New York, NY 10158, ATTN: Electronic Services Department**. Include your name and full address. Disks will take 4–6 weeks for delivery.

5.2.4. How to Make a Backup Disk

Before working with the enclosed disk, we recommend that your first make a backup copy. Making a backup copy of the original disk allows you to have a clean copy in case you make a mistake while working with the files. To make a backup copy on a system with two floppy disk drives follow these steps:

- Insert your DOS 5.0 disk into drive A.
- Insert a blank floppy disk into drive B.
- At the DOS prompt for your floppy drive, type `DISKCOPY A: B:` and press ↵.
- Follow the instructions on-screen, inserting the original *Statistical Methods in Analytical Chemistry* disk into drive A when prompted.

After the copy is completed, store your original disk in a safe place and use the backup copy to run the programs.

If you only have one disk drive and a hard disk, you can install the files to your hard disk in lieu of making a backup copy. To install the SMAC disk, see Section 5.2.5.

5.2.5. How to Install the SMAC Disk on Your Hard Disk

In lieu of making a backup copy, you can install the SMAC disk on your hard disk. After you complete installation, store the original disk in a safe place.

To install the program, follow these steps:

- Insert the original disk for *Statistical Methods in Analytical Chemistry* into your floppy disk drive.
- At the DOS prompt for your floppy drive, type `INSTALL` and press ↵.
- Follow the instructions on-screen.

The on-screen instructions give you the procedure for naming your source drive (usually A or B) and the target drive (usually C or D). For instance, to install from A: to C, you would do the following:

- Type `SMACINST A: C:` ↵

5.2.6. How to Load the Main Menu

To load the Main Menu for *Statistical Methods of Analytical Chemistry*:

- Move to the \SMAC subdirectory by typing `CD\SMAC` ↵
- At the DOS prompt, type `SMACMENU` and press ↵.

Once the Main Menu appears (Fig. 5.1), you can select an option by typing in the appropriate letter and pressing ↵. From this menu, you can start any of the programs in the package.

Remember, however, that the programs require QBASIC in order to run. To use the menu system, you must have QBASIC on your disk. If you are using a floppy disk system, QBASIC must be available in one of your floppy drives. If you have a hard disk system, you must have QBASIC in a place that can be reached by your PATH statement.

The Main Menu includes three types of QBASIC programs: utilities, statistical programs, and various others. The utility programs allow you to perform different functions on the data and views. These include `DATA`, `IMPORT`, `VIEW_XY`, and `XYZ`, which are listed as options A, B, P, and Q, respectively, on the menu. In addition to these programs, there are some utility programs for using the software. These include `PRI` for selecting your printer type and option Y from the Main Menu, which lets you change the path statement to locate your data and program files. This option does not change the path statements that are already part of your computer configuration.

The bulk of the menu contains the statistical programs. These include SIMILAR, SIMGAUSS, MSD, HISTO, HUBER, TTEST, TESTFIT, SMOOTH, HYPOTHES, MULTI, CORREL, FACTORS, EUCLID, LR, VALID, VALIDLL, WLR,

```
STATISTICAL METHODS          P R O G R A M S                      Peter C. Meier
        IN                         for                          Richard E. Zund
ANALYTICAL CHEMISTRY       QBASIC under DOS 5.0       John Wiley & Sons ©1993

create/import/generate data sets        evaluate/display/compare data sets
A : Data set entry/editing     DATA     M : Multiple data sets              MULTI
B : Import spread-sheet      IMPORT     N : Correlations                   CORREL
                                        O : Factor optimization           FACTOR8
D : Generate similar data   SIMILAR     P : 2-dim display                 VIEW_XY
E : Simulate noisy data    SIMGAUSS     Q : 3-dim display                     XYZ
                                        R : evaluate N-dim distance        EUCLID
F : Mean, Std.Dev.              MSD     S : Standard Lin. Reg.                 LR
G : Histogram                 HISTO     T : Validate calibration            VALID
H : Outlier detection         HUBER     U : Validate log/log calib.       VALIDLL
I : t-test                    TTEST     V : Weighed Lin.Reg.                  WLR
J : Test fit of function    TESTFIT     W : Determine shelf-life         SHELFLIF
K : Smooth time series       SMOOTH     X : Arrhenius act.energy         ARRHENIU
L : Test hypotheses        HYPOTHES     Y : Change default data path
ZVAL CPVAL PDVAL TVAL PVAL NVAL FVAL CHI CONV   PRInter emulat.   Z : quit
```

Figure 5.1. *Main Menu.*

SHELFLIF, and ARRHENIU, which are listed as options D–O and R–X on the menu, respectively.

The remaining programs allow you to display, plot, and print approximations to statistical tables. They are listed in the box on the bottom left of the menu, and include ZVAL, CPVAL, PDVAL, TVAL, PVAL, NVAL, FVAL, CHI, and CONV.

5.2.7. How to Change the Default Drive and Subdirectories

Upon startup, the programs are set for being used on C drive. They look for data in the subdirectory \SMAC \ BD on drive C.

If you would like to change these options to fit your system, you need to select option Y from the Main Menu to change the selected paths. After your choose Y, you will be given instructions on-screen for changing the drive designation for the data and program files and the subdirectory name for the data files.

5.2.8. How to Use the Program Files

These programs were specifically designed to allow you to customize them to your specific use. Customization requires that you go into the QBASIC programming language and alter the code for a particular program. Instructions about making code modifications are included in Section 5.2.9.

This section gives you general information about running the programs. For simplicity, many of the programs have common features to allow you to move easily between them. Common features of all the programs are listed next.

Program Structure

Each program on this disk performs the same fundamental functions:

- Displays the name, version, and features of the program on a title screen.
- Shows the available data files (for more information on the use of data files with particular programs, refer to Section 5.2.10), lets you select a file for use, and displays the file that you choose.
- Allows you to select one or more columns (vectors) for analysis.
- Performs the basic data analysis and displays the results.

- Includes additional options from an abbreviated menu. Common menu options include printing data tables and graphics, changing the scale for a graph, and adding data to a graph.
- Allows you to quit the program at any time by pressing `Q`.

Program Commands

In all of the programs, the following commands are available by pressing the keys indicated:

`Q`	Lets you quit from the program at any time.
`CTRL-BREAK`	Lets you stop a program in mid-run. You are returned to the QBASIC editor and must select `EXIT` from the FILE menu to return to the SMAC menu.
`PRINT SCREEN`	Allows you to print graphics. This is fully discussed in the next section.

Printing Graphs and Tables

The programs will display and print your results in both graphical and tabular formats. When a printer is required for a particular function, you will see a boxed reminder on screen to turn on your printer. If a printer is not available during these times, the program will display an error message that reads `DEVICE TIMEOUT`.

When you select the `Graph` option from a program, you will get a full-screen image of the graph. To print it, merely press the `PRINT SCREEN` key on your computer (some systems may require that you press `SHIFT` and the `Print Screen` key together). Depending on the complexity of the graph, the printout can take up to several minutes.

To view a graph without printing it, just press ↵ after the graph appears on screen.

Setting the Page Length

When a potentially long list must be printed, you are asked for the maximum number of lines per page for your printer. This number represents the maximum number of lines that will fit on one sheet of paper; for most printers, this is a number between 66 and 72. To find this number for your particular printer, refer to your user manual. Alternately, you can merely print out a table that is very long and count the number of lines that fit on the page before it ran out of room.

The page length is defined to be no less than 50 lines, to avoid unnecessary waste of paper in case of an accidental entry.

5.2.9. Customizing the Programs

These programs have been specifically created to allow you to customize them to your applications. You can customize the programs in a number of ways, from changing the number of significant digits to modifying equations. To make any modifications to a program, you must first load it in QBASIC and then change the program code. If you do not need to make any of the changes listed in this section, you can skip to Section 5.2.10, which discusses the data files.

Before you make changes to the code of any program, be sure that you have made a backup copy of your original disk to avoid overwriting the original program (see Section 5.2.4).

To load the QBASIC program, at the DOS prompt type QBASIC and press ↵. Once QBASIC is running, open the particular program by selecting `OPEN` from the FILE menu and choosing the appropriate program file.

Once you have loaded the program file, you will see the program code. Each program contains a `REM` statement (which stands for Remark) on line 10 that gives the program name, size, and revision date. In addition, the first few lines of the program code include any notes about revising the file.

General QBASIC Commands

For a full listing of QBASIC commands, refer to your QBASIC manual. Since the QBASIC environment is menu driven, it is quite easy to move around the menus to perform functions. Alternately, there are a series of quick keys available for experienced users. Some common commands include the following:

Select	From	To Do
`Open`	The File Menu	Open a program file
`Print`	The File Menu	Print out the program code
`Quit`	The File Menu	Exit form QBASIC
`Save As`	The File Menu	Save a program under a new name
`Start`	The Run Menu	Start running a program that you have modified

Displaying Numbers

The programs are written with certain numerical input in mind, for instance, from 0.1 to several thousand. Numbers in this range are presented with a reasonable number of significant digits. The statement that defines the format for numbers can be found just above the print statements in the program code. The standard format for displaying numbers in each program is defined in the `R0 = sd.cw` statement, where `cw` represents the column width. Most programs have a column width of 7 to 9 digits; however, if a column width is not defined, the program will default to a width of 9 digits.

Output that cannot be presented without loss of significant digits is converted to scientific format with as many significant digits as the column width that is defined will allow. The exponent sign `E` is dropped to gain an additional significant digit; for example, 1.2345E ± 03 is displayed as `1.2345 ± 3`. If you have very small or very large numbers, you should introduce a power of 10 to bring your raw data in line for the programs. For instance, 0.0000649 g/L could be changed to 64.9 μg/L for the concentration axis.

Displaying Significant Digits

Like displaying numbers above, the number of significant digits is governed by the statement `R0 = sd.cw`, where `sd` represents the number of significant digits.

Unformatted Results

If you want an unformatted result, stop the program by pressing `CTRL-BREAK` and switch to the editor mode by pressing `F6.Ask?` and enter your variable name. Then you can press `ALT-F5` to restart the program or select `EXIT` from the FILE menu to return to the SMAC menu. If the variable name is unknown, use the `PAGEUP` or `PAGEDOWN` keys to find it.

Viewing Graphics

The standard display uses QBASIC mode 9, which yields 640 × 350 pixels on-screen. The screen window is defined by the statement **VIEW (240,0)-(639,285)** and gives a 400 × 285 pixel EGA graph, which measures approximately 14 × 14 cm on a 32-cm screen. The remaining screen is used to display relevant results or status information.

Setting the Page Format

The pages are formatted to include one graph plus comments or one table per page. If you want to print many short tables on one page, you should run the program until you see the query "How many lines to a page?" At this point, press CTRL- BREAK to leave the program run and move to the program code. In the code, find the statement **LPRINT CHR$(12)** and remove it. The command CHR$(12) gives the instruction to insert a page break (also called a form feed). After you have removed the statement, press ALT- F5 to restart the program.

Changing the Equations

Some programs require that you modify the equations to meet your needs (for instance, the fitted equation in TESTFIT). For these programs, the title screen when you run the program will give the appropriate line number for the equations that need to be modified.

5.2.10. Using the Data Files

The enclosed disk includes sample data files that can be used with the programs. The sample files work with certain programs, as listed in the following table. All program files have the extension BAS (for BASIC), while the data files use the extensions 001, 002, and 003 and can be found in the subdirectory \SMAC \ BD (which stands for basic data).

Option	Program	Applicable Data File(s)	# Columns M
A	DATA	Any data file	
B	IMPORT	Any data file from Lotus 1-2-3, SuperCalc, Microsoft Excel, Symphony or dBASE III + or IV	
D	SIMILAR	CYANIDE (equation programmed); any other file	$M \geq 2$
E	SIMGAUSS	Generates synthetic data sets using Gaussian noise, 5 models programmed	
F	MSD	MSD, MOISTURE, JUNGLE, VOLUME, HISTO	$M \geq 1$
G	HISTO	HISTO, SIM1, SIM2, SIM3	$M \geq 1$
H	HUBER	MOISTURE, JUNGLE, HISTO, FILLTUBE, TABLET__C, TABLET__W	$M \geq 1$
I	TTEST	Data entered via keyboard	
J	TESTFIT	PARABOLA (equation programmed), CYANIDE	$M \geq 2$
K	SMOOTH	SMOOTH, PARABOLA, SIM1, SIM2, CYANIDE, PARABOLA, JUNGLE	$M \geq 1$

Option	Program	Applicable Data File(s)	# Columns M
L	HYPOTHES	Data entered via keyboard	
M	MULTI	MOISTURE, SIM2 etc.	$M \geq 2$
N	CORREL	MOISTURE, JUNGLE, SIM2, etc.	$M \geq 2$
O	FACTOR8	Data entered via keyboard	
P	VIEW_XY	Any data file	$M \geq 2$
Q	XYZ	Any data file	$M \geq 3$
R	EUCLID	SIEVE, HPLC2	$M \geq 4$
S	LR	LRTEST, CALIB1, CALIB2, CALIB3, PARABOLA, JUNGLE, HPLC, CYANIDE, (but not VALID!)	$M \geq 2$
T	VALID	VALID (Note: the files VALID.OON can only be used with the programs VALID and VALIDLL,	$M \geq 2$
U	VALIDLL	VALID	$M \geq 2$
V	WLR	see program LR (option S above)	
W	SHELFLIF	SHELFLIF, CREAM, GLASS, ARRHEN	$M \geq 2$
X	ARRHENIUS	ARRHEN	$M \geq 4$

Data Format

The data format is identical for all programs, as follows:

`N, M, CW` Represents the number of rows (N), the number of columns (M) and the column width (CW) in characters, where appropriate. These three integer numbers can be all in one record, separated by commas, or in three different records.

`AAAA, [mg]` Represents M records containing column heading (AAAA) and column dimension ([mg]). Headings and [dimensions] can be any length, but only the leading 8 characters are displayed. The total record length may not exceed 256 characters.

`R(i,j)` Represents table entries, where $i = 1, \ldots, N$ rows and $j = 1, \ldots, M$ items/row. Note: The array $R(i, j)$ contains the values; "i" is the row number, that is, the ith measurement; "j" is the item or dimension number. For a program based on linear regression (LR, WLR, VALID, or SHELFLIF), since the array $R(\, , \,)$

has M columns, it is up the user to decide which column will be identified with Abscissa X (Index K), and which with Ordinate Y (Index L); R(I, K) is the independent variable X, R(I, L) is the dependent variable Y. K and L (if necessary) are requested in each program after the file has been selected. When any program is started, the available data in the chosen file will be shown in an abbreviated format for review.

Using Data from Lotus® 1-2-3, dBASE® III + and IV, Microsoft® Excel®, SuperCalc®, and Symphony®

You can use data from Lotus® 1-2-3, dBASE® III + and IV, Microsoft® Excel®, SuperCalc®, and Symphony® with the SMAC programs included on this disk. For your reference, there are sample data files included which illustrate the proper format for data files in these popular packages. The sample data files are available in the following subdirectories:

\SMAC\BD\123	Lotus® 1-2-3 spreadsheet files
\SMAC\BD\DB3	dBASE® III + data files
\SMAC\BD\DB4	dBASE® IV data files
\SMAC\BD\EXL	Microsoft® Excel® spreadsheet files
\SMAC\BD\SC	SuperCalc® spreadsheet files
\SMAC\BD\SYMDATA	Symphony® spreadsheet files

5.2.11. Definitions of Some Statistical Terms

f number of DEGREES OF FREEDOM.

z NORMALIZED DEVIATE; in the Normal (Gaussian) Distribution function ND(0, 1), z is defined by $z = (x - \text{mean})/\sigma$ for any given, positive or negative, deviation of an observed x from the population mean μ, where mean = 0 and $\sigma = 1$. Range: $-\infty < z < +\infty$. The range $-1.96 \le z \le +1.96$ contains approximately 95% of the area under the ND(0, 1) curve.

The function ND(0, 1) is valid for "large n," a requirement that is well fulfilled for $n > 20$. The deviate z is used to test the difference between two means for significance. Since the ND(0, 1) function is symmetrical about μ, $|-z| = |+z|$ for a given deviation ABS($x - \mu$).

CP CUMULATIVE PROBABILITY; CP $= f(x)$ is obtained as the integral of the function ND(0, 1) over the range $-\infty$ to x. The form of the curve is sigmoidal and is bounded by the values $0 \leq \text{CP} \leq 1$.

t STUDENT's t; formally equivalent to the normalized deviate z, but instead of being defined for ND(0, 1), there is a separate t Distribution for every degree of freedom. Student's t, in contrast to z, takes account of the fact that for a sample containing a small number of repeats n the estimated `xbar` and `sx` have nonzero confidence intervals, whereas the theoretical μ and σ used in the Normal Distribution are assumed to be error-free. The t Distributions converge toward ND(0, 1) for large n; for $n > 20$ there is no practical difference between the two (for this reason program HYPOTHES uses ND(0, 1), which is much easier to program). Range: $-\infty < t < +\infty$. In statistical tables the probability density PD(x) is not given, instead, the t-values indicate at what deviation $(x - \mu)$ the cumulative probability for a specific f. Interpret CP as the area under the distribution curve to the left of x; alternatively, interpret α, the probability of error under hypothesis H0 (one-sided case), as the area under the CP curve to the right of x. The Student's t is used to test the difference between two means for significance. Since the t Distributions are symmetrical about μ, $|-t| = |+t|$ for a given deviation ABS($X - \mu$).

p PROBABILITY OF ERROR as defined for H0 under the one-sided hypothesis H1: $\{x\text{-test} > \mu\}$, also given by $\alpha = (1 - \text{CP}(x))$.

If the two-sided hypothesis H1: $\{x\text{-test} <> \mu\}$ is used, p is the sum of the (usually equal) areas (lower alpha, or CP(at $-z$ resp. $-t$)) and (upper alpha, or (1 − CP(at $+z$ resp. $+t$))).

F FISHER's F, the quotient $V1/V2 = (s1/s2)^2$, for $V1 \geq V2$; thus the degrees of freedom are assigned as $V1 \rightarrow f1$, respectively $V2 \rightarrow f2$.

`F(f1,f2,alpha)` is used to test for differences between two variances (see F-test).

χ^2 CHI-SQUARE; a family of curves that provide upper and lower critical limits to the χ^2 Distributions for given α. As for the t Distributions, there is a χ^2 curve for every degree of freedom. Since the χ^2 Distributions are asymmetrical, the upper and lower critical levels have to be calculated separately. The χ^2 values are used in setting confidence limits to standard deviations (program MSD) and in determining the significance of differences between the profiles for a test and a reference distribution (see program HISTO).

N NUMBER OF REPEAT SAMPLES N; in essence, program T-val is inverted, so that when left and/or right Specification Limits (SL1, SL2), a probability of error p, and a standard deviation sx are given, the program calculates the number of repeat measurements N that are needed to satisfy the distance equation `[SL-xbar] ≥ t*sx / SQR(N)`, where `xbar` is the average found for the N repeats.

5.2.12. For More Information

The following can be found in Sections 5.3 and 5.4 of the Appendix:

Section 5.3: Detailed reference information on each program included in the package, in order of appearance of the program on the Main Menu.

Section 5.4: Technical information on the structure and running of the programs, as well as troubleshooting help.

5.3. SOFTWARE REFERENCE

This section gives detailed reference information for each of the programs and data files included on the enclosed disk. The program reference is listed by program name in order of appearance on the Main Menu.

5.3.1. DATA (Option A)

Purpose. To create a new SMAC data file or edit/modify an existing one.

Features

INPUT

Allows you to perform the following steps:

- Define the number of columns M and the number of rows N.
- Define a column heading and dimension (the square brackets as in [mmol/L] are automatically added).
- Input data in a row-by-row mode (R) or in a column-by-column mode (C).
- Empty cells in the data table are skipped by pressing ↵, which results in an entry of "9999". The data "9999" is recognized by all programs and handled as follows:
  (1) In the review mode before data is selected for analysis, 9999 entries are displayed as blanks.
  (2) In the analysis mode, $x = 9999$ and/or $y = 9999$ result in a skip; this data point is ignored.
  (3) In the edit mode, 9999 is displayed as the entry.

EDIT

Allows data entries to be corrected, using the same display of data as in review mode in other programs. Also allows column headings and dimensions to be corrected.

Editing is done in 7-columns-by-16-row frames; as soon as the next frame is requested with [N], the next 16 rows are shown (1 row overlap), until up to 7 columns have been viewed. Then the next 7 rows are displayed. The [Q] entry is used to skip to the end. At an entry of [C] and then [row number.column number] ↵ highlights the corresponding elements. If the chosen entry is off-screen, the frame is shifted so that the entry is then near the middle of the screen.

APPEND

Allows you to append a row of new data to an existing data file (update mode). If more data is to be added, repeat the APPEND procedure. New columns can be chosen to remain empty for filling with the MODIFY routine (see below), or can be manually filled with data. Entry of a # character into any element is interpreted as an "end of new data" command.

DELETE

Deletes entire columns and/or groups of rows from an existing data file. The rest of the data is shifted to eliminate empty rows or columns.

MODIFY

Reads an existing data file and modifies its contents according to one or more equations. A number of commonly used transformations $v = f(u)$ are predefined and can be chosen from the following menu:

Option	Menu Choice
A	Logarithm to base 10
B	Antilog to base 10 (10^ u)
C	Natural log
D	Exponential
E	Square root
F	Square
G	Reciprocal
H	Addition
I	Multiplication
J	Clipping
K	Adding Gaussian noise (ND(0, 1))
L	Normalize the array
M	User-defined function
N	Transpose

The independent variable u can be defined to be any of the M columns, and the same is true for the dependent variable v; thus it becomes possible to enter one or more columns of real data in module INPUT, then create one or more additional columns with APPEND, and either fill this new column with data on the spot, or have it filled with 9999. Data is transferred/transformed from one column to the other by using the appropriate predefined functions. Combination of entries as in defining a quotient such as $R(i,3) = R(i,2)/R(i,1)$ is possible by using option N for the user-defined option.

Normalize the Array (Option L). The data array can be normalized by selecting this option. Normalization entails taking the average and standard deviation over one or a group of or all column(s) and subtracting the found mean from each individual entry to obtain a residual, and then dividing each residual by the SD, so as to obtain data with mean = 0, SD = 1.

Four options are available for normalizing the array:

(1) The normalization of one given column J, i.e., `R(i,j)' = (R(i,j)-Xbar(j))/Sx(j)`
(2) The normalization of all columns at once (same as above)
(3) The normalization of a group of columns K...L relative to their common average `Xbar(K...L)` and standard deviation `Sx(K...L)`, i.e., `R(i,j)' = (R(i,j)-Xbar(K...L))/Sx(K...L)`
(4) The same as option 3, except that the global average $\bar{\bar{x}}$`-Xbar(1...M)` and the global standard deviation $\overline{\overline{Sx}}$ `= Sx(1...M)` are used; all columns $J = 1, \ldots, M$ are processed at once.

In conjunction with option transpose, applied before and after normalization, an array can be normalized row-wise.

Transpose (Option M)
A new array $T(\ ,\ )$ is created, into which the array $R(i,j)$ is copied; the indicies are exchanged, i.e., $R(i,j)$ becomes $T(j,i)$ and the size of the array $T(\ ,\ )$ is now N columns of M rows each. New column headers `Col_#` and dimensions `[-]` are created. Array $T(\ ,\ )$ is saved. This allows normal BASIC data files to be transposed for use with the program EUCLID, as well as the row-wise normalization of an array in conjunction with the normalize function.

User-Defined Function (Option N)
The user can define a special function on lines LLLL. To do this, run the program and stop it with a `CTRL-BREAK`; access line LLLL using the arrow keys; modify the equation; restart the program by pressing `ALT-F5`. Line number LLLL is given on the title screen for each program.

PRINT

Prints the data file in the same 7-column format that is used for display purposes.

SAVE

Saves the data file. The filename is requested for a new file. For an already existing file that is being edited, simply hit the ↵ key to send the data to the same filename or enter a new name (with the appropriate extension) to create a new file. The default data path is C:\SMAC\BD, this can be changed by using option Y from the SMAC menu. Any extension beginning with 0 (zero) can be used for data files.

The data format is the following:

```
N
M
W                                  W = column width in bytes
H$(1)D$(1)
.
H$(M)D$(M)
X(1,1)X(1,2)...X(1,M)
.
X(N,1)X(N,2)...X(N,M)
```

Notice that the column headings and dimensions are arranged differently than in the spreadsheets that can be read using IMPORT (see Section 5.3.2 for information about the IMPORT function).

5.3.2. IMPORT (Option B)

Purpose. To transform an existing spreadsheet file into a data file for use with SMAC.

Legend

- N is the number of records (starting with $X(1, 1)$)
- M is the number of columns
- W is the column width in bytes (= characters)
- H$(J) is a column heading (text)
- D$(J) is an associated dimension (square brackets are added automatically)

Features

Scientific Format Symphony® or Lotus 1-2-3® Files

A Symphony spreadsheet created in the same format as given in the sample file \SYMDATA\GLASS_S.WR1 and below:

N				
M				
W				W = column width in bytes,
H$(1)	H$(2)	...	H$(M)	must be the same for
D$(1)	D$(2)	...	D$(M)	all columns
X(1, 1)	X(1, 2)	...	X(1, M)	
.	.	.		
X(N, 1)	X(N, 2)	...	X(N, M)	

Example:

3	
2	
12	
COLUMN_1	COLUMN_2
[mg]	[%]
1.34E + 00	1.79E + 01
6.92E + 00	2.36E + 01
1.50E + 01	−9.26E − 02

Lotus 1-2-3 files are handled analogously to the Symphony case presented here. The ∗.WR1 file can contain comment or other information outside the block of data indicated here.

Necessary steps to transform the data to a ∗.PRN file:

- [F10][Format][Scientific] to give the entries a common format; if the column width is insufficient, change this by selecting a particular column with the cursor, press [F10][Width][Set] number ↵; instead of "number ↵", the cursor keys can be used to change the column width. *Note*: all columns must have the SAME width!

- The print format ([F9][P][S][M][T]) should have a top margin of zero and the print format ([F9][P][S][M][L]) should have a left margin of zero.
- [F9][Print][Settings][Source][Range] use the cursor keys to define one corner of the array, press the period [.] key to secure the corner, and draw the cursor to the opposite corner, press ↵.
- [Destination][File], enter a file name, [Quit][Go].
- Access QBASIC, choose IMPORT (item [B]) from the menu, choose the name of a source file, and chose a target file name.

Comma-Separated Symphony® or Lotus 1-2-3® Files

Similar to above, except that, instead of forcing a scientific number format, any format(s) can be used. To enable correct transformation, a reading frame must be established; the spreadsheet is modified so that one has alternating wide and narrow columns, e.g., columns A, C, E, ... have widths X, Y, Z, and columns B, D, ... have width 1. The narrow columns are filled with commas, see GLASS_C.WR1. The number entered for item W is read but not used, so any number can be written.

```
N
M
W                                          W = column width in bytes
H$(1)   ,  H$(2)   , ..., H$(M)                can be any number
D$(1)   ,  D$(2)   , ..., D$(M)
X(1,1)  ,  X(1,2)  , ..., X(1,M)
  .          .        .
X(N,1)  ,  X(N,2)  , ..., X(N,M)
```

Example:

```
3
2
0
COL_1 , COL_2
[mg]  , [%]
1.34  , 1.79E + 01
6.92  , 23.6
15.01 , -0.0926
```

dBASE® III + and dBASE® IV Files

Transfer data files from dBASE III + *.DBF or dBASE IV *.DBF files to Basic data files. There is no provision for writing an ASCII file from dBASE, so the normal .dbf format is read. All columns of a dBASE file must have the same width; the first three rows in columns 2 and higher remain empty.

W			
N			
M			
H$(1)	H$(2)	...	H$(M)
D$(1)	D$(2)	...	D$(M)
X(1, 1)	X(1, 2)	...	X(1, M)
.	.	.	
X(N, 1)	X(N, 2)	... X(N, M)	

SuperCalc® Files

The SuperCalc spreadsheet has a built-in output option for ASCII files. A .CSV extension is appended. The structure of the data array is the same as for Symphony or Lotus (cf. sample file \bd\sc\glass.csv).

Microsoft® Excel Files

Excel has a built-in facility for generating an ASCII output with the .ASC extension. The delimiter must be defined as being a comma. For Excel for Windows files, the option export-to-ASCII-file "*.asc" no longer exists. The way around this is to use the export-to-SuperCalc function that results in a .TXT file; change the extension from .TXT to .CSV; read as a SuperCalc file.

ASCII Files

Two types of straight ASCII files are allowed, with the individual items separated by commas or of constant width. It is assumed that the ASCII files contain only the body of the table $R(1, 1) \ldots R(N, M)$. The user supplies the exact number of columns M and, in case of constant-width columns, an exact number of rows N, an exact column width W (in [bytes]) and an offset (in [bytes]). The offset might have to be experimentally determined by typing various numbers, starting with zero.

Dummy column headings and dimensions are assigned; these can be changed using program DATA, option EDIT (see Section 5.3.1).

Storing BD-Format Data Files

Notice that the BD-format data files have a different arrangement for the column headings and dimensions, see above.

5.3.3. SIMILAR (SIMulate statistically simILAR data sets) {3.4} (Option D)

Purpose. To take an existing data file that comprises at least a column X (independent variable) and a column Y (dependent variable). Call up the option 2 or 3 to edit the equations. Restart the program with [Alt][F5] and then select option LOAD. For each x-value a new y-value is generated in column M + 1, which is automatically appended using the same column heading (modified in position 8) and dimension; the resulting simulated data set is stored under the same file name. A new column Y can then be chosen, and the process repeated. Create a number of such files $(X, Y, Z, \ldots, Y', Z' \ldots)$. Use the modified data files as if several series of measurements had been acquired and test the intended statistical program and the interpretation scheme for robustness (ruggedness): if, despite the stochastic variability, always similar results and the same interpretation are found, the evaluation procedure is robust. A second set of measured values that cannot be distinguished from the first set, or its simulated statistical look-alikes cannot be interpreted to be different.

Features

- Defines a model for the standard deviation SD = f(x) on line LLLL; for example (the coefficients A and B have to be defined as numbers, of course):
  (1) `SDFX = A + B*X`, as for FID detectors in gas chromatography.
  (2) `IF X < 10 THEN SDFX = A ELSE SDFX = B`, i.e., where a detector's sensitivity is automatically switched at a certain point, say X = 10, with a concommitant change in the signal-to-noise ratio.
- Optionally defines a Y-model YM = f(x) on line LLLL; for example:
  (1) `YM(I,L) = A*X`, for a conventional linear detector.
  (2) `YM(I,L) = A + B*LOG(C + D*X)`, for a logarithmic detector with a constant interference (background) C, as for ion-selective electrodes.

- The complete model, the sum of the deterministic and stochastic components, is either `Y' = Y + SDFX * RN` or `Y' = Y(x) + SDFX * RN`, where Y are the y-values R(I, L) already in the file, Y(x) are values calculated from a model YM provided by the user, SDFX = SD(x) is a user-provided model describing the dependency of the standard deviation on x, and RN = ND(0, 1) is a Gaussian (normally distributed) stochastic variable with zero mean and unity standard deviation.
- Uses a preexisting data file comprising at least columns X and Y. Simulated data are added as extra columns to this file; an additional column carries the same header as the column from which the ordinate data were taken, but the 8th character of the header is changed to a tilde [~] (see data file LRTEST.001).
- Using a Monte Carlo simulation, simulate "measured" values that are similar to and of statistically indistinguishable properties to those provided in the original data file (provided that the SD = f(x) model is appropriate).
- View the original X and Y values, together with the newly created Y′, the residual in its absolute $Y - Y'$ and its relative $100*(Y - Y')/Y$ [%] form.
- Since the seed for the random number generator is reset using the internal clock every time the program is started, no two simulated data files thus created are identical.
- Empty cells "9999" are ignored, and if a column Y should contain none, one, or several values so that Ymin = Ymax, the model is automatically substituted. If Y contains actual data, the user is free to substitute the model.

5.3.4. SIMGAUSS {1.4} (Option E)

Purpose. Generate data sets using mixed deterministic/stochastic models. These data sets can be used to test programs or to do Monte Carlo studies. For purposes of demonstration, five different models are predefined: sine wave, sawtooth, baseline, GC-peaks, and step functions. Data file SIM1.001 was generated for $N = 200$.

Features

- Column headings and dimensions can be defined.
- The set of 5 equations on lines LLLL can be changed.
- The length of the generated file can be defined.

5.3.5. MSD (Mean, Standard Deviation) {1.1, 1.3.2, 1.7.2} (Option F)

Purpose. Calculates the mean and standard deviations, and their confidence limits, for a vector; compares with specification limits.

Features

- Calculates x-bar and the (symmetrical) confidence limits
- Approximates Student's t for $N - 1$ degrees of freedom and various p (0.0001, 0.001, 0.002, 0.01, 0.02, 0.05, 0.1, 0.2, and 0.5).
- Calculates and displays the standard deviation and the (asymmetrical) confidence limits.
- Approximates the Chi-square tables for $N - 1$ degrees of freedom and various p (0.001/0.999, 0.005/0.995, 0.01/0.99, 0.025/0.975, 0.05/0.95, and 0.1/0.9; this corresponds to two-tailed confidence limits for $\alpha = 0.002$, 0.01, 0.02, 0.05, 0.1, resp. 0.2)
- Displays (graphically and as a table) the mean and its confidence limits as a function of p.
- Displays (graphically and as a table) the standard deviation and its confidence limits as a function of p.
- Prints the same information
- Options "display S.D." and "print S.D." feature two extra rows, which give the factors that link the lower and upper confidence limits CL(s) to s.

5.3.6. HISTO {1.8} (Option G)

Purpose. Determines the distribution of many repeat measurements and compares this distribution to the Normal Distribution.

Features

- The overall x range (which scales the plot) and that part of the x range that is to be subdivided into D classes (bins) can be individually defined; essentially, this means that the plotted window can be adjusted to be the same for comparing several histograms, while bins need only be defined in that part of the x axis where the measurements are concentrated.
- No autoscaling is available, which, while convenient, exposes the individual plot limits and bin boundaries to the vagarites of measurement and sampling noise.

- Bin numbers and x range can be varied.
- An equivalent-area Gaussian ("Normal") Distribution curve can be superimposed.
- Displays and/or prints tables that contain bin number, number of events, individual, cumulative percentage population, and the normalized bin boundaries ($z = (bb - xb)/sx$), Chi-square components, as well as various statistical indicators (extreme values, number of events outside bin range, mean, SD). Option TABLE gives the z-values, the delta CP, the expected frequencies, and the associated χ^2 contributions for the two pseudoclasses "0" and "D + 1", i.e, for the areas of the distribution function ND(x-bar, S.D.) between $-\infty$ and the lower bin boundary of class 1, respectively between the upper bin boundary of class D and $+\infty$. Thus the delta CP adds up to 1.000, and the expected frequency equals the number of events.
- The cumulative percentage points can be plotted on a distorted %-axis (so-called Normal Probability Scale) that yields a straight line for perfectly ND data.

5.3.7. HUBER {1.5.5} (Option H)

Purpose. Checks a vector for outliers.

Features

- Selects a column from a data file and calculates mean, median, and standard deviation. Displays the original data together with x-bar, xm, and x-bar $\pm$ sx.
- For a better overview, a large vector is displayed in groups of 10, every group being plotted a bit higher than the previous group.
- Calculates all deviations ($x(i) - xm$), and sorts according to absolute size; calculates the Mean Average Deviation MAD; calculates cutoff limits for outliers according to Huber [see P. L. Davies, *Fresenius Z. Anal. Chem.* (1988) **331**, 513–519] by assuming the recommended value for Huber's k (3.5). Different values for this multiplier can be selected.
- Displays the cutoff limits and the clipped data set.
- Displays x-bar, xm, and sx before and after elimination.
- The ± 2- and ± 3-sigma cutoff limits of the conventional outlier-detection models are plotted (two tic marks connected by a dotted line).

- Uses the sorted absolute deviations above and plots the critical Huber's k for each $x(i)$ versus the percentage of points retained (cumulative number of ordered absolute deviations); this yields insight into the sensitivity of the clipping process to changes in Huber's k (full line, points). Note that a scale for $k = 1, 2, \ldots$ is given on the left side.
- Plots analogous critical z-values for the mean/SD before or after elimination of points (dotted lines). Since the standard deviation will decrease on elimination of suspected outliers, the dotted sensitivity curve for "after elimination" will be higher than the one for "before".
- The largest (absolute) deviate from the median (LADM), $|x(i) - xm|$ is compared to the largest (absolute) deviate from X-bar, LADX, |x(i) − xb|. Since these deviations are measured relative to the median and the mean, respectively, which do not always closely coincide, the first point to be eliminated need not be the same for the two rules. An equivalent critical signal-to-noise outlier rejection rule is given in the form `Limit, L,` ``L = ±z*s'' through use of a proportionality: LADM$/k *$MAD = LADX$/z *$SD. If Huber's k is chosen to be just a bit higher than the critical value for the point to be first eliminated, the corresponding mark on the right-hand side will be at the height z of the last point on the dotted curve.

Conventionally, outliers are identified as those points that have residuals larger than 2–3 times the standard deviation (this z could be interpreted as a Student's t factor of 2–3). Because the SD is sensitive toward the presence of outliers, outlier rejection schemes based on it are inferior to the one presented here, which uses medians. Huber's rule can be used with asymmetrical distributions.

5.3.8. TTEST {1.5.2} (Option I)

Purpose. For a given set of results (two means, their standard deviations and numbers of determination: x1, x2, sx1, sx2, n1, n2) the F-test is performed; depending on the outcome (sx1 significantly different form sx2, or indistinguishable) and the identity or nonidentity of n1 and n2, different forms of the t-test are carried out. Because cases exist where different authors propose similar but not identical equations for treating the same data, ambiguous situations can arise; usually, this is due to varying and sometimes unstated assumptions as to what represents a large number ($n > 200$, for example), whereupon equations are simplified by replacing $(n - 1)$ by (n). One variation of an equation that is given in more than one textbook is included despite the fact that it probably contains a transcription error ("$n + 1$" instead of "$n - 1$").

Features

- Equations and references can be requested to appear on the screen (use PRINT-SCREEN to get a hard copy).
- The estimated probabilities of error for both the one-sided and the two-sided cases are given, along with the appropriate critical values (0.025 or 0.05).
- Of the six inputs, one at a time can be changed to evaluate the sensitivity of the t-test decision relative to changes in N, x-bar, or sx.

5.3.9. TESTFIT {4.13} (Option J)

Purpose. Quantitates the goodness of fit between a model and a data set. The model must be provided together with a set of estimated default coefficients on lines LLLL-; up to 5 coefficients A1 ... A5 can be used.

Features

- The program is stopped by selecting menu option 2; an equation is entered on line(s) "LLLL" if the program code that describes the dependency of the measured variable $Y(i)$ on the settings of the independent (controlled) variable $X(i)$, that is, $Y(i) = f(X(i))$ with model coefficients a1, ..., a5 (up to 5 coefficients can be used in the equation). A string is available so the equation and/or comment will appear on the screen and printer output to clearly define the conditions, e.g., `T$(23) = ``Smith's Eq. with quadr. term: A4*y*y; A4:-3.75''`. The program can easily be amended to provide for more than one independent variable, e.g., $Y = f(x1, x2, \ldots)$. The program is started with [Alt][F5].
- After the data file is entered, the columns are chosen that are to correspond to the independent variable, i.e., the abscissa X, respectively the dependent and measured variable (ordinate Y).
- The above user-supplied equation is evaluated for every point $X(i)$, and the residual $Y(l) - Y$, where $Y = f(x)$, is calculated.
- The extreme values for both x and y ranges are determined, and autoscaling of the display is performed (this can later be changed by using the SCALE module).

- The residuals are initially plotted across the bottom of the display using an expansion factor of 1; the expansion factor can be influenced.
- If the fit is poor, the coefficients A1 ... A5 can be modified without stopping the program.
- Model interpolations $Y = f(x)$ can be requested for a given x.
- Model interpolations $X = f(y)$ can be requested for a given y and an estimated x; interpolation is done using two closely spaced points (x or $x + 2\%$ of the x range); iterations are continued until the calculated Y changes by less than 0.1%.
- The displayed graph can be output on the line printer.
- A table of $x(i)$, $y(i)$, the estimate $Y(x(i))$, and the residual can be printed; the averages for x and y, as well as the residual standard deviation are printed.

Interpretation. The model can only be improved upon if the residual standard deviation remains significantly larger (F-test!) than the experimental repeatability (standard deviation over many repeat measurements under constant conditions, which usually implies "within a short period of time"). Goodness of fit can also be judged by glancing along the horizontal (residual = 0) and looking for systematic curvature.

5.3.10. SMOOTH {3.5} (Option K)

Purpose. Constructs a smoothed trace over a (large) series of observations to improve presentability or to extract a "trend".

Features

- Ordinate values are assigned by choosing a column number in the data table; abscissa values can either be chosen similarly or can be assigned sequential numbers ("index", which makes them equidistant on the x scale). The Moving Average and Savitzki–Golay filters assume equidistance for correct application; *note*: if this condition is only mildly violated, not much harm is done. The residuals and the residual standard deviation are strongly affected if the abscissa distances between successive points ("delta x") changes appreciably from one interval to the next.
- Error bars of constant or signal-proportional length can be assigned to the points.
- The coordinates are automatically sorted according to ascending x if x is not an index.

- Three smoothing algorithms are incorporated:
  (1) BOX-CAR averaging with a box width (filter width, window) in the range $1 \leq W \leq N$; $W = 1$ just connects the points, while $W = N$ calculates the overall average x-bar. The average for each box will be displayed as a horizontal line.
  (2) MOVING AVERAGE; a filter width in the range $1 \leq W \leq N$ is possible, W being restricted to odd values. Both the x- and the y-values are averaged in the filter window and the averages are assigned to $u(i)$, respectively $v(i)$; the trace is plotted using these smoothed coordinates u/v. The residuals, however, are calculated as $r = y(i) - v(i)$ without interpolation, under the assumption that the filtered $v(i)$ does not appreciably vary locally; if it is suspected that this condition is not met, it is suggested that the filtering be done using the index $i = 1, \ldots, N$ instead of a column from data table (option "column = 0 for abscissa").
  (3) SAVITZKY–GOLAY smoothing; filter widths in the range $W =$ 5, 7, 9 or 11 are possible; smoothing can be carried out using polynomial filters of order 2 (square parabola) or 3 (cubic parabola). Differentiating filters and filter widths above 11 were not implemented because such applications only make sense if very large numbers of measurements are available; if such data series are acquired on EDP equipment, it is recommended that dedicated programs optimized for speed be used. Since the SG filter coefficients are generated in situ according to P. A. Gorry [*Anal. Chem.* **62**, 570–573 (1990)], much time would be wasted if the filter type and/or width were often changed in order to find the optimal combination. Whereas the usual implementation of the SG-filtering algorithm allows only for a smoothed trace in the range $x(\text{INT}(W/2) + 1) \ldots x(N - \text{INT}(W/2) - 1)$ (the ordinate in the center of the filter is estimated), analogous filters are also calculated here that allow the estimation of the smoothed trace at the ends of the range, i.e., $x(1) \ldots x(\text{INT}(W/2))$. Thus the smoothed trace includes all N points.
- A technique to detect deviations from random scatter in the residuals (symmetrical about 0, frequent change of sign) is included: CUMSUM (Cumulative Sum of residuals) detection of changes in trend or average. Here, an average is subtracted to yield residuals; these residuals are then summed over points $1 \ldots k \ldots N$, with the sum being plotted at every point $x(k)$. Two uses are possible:
  (1) No averaging has taken place (options 5 in the menu): the individual average is equal to the overall mean y-bar, which is displayed as a horizontal line; this corresponds to the classical use of the CUMSUM technique. By this means, slight shifts in

the average (e.g., when plotting process parameters on control charts) can be detected even when the shift is much smaller than the process dispersion, because the CUMSUM trace changes slope.

(2) If smoothing has taken place, the individual "average" associated with every point will equal the local estimate for the smoothed trace; essentially, the CUMSUM now detects how well the smoothed trace represents the measurements. For example, if peak shapes are to be filtered (see data file SIM1.001) and too wide a filter is used, the smoothed trace might "cut corners"; as a result, the CUMSUM trace will change slope twice.

The CUMSUM trace can be shifted vertically, and an expansion factor can be chosen. Ordinate rescaling is done automatically.

- The original and the filtered data, together with the appropriate residuals, are displayed and printed.
- The residuals relative to the smoothed trace (to y-bar, if no smoothing has been done) are plotted; a vertical shift and an expansion factor can be chosen.
- A linear regression can be carried out over a restricted range; the regression line is displayed with 95% confidence limits.
- Interpolation on the smoothed trace can be carried out for any x within the bounds of the trace; the result of the interpolation is displayed and printed in numerical format and is indicated by cross hairs on the screen.
- Interpolation on the smoothed trace can be carried out for any y; multiple intersections with the trace are possible; the results of the interpolation are displayed and printed in numerical format (list) and are indicated by cross hairs on the screen.
- Without restarting the program, a different filter type and/or width can be chosen; this results in recalculation of the SG filter coefficients.
- Without restarting the program, new data can be chosen for analysis: either a different column from the same data set, or a new data file. If an SG filter has already been calculated and is again selected, the filter coefficients are not recalculated. A recalculation must be forced by selecting option 7 (different filter).

5.3.11. HYPOTHES {1.9} (Option L)

Purpose. Displays the type I error (alpha) and the type II error (beta) both as (colored) areas on the ND(μREF, σREF) and the ND(μTEST, σTEST) distribution functions and as lines in the corresponding cumula-

sponding cumulative probability curves. Since a Normal Distribution is used instead of t Distributions, the results are, strictly speaking, only valid for Nref and Ntest larger than about 20.

Features

- Freely chooses alpha.
- The means and standard deviations can be freely chosen to represent any combination of reference measurement and test measurements. The hypothesis H0 applies if the REF and TEST measurements are indistinguishable, H1 if they can be distinguished.
- The following tests can be conducted: REF < TEST, REF < > TEST, REF > TEST.
- Displays and prints graphics
- Displays the estimated error beta numerically.

5.3.12. MULTI {1.5.3, 4.4} (Option M)

Purpose. Tests several sets of data for deviations from the null hypothesis H0 ("all data sets belong to the same population, that is, the individual standard deviations do not differ from the overall standard deviation"). It is assumed that the individual data sets consist of a number of repeat measurements of one observable, such as: repeat moisture or temperature measurements at several locations; hourly concentration determinations of a given chemical species in several reaction vessels that are being run in parallel; or the same reaction in the same vessel at different times of the week. If the standard deviations do not significantly differ (Bartlett Test), continue with the Simple ANOVA Test to find whether the data set averages belong to one or more populations. If more than one population is involved, find which data set averages can be grouped into homogenous subpopulations (Multiple Range Test). It is possible that one and the same data set could be grouped into two or more subpopulations that partially overlap.

Features

- Each data set fills a column; if data sets are of unequal length, N is determined by the largest set (program DATA; nonexistent entries are entered as "9999" in DATA by just pressing ↵ and are ignored by MULTI).

- After the chosen data file is read into MULTI, sequential columns K through L are selected for analysis; K and L must be within the bounds $1, \ldots, M$, and L must be larger than K.
- The heading, dimension, average x-bar, standard deviation sx, RSD (or c.o.v.) $100 * sx/x$-bar, and number of determinations Nj are displayed for every column K...L.
- The following three tests can be individually called, but execution is blocked if the preceding test has not been performed.
- The Bartlett Test yields an uncorrected and a corrected Chi-square value, which are compared to the critical χ^2 for $f1$ degrees of freedom and $p = 0.05$ (calculated with a relative accuracy of typically 0.5–1%) using the algorithm from Table 5.2. The interpretation is given. If at least one standard deviation is significantly different from the others, the program stops here.
- If the standard deviations are indistinguishable, an ANOVA Test is carried out (simple ANOVA, one-parameter additivity model) to detect the presence of significant differences in data set averages. The interpretation of the F-test is given (the critical F-value for $p = 0.05$, one-sided test, is calculated using the algorithm from Section 5.1.3).
- The Multiple Range Test yields a triangular matrix of differences delta-x-bar-ij (difference in x-bar for every possible combination of x-bar-j with x-bar-j).
- The triangular matrix of differences delta-x-bar-ij is converted into a triangular matrix of q-values using the values x-bar-i, x-bar-j, $N - i$, and $N - j$; since tables of q-values are rarely given in statistics textbooks, especially for p-values other than 0.05, the q-values are converted to "reduced" q-values by division through the appropriate Student's t and square root of 2. This permits a delineation of subpopulations on the basis of a critical reduced q of 1.1 with an only small chance of misinterpretation, even if the statistical probability of error p is changed from $p = 0.05$ to some other value (0.02, 0.1, etc.) by exchanging the algorithm for $t(f, p = 0.05)$ (Table 5.1).
- The subpopulations of data sets that can be distinguished are given as lists of x-bar values; the number of lines corresponds to the number of subpopulations.
- A graph is displayed that contains the individual data points, and the associated averages x-bar and standard deviations sx; the data sets are arranged left-to-right in the same order as they appear in the data file.
- A second graph, with the data sets ordered according to increasing x-bar, depicts the averages and standard deviations, and, as stacked

horizontal lines, the range of averages spanned by the individual subpopulations.

- The tables of differences-of-averages, respectively of reduced-q-values are displayed with the corresponding ordered averages x-bar arranged at the top and down the left margin, 7 columns and 15 rows at a time; by pressing ↵, the next frame is shown. The last frames are arranged to have column L appear at the right of the screen, so that a maximum overlap is achieved.
- All results shown on the screen are printed on demand.

5.3.13. CORREL {4.11} (Option N)

Purpose. Finds correlations in a data table; the data table is organized into M columns, each of which corresponds to a dimension, e.g., concentrations of impurities, pH, absorbance at various wavelengths, etc. Each row corresponds to a sample, e.g., a batch of material analyzed according to M methods.

Features

- Calculates mean and standard deviation for every column.
- Calculates the correlation coefficient r for every combination of columns, and displays the results in a triangular matrix (an absolute value just under 1.00 indicates a strong correlation between the measurements in columns i and j).
- Using the correlation coefficients, calculates the Student's t factors and displays these (the larger this value, the more probable the correlation; values below about 2 are insignificant).
- Using the Student's t factors and the number of degrees of freedom, calculates the probabilities p that discerned correlations are due to chance alone (error probabilities); these are interpreted as follows:

p above about 0.1–0.2:	insignificant
p in the range 0.05–0.1:	weak
p in the range 0.02–0.05:	significant
p below about 0.02:	highly significant

Negative slopes are flagged: an "n" is to be interpreted in the sense that the slope is negative, and J decreases as I increases; positive slopes remain unmarked.

- Suspected correlations can be viewed as scatterplots; a scan-up/down feature allows all combinations of vectors to be efficiently inspected.

Any interpretation must take the physical and/or chemical situation into due consideration:

(1) A correlation can indicate a mechanism that links I and J; if I is, say, a concentration (the independent variable), the absorbance J is the dependent variable, but not vice versa!

(2) A correlation could also be due to a mechanism that links a third, known or unknown, factor to the two observables I and J; an increase in concentration of a complexing agent, say, could lead to increased solubility (I), and at the same time shift a UV-absorption feature by a few nanometers (J).

(3) Spurious correlations are often observed, e.g., when only a small number of observations N is available; if uncontrolled forces are at work, or important data has not been collected; or when coincidences apply, for example, when the time between injections is such that the (saturation-broadened) main peak coelutes with a late peak from a previous injection.

(4) Lack of correlation can mean just that, or could be due to the fact that the observations did not span a sufficiently broad range of (controlled) experimental conditions.

5.3.14. FACTOR8 {2.4.2} (Option O)

Purpose. To determine, from eight initial experiments performed under certain conditions, whether the three controlled parameters have an effect on the measurement, and which model is to be used. This factorial approach to optimization is an alternative to the use of multidimensional SIMPLEX algorithms; it has the advantage of remaining transparent to the user.

Features

- The low/high values for the three parameters are entered.
- The 8 measurements corresponding to the experimental conditions 1, a, b, c, ab, ac, and abc are entered. "1" means all three parameters are set "low", while "abc" connotes the opposite.
- The model is `Y = m(1) + m(2)*a + m(3)*b + ... m(7)*a*b*c`.
- The model is fit; effects, specific effects, model coefficients, and residuals are displayed.

- The assumed residual standard deviation, i.e., the precision of measurement, can be varied to study its effect.
- Option 6, $s = f(\mathrm{SD})$ sweeps through a range of assumed SDs (start with a relatively high value, for instance, 20 times the experimentally determined sx) to detect the critical points where the 7 model coefficients $m(i)$ are set to zero for lack of statistical significance.
- By brute-force iteration, the highest Y within the cube spanned in 3-space is located with a resolution of a few percent of the parameter ranges. The cube's center is accordingly moved, and the model is reevaluated; in this fashion a track of steadily higher Y-values in the immediate vicinity of the initial cube is displayed. Thus the direction to be taken for further experiments is indicated.
- The model can be evaluated for any combination of a, b, and c.
- A "Change Data" option allows the previously input values x1, x2, s1, s2, n1, or n2 to be individually changed for a new calculation in order to study the ruggedness of the t-test.

5.3.15. VIEW_XY {2.1, 2.4.1} (Option P)

Purpose. Displays x/y coordinates (a so-called scattergram); the coordinates are not marked by small squares, as in all other programs, but with the appropriate index numbers, so that "outliers" can be readily identified.

Features

- Vector selection and many other functions are similar to those provided in program LR.
- Once a graph is displayed, functions "4: ≪ Scan" and "5: ≫ Scan" advance the indices appropriately, so one combination can be viewed after another.
- Routines are incorporated that allow fairly readable presentations of the digits $0, \ldots, 9$ to be drawn in the graphics mode.
- The program centers 1-, 2-, and 3-digit numbers (index $1, \ldots, 999$) on the coordinate.

5.3.16. XYZ {2.4} (Option Q)

Purpose. Plots pseudo-three-dimensional (isometric) presentations in order that complex relationships can be studied.

Features

- Select any vector to be the left-to-right axis (X axis).
- Select any vector to be the front-to-back axis (Y axis).
- Select any vector to be the vertical axis (Z axis).
- Autonormalize every vector to the range $0 \ldots 100$ so that a cube results.
- Depict every coordinate by a small square.
- Mark the footprint of every coordinate on the X–Y plane.
- Connect $z > 0$ points to the corresponding footprints with full lines.
- Connect $z < 0$ points to the corresponding footprints with dotted lines.
- Mark every axis with the corresponding header.
- Rotate around the Z axis in a range $-90 \ldots +90°$ for better view.
- The displayed X-, Y-, and Z-ranges can be scaled; the projected cube can be rotated about its z axis in the range $-90 \ldots +90°$.

5.3.17. EUCLID {4.12} (Option R)

Purpose, Scheme of Calculation. Compares two sets of objects (groups A and B, for a total of $K + L \leq M$ objects); each object is profiled by determining its response to measurements in N dimensions. These measurements are averaged within each group to establish typical profiles. A "distance" is established between the A and the B averages by calculating the Euclidian distance in N-space:

$$D = \mathrm{SQR}\left((xa - xb)^2 + (ya - yb)^2 + \cdots\right),$$

where x, y, ... are the results for dimensions $1, \ldots, N$; for example, a concentration, a pH, an absorbance. The Euclidian distances between every object and the group averages A and B are calculated analogously; the triangle A–B–O is projected onto the screen, the side A–B being the base. The vertices are marked with filled (Group A) and open (Group B) squares. The scatter and clustering of the upper vertices (O) is analyzed: if two distinct groups are formed, then the hypothesis that the objects can be separated into two groups gains credence.

Features

- A table of M columns (M objects) and N rows (dimensions) is split into two groups, i.e., objects 1 to K belong to group A, objects K + 1 to K + L ≤ M belong to group B.

- The group averages A and B are formed for every dimension.
- The differences between the group averages are calculated and displayed for every dimension.
- The N-space Euclidian distances for every object relative to the group averages A and B are calculated and plotted.
- The contribution every dimension makes towards the total distances A-B, O-A, or O-B is plotted.

Note: If the N dimensions yield very different numerical values, such as 105 ± 3 mmol/L, 0.0034 ± 0.02 m, and $13{,}200 \pm 600$ pg/ml, the Euclidian distances are dominated by the contributions due to those dimensions for which the differences $A - B$, $A - O$, or $B - O$ are numerically large. In such cases it is recommended that the individual results are first normalized, i.e., $x' = (x - xb)/sx$, where xb and sx are the mean and standard deviation over all objects for that particular dimension X, i.e., by using option Modify/Normalize in program DATA. The case presented in sample file SIEVE.001 is different: the individual results are wt-% material in a given size class, so that the physical dimension is the same for all rows, namely, [wt-%]; since the question asked is "are there differences in size distribution?", normalization as suggested above would distort the information and statistics-of-small-numbers artifacts in the poorly populated size classes would become overemphasized. The lower graph gives the contributions by dimension; if some dimensions contribute virtually nothing to the Euclidian distance, e.g., dimensions (rows) 10–15 in SIEVE.001, these rows can be eliminated using program DATA/option DELete, and a reanalysis is done in a lower-dimensional space. Use the "Transpose" feature in program DATA/option "MODify" to exchange rows and columns, if necessary. (See 5.3.1 for information on the DATA program.)

5.3.18. LR (Standard Linear Regression) {2.2} (Option S)

Purpose. Performs a linear regression analysis over the selected data points; displays and prints results, does interpolations, determines limits of detection.

Features

- Chooses any two columns from the M-column data file to represent the abscissa X respectively the ordinate Y.
- Performs the LR.

- Approximates Student's t for $N - 2$ degrees of freedom and $p = 0.05$. (Alternatively, coefficients for the algorithm for $p = 0.0001$, 0.001, 0.002, 0.01, 0.02, 0.1, 0.2, and 0.5 are given in Table 5.1 or can be looked up in program MSD).
- Plots the regression line and its confidence limits.
- Interpolates $Y = f(X)$ and $X = f(y)$, including confidence limits on the results; displays interpolation result.
- Calculates and displays the limits of detection and quantitation LOD and LOQ according to Luthardt et al. [*Fresenius Z. Anal. Chem.* **326**, 331–339 (1987)] resp. Oppenheimer et al. [*Anal. Chem.* **55** 638–643 (1983)]. [*Note*: This form of calculating the LOD or LOQ was chosen because the results are influenced not only by the noise on the baseline, but also by the calibration scheme; from the educational point of view this is more important than the consideration whether any agency has officially adopted this or that LOD model. For a comparison, see Fig. 4.29.
- Calculates and displays the residuals $(y(l) - Y)$, where $Y = A + B * x(l)$.
- Selects and displays specification limits on the acceptable X-range, as when doing calibration/assay_of_sample work.

5.3.19. VALID {4.23} (Option T)

Purpose. Same as program LR; it is assumed that repeat determinations were performed for most concentrations and that the results are grouped; a reduced input format is used, see sample file VALID.001.

Features

- Since the determinations are grouped by concentration, repeat measurements may be entered by just hitting ↵ in the concentration column in program DATA (this is equivalent to a "9999" value, which is displayed as a blank); "9999" will be converted to the correct value after the columns are selected.
- The average, the standard deviation, and the confidence limits of the population at each concentration with multiple measurements will be calculated and tabulated.
- Option VALID presents a graph of relative standard deviation (c.o.v.) versus concentration, with the relative residuals superimposed. This gives a clear overview over the performance to be expected from a

linear calibration Signal $= A + B *$Concentration, both in terms of (relative) precision and of accuracy, because only a well-behaved analytical method will show most of the residuals to be inside a narrow "trumpet"-like curve; this trumpet is wide at low concentrations and should narrow down to c.o.v. = ±5% and rel.CL = ±10%, or thereabouts, at medium to high concentrations. Residuals that are not randomly distributed about the horizontal axis point either to the presence of outliers, nonlinearity, or errors in the preparation of standards.

- The back-calculation feature in option "Table" gives each measurement as the estimate $X(y(i))$ normalized to the nominal concentration; the results should all be around 100%. The symmetrical limits ±SD are also given.
- The other features are identical to those of program LR.

5.3.20. VALIDLL {4.23} (Option U)

Purpose. VALIDLL is identical in concept and features to VALID, the difference being that the use of a log-log depiction is assumed. Linear (i.e., nontransformed) data are read, a linear regression is calculated as in VALID, and the limits of detection/quantitation are determined. Thereafter, the data set is logarithmized and displayed.

Features

- Interpolations are done by entering nontransformed values, the results being back-transformed; confidence intervals and back-calculated estimates are now nonsymmetrical.
- LOD and LOQ are displayed in both graphs.

5.3.21. WLR (Weighted Linear Regression) {2.2.10} (Option V)

Purpose. Same as program LR; the user has the option of defining a functional dependence of the repeatability (defined as a standard deviation) of the measurements Sy on the independent variable x, e.g., $Sy = a + b * x$ (typical for gas chromatography, where the relative standard deviation (in [%]) of the measured peak area is often constant over a very large concentration range; the constant "a" represents the intrinsic, concentration-independent repeatability of the instrument). Appropriate functions can be fitted using the output of the results table of programs VALID and TESTFIT.

Features

- Practically identical with program LR.
- $Sy = f(x)$ function is defined on lines LLLL and is displayed and printed in the protocols in the form $y(i) \pm$ SD. The function is changed by stopping the program with `Ctrl-Break`, using the arrow keys to find line LLLL, modifying the equation, and restarting the program with `Alt-F5`. A string function is provided to add a note that will be displayed on the screen and printed on protocols.
- In the graphs each point is marked with a square and attached error bars that correspond to $yi \pm \mathrm{SD}(xi)$.
- The statistical weight attached to each point is given in option "Table".

5.3.22. SHELFLIF {4.20} (Option W)

Purpose. Determines the shelf life of a product (e.g., a pharmaceutical) by evaluating analysis results as a function of storage time. The points at which the lower 95% confidence limit of the population and a horizontal at 90 or 95% of the nominal content intersect determines the acceptable shelf life as promulgated by the FDA in their "Guidelines for Submitting Documentation for the Stability of Human Drugs and Biologics, February 1987". The program is a modification of the LR program.

Features

- The x-range is fixed to show only the 0–60-month interval because shelf lifes above 5 years are uncustomary in the pharmaceutical industry.
- The y-range is initially set to 80–110% of nominal, but can be changed; data input is in the form x = time, y = %-of-nominal concentration or amount.
- An algorithm for calculating the symmetrical (two-tailed) t-factors for $p = 0.1$ is incorporated; its use corresponds to the statement that "the probability that measurements on a future batch, given the linear trend already established, will inadvertently be found to be below the specification limits of either 90 or 95% of nominal, at a shelf life that would lead one to expect a residual content at or above the specification limit, is $p = 0.05$." This particular model for calculating the shelf life is accepted by the FDA.

- The equation for the lower confidence limit is $Y = A + B * x - t * \mathrm{SQR}(V\,\mathrm{res} * (1/N + 1 + ((x - xb)^2)/Sxx)$, where time x is measured in months.

5.3.23. ARRHENIUS {4.21} (Option X)

Purpose. From a series of assays done on samples stored at different temperatures over various lengths of time, the assay-vs.-time trend is calculated for every temperature. These slopes and the actual storage temperatures [deg.C] are used to construct an Arrhenius activation-energy diagram, from which the decomposition rate at any temperature within the investigated interval can be estimated, and a shelf life can be assigned. Note that zero-order decomposition kinetics are assumed (a zero-order reaction proceeds at a constant rate, i.e., independent of the remaining concentration); when this assumption is violated, the activation energy changes with temperature, and the Arrhenius diagram becomes nonlinear. The data format is demonstrated in file ARRHEN1.001, which was taken from a Ph.D. thesis (Ref. 146).

Features

- Performs linear regression for every data set (= storage temperature).
- Plots Arrhenius diagram: slopes $\pm$CL versus $1/T$.
- Tabulates assay-vs.-LR residuals.
- Tabulates LR statistics and activation energies.
- Estimates shelf life for a given temperature.

5.3.24. Other Programs

Nine programs featuring the approximations to the ND, the t, and the χ^2 distributions appear as nine abbreviations on the last line of the basic menu BASICM.ML. To call up one of these programs, type, e.g., [zval] ↵. Appearance of, use of, and escape from these 9 programs are analogous to what the other 23 programs offer.

The nine programs are called ZVAL, CPVAL, PDVAL, TVAL, PVAL, FVAL, NVAL, CHI, and CONVerge; they provide the algorithms used in other programs in a look-up form that allows direct comparison with tables.

ZVAL

Purpose. Tabulates and plots $z = f(\mathrm{CP})$.

Features

- Approximates $z = f(\text{CP})$.
- Prints and plots individual CP/z coordinates for a given CP.
- Lists accuracy and precision information, reference.

CPVAL

Purpose. Tabulates and plots $\text{CP} = f(z)$.

Features

- Approximates $\text{CP} = f(z)$.
- Prints and plots individual z/CP coordinates for a given z.
- Lists accuracy and precision information, reference.

PDVAL

Purpose. Tabulates and plots $\text{PD} = f(z)$.

Features

- Approximates $\text{PD} = f(z)$.
- Prints and plots individual z/PD coordinates for a given z.
- Lists accuracy and precision information.

TVAL

Purpose. Tabulates and plots $t = f(f, p)$.

Features

- Approximates $t = f(f, p)$.
- Prints and plots individual $\text{lgt}(f)/t$ coordinates for a given f.
- Displays linearity of $t(f, p)$ vs. $\text{lgt}(p)$.
- Lists accuracy and precision information, reference.

PVAL

Purpose. Tabulates and plots $p = f(f, t)$.

Features

- Approximates $p = f(f, t)$.
- Prints and plots individual lgt$(f)/p$ coordinates for a given f.
- Lists accuracy and precision information, reference.

NVAL

Purpose. Tabulates and plots the necessary number of repeat determinations $N = f(x, p, sx)$ such that for a fixed probability p the calculated x-bar will be within specification limits.

Features

- Approximates $t = f(f, p)$ for given SL, p, and sx.
- Calculates $N = f(f, p, \mathrm{SL})$.
- Prints and plots individual N/x coordinates.
- A one-sided t-test performed at $p = \alpha/2$ will indicate that x-bar does not violate the SL, or, inversely, that the chances that a series of N repeat measurements on one sample that yield the average x-bar will result in a statement "H1: the sample does not comply with the Specification Limit" do not exceed $100 - p\%$. This procedure is applied to both SL1 and SL2 and gives the requisite value N for all x-bar SL1 $\leq x$-bar $\leq$ SL2. The results $N = f(x)$ are given in tabular and graphical format. The probability p can be chosen from the list $p = 0.00005, 0.0005, 0.001, 0.005, 0.01, 0.025, 0.05, 0.1$, and 0.25. The limits SL1 and SL2 can be chosen at will, 90 and 110% being the default values.

FVAL

Purpose. Tabulates and plots $F = f(f, p)$.

Features

- Approximates $F = f(f, p)$.
- Prints and plots individual lgt$(p)/F$ coordinates for $p = 0.025$ and $p = 0.05$.
- Lists accuracy and precision information, reference.

CHI

Purpose. Tabulates and plots $\mathrm{Chi}^2 = f(f, p)$.

Features

- Approximates $\text{Chi}^2 = f(f, p)$
- Prints and plots individual $\text{lgt}(f)/\text{lgt}(\text{Chi}^2)$ coordinates for a given f.
- Displays linearity of $\text{Chi}^2(f, p)$ vs. $\text{lgt}(p)$.
- Lists accuracy and precision information, reference.

CONV

Purpose. Demonstrates how a calculated average and standard deviation converge on the expected values $\mu = 0$ and $\sigma = 1$ as more ND(0, 1) distributed measurements are added to the calculation. Accuracy and precision are plotted. The evaluation is for a probability of error (two-sided) $p = 0.002$, 0.01, 0.02, 0.05, 0.1, or 0.2.

Features

- Generates new data set of length N.
- Set N.
- Sets p.
- Plots and tabulates results

5.3.25. Utility Programs

PRI (Printer)

Purpose. Defines the printer.

Features

- Any of the following three printer types can be directly used, and other printer models capable of emulating one of them can be addressed:
  (1) IBM PROPRINTER
  (2) any HP DESKJET printer
  (3) HP LASERJET printer, model II or higher

Option Y

Purpose. Allows you to save data files in and retrieve data files from directories other than \BD. Note that subdirectories to the chosen director (e.g., bd\symdata) can be accessed without changing the data path, e.g., by typing [symdata\glass_s.001] in order to use \BD\SYMDATA\GLASS_S.001.

Features

- The program opens basicsm.bat in \smac\basicm5 in the edit mode.
- The "set datadir = smac\bd" instruction is visible, and instructions how to change this are given.
- Instructions for either saving the change/returning or just returning without saving are given on-screen.

Option Z

Purpose. Exit from QBASIC menu. Type Z, hit ↵ to leave the BASIC-Menu and return to the AUTOEXEC.BAT file, so the configuration as it existed before QBASIC was accessed is again reinstated. (For more information, see the Technical Notes Section 5.4.)

COLOR

Purpose. Displays the possible foreground/background color combinations to check for contrast, hue, and saturation. This program was used to determine the colors that would work both on color and b/w screens. One and the same color can look different if viewed on two different color screens side by side.

5.3.26. Data Files

The following files are provided as instruction aids and examples.

ARRHEN.001

A product was put on stability at 25, 30, and 40 deg.C. for 24, 24, or 3 months; see Ref. 146.

ARRHEN.002

A peptide solution was put on stability at 30, 40, 50, 60, 73, and 80 deg.C for between 10 and 298 days; see Ref. 146.

ARRHEN.003

Another peptide solution was also put on stability at 30, 40, 50, 60, 73, and 80 deg.C. for between 10 and 298 days; see Ref. 146.

CALIB.001

Calibration measurements at 8 concentrations (double determinations) using a GC; peak area measurements in [mV * sec] vs. weight in [mg].

CREAM.001

Two active components of a cream were measured during a stability monitoring program; data for $0, \ldots, 24$ months. One component decomposes faster than the other.

CYANIDE.001 {4.13}

Two calibration series over the same range, and one over a short range (3 groups of columns Concentration/Signal), and a fourth group that combines all the above data; the data can be fitted to a parabola $Y = -0.002125 + 0.005211 * X - 0.0000009126 * X^2$ with a residual standard deviation of ± 4.5 mAU. Use with LR, TESTFIT.

FILLTUBE {4.19}

Tubes must be filled to a nominal 20 g; 10 simulations are provided each for the EEC and the Swiss Guidelines; the average fill weights are 19.7 g and higher; $n = 50$ tubes per sample.

GLASS_.001

Data from a degradation study of a pharmaceutical dosage form (tablets) in two different types of glass container (signal versus time). The underscore character represents the digits 1, 2, and 3; the same data is provided in Symphony .WR1 or .PRN, in LOTUS 1-2-3, and in dBASE .DBF files, to show how these files must be formatted to make the contents transferable to Basic .001 files using program IMPORT. Use also with SHELFLIF to determine effect of differences between containers on potential shelf life.

HISTO.001 {1.8}

Nineteen repeat measurements of a normally distributed (ND) signal to be used for programs HUBER, MSD, and HISTO.

HPLC1.001

Eight impurities were measured by the area-% technique; nine batches of a raw material were tested; file can be used with any program except EUCLID.

HPLC2.001

Eight impurities were tested over 14 runs; file is to be used with program EUCLID to determine whether the 14 samples belong to two different groups; see also file SIEVE.001.

JUNGLE.001 {4.11, 5.2.4}

Three parameters (impurity content, HPLC-assay, and titration assay) were measured on five batches of a raw material.

JUNGLE.002 {4.22}

Eight parameters are measured per batch; the parameters are partially linked. $N = 46$ batches are simulated. Use with CORREL and VIEW_XY to find these links.

JUNGLE.003 {4.22}

Same as JUNGLE.002, but with very specific artifacts added; see text.

LRTEST.001

Synthetic data to serve as a test file for programs LR and WLR.

MOISTURE.001 {4.4}

At 10 selected locations inside a dryer, samples of 8 tablets each were drawn to determine water content by the Karl Fischer method; using MULTI, the hypothesis H0 is tested that all 10 sample averages and standard deviations are indistinguishable.

MSD.001

Test file for MSD (HISTO, HUBER) that contains four data sets of different size and distribution.

PARABOLA.001

Thirteen EMF vs. temperature measurements that conform to the equation $Y = -20.63 + 0.6395 * X - 0.007295 * (X - 31.09)^2$, to be used with TESTFIT, LR, and WLR.

SHELFLIF.001

The content (% of nominal) of two active components in a dosage form was assayed at various times (0–60 months) during a pharmaceutical stability trial to determine the acceptable shelf life of the formulation; the point at which the lower 90% confidence limit of the linear regression model intersects the 90%-of-nominal line gives the answer. Since there is a trend for health authorities to demand shelf life determinations based on the 95%-of-nominal level for certain products, the corresponding calculation is also included.

SIEVE.001 {4.12}

A crystalline raw material is purchased from two different suppliers on the basis of the same specifications; crystal size distribution was relatively loosely defined, so that both vendors' materials passed specs. Production trials with 7 batches from each vendor resulted in products of unequal properties: sieve analysis was carried out on retained samples using a laser light-scattering technique, yielding %-content for each of 15 classes. Analysis of the vendor-averaged sets by the conventional Chi-Square Test yielded no conclusive answer due to the high within-vendor-group variability. Using EUCLID, the 15-point data set for each sample was projected from 15-space into the plane defined by the three Euclidian distances $A - B$, $A - S$, and $S - B$, where A, B, and S are the coordinates in 15-space of the vendor-averages A and B respectively the individual sample S. Two nonoverlapping groups of points could be distinguished that confirmed the impression gained during the casual inspection of the 14-column–by–15-row table.

SIM1.001 {1.4}

Five data sets of 200 points each generated by SIMGAUSS; the deterministic time series Sine Wave, Saw Tooth, Base Line, GC-Peak, and Step Function have stochastic (normally distributed) noise superimposed; use with SMOOTH to test different filter functions (filter type, window). A comparison between the (residual) standard deviations obtained using SMOOTH, respectively HISTO (or MSD), demonstrates that the straight application of the Mean/SD concept to a fundamentally unstable signal gives the wrong impression.

SIM2.001

A 25-row–by–25-column table of random numbers; to be used with various programs.

SIM3.001

A 6-row–by–25-column table containing integer numbers in random and not-so-random sequences; a few empty cells (coded by "9999") are included. Use this file to play with the editing functions contained in DATA.

SMOOTH.001

A 26-point table of values interpolated from a figure in P. A. Gorry *Anal. Chem.* **62**, 570–573 (1990), to demonstrate the capability of the discussed Extended Savitzky–Golay filter to provide a smoothed trace from the first to the last point in the time series.

TABLET_C.001 {4.18}

Simulated drug content uniformity measurements; 10 different means, starting from 46 mg, with 2 samples of 10 tablets each at every weight. $N = 10$, $M = 20$. To be used with HUBER, HISTO, but also CORREL to test for spurious correlations in table of random numbers.

TABLET_W.001 {4.18}

Similar to TABLET_C.001, but with tablet weights, starting at 330 mg, and $N = 20$ tablets per sample. $N = 20$, $M = 20$.

UV.001 {2.2}

A set of five calibration points (absorbance vs. %-of-nominal concentration) to be used with LR. This file was used for many of the numerical examples of Chapter 2.

VALID.001 {4.23}

A set of repeat determinations ($m = 10$) at concentrations that are logarithmically spaced (10, 25, 50, 100, 250, 500 ng/ml); the first of every set of concentrations is entered as given, the second is skipped (display; 2, ,5, ,10, ,25, etc.; code: 2, 9999, 5, 9999, 10, 9999, 25, etc.); the measurements are all entered, of course. To be used with programs VALID and VALIDLL (programs LR and WLR recognize only those entries that have $x <> 9999$, i.e., $n = 9$).

VALID.002 {4.23}

A set of duplicate determinations carried out on each of eight successive days, with logarithmically spaced concentrations (1, 2, 5, 10, 20, 50, 100, 200, and 500 ng/ml). $M = 7$, $N = 18$.

VALID.003 {4.23}

A set of replicate determinations on each concentration (1, 2, 5, 10, 20, 50, 100, 200, and 500 ng/ml), and multiple determinations at 35 and 350 ng/ml. $N = 46$, $M = 2$; all measurements were carried out on one day.

VOLUME.001 {1.1.2}

A set of five precision weighings of a water-filled 100-ml flask; the weights in grams were converted to milliliters using the standard density-vs.-temperature tables. Use with MSD to test the effect of truncation errors on the calculation of the standard deviation; the true result should be SD = ± 0.004767284342265; see also Table 1.1.

WLR.001 {2.2.10}

A set of peak area vs. concentration results of a gas chromatography calibration. Use with LR and WLR to test the effect of a weighing scheme. The originally estimated dependence of the standard deviation of determination vs. concentration is described by the equation SD $= 100 + 5 * x$; see lines LLLL.

QRED.TBL {Table 1.7}

Reduced critical q-values (division of q-values for $p = 0.05$ by the appropriate Student's t factor and SQR(2)), for use with the multiple range test.

5.4. TECHNICAL NOTES

For your reference, this section includes additional notes on the technical aspects of the software to accompany this book.

5.4.1. Program Structure

Once the programs are installed, they will be included in the subdirectory structure listed below. For more information about using subdirectories on your computer, refer to your DOS manual.

\SMAC\	This is the main subdirectory, which includes all the program, data, and menu files. This subdirectory is directly attached to your root directory.
\SMAC\BP5	This subdirectory includes 35 programs which run under QBASIC
\SMAC\BASICM5	This subdirectory includes the files which create the user menu.
\SMAC\BD	This subdirectory include 49 data files for use with the SMAC programs in the subdirectory listed above. In addition, it includes 6 subdirectories that contain sample data files in different formats from popular spreadsheet and database programs. The 6 subdirectories include data files in the following formats:

\SMAC\BD\123	Lotus® 1-2-3 spreadsheet files
\SMAC\BD\DB3	dBASE® III + data files
\SMAC\BD\DB4	dBASE® IV data files
\SMAC\BD\EXL	Microsoft® Excel® spreadsheet files
\SMAC\BD\SC	SuperCalc® spreadsheet files
\SMAC\BD\SYMDATA	Symphony® spreadsheet files

These data files are included to illustrate the proper data format for popular numerical applications.

5.4.2. BASIC Statements

There follows a list of QBASIC statements/functions that were used in the programs described above:

ABS, ATN, CHR$, CLOSE, COLOR, COS, CLS, CSNG, DATA, DEFINT, DEFDBL, DEFSNG, DELETE, DIM, EDIT, END, ERASE, EXP, FILES, FIX, FOR..NEXT, GOSUB..RETURN, GOTO, IF..THEN.. .ELSE..GOTO, INKEY$, INPUT, INPUT#, INSTR, INT, KEY ON/ OFF, KILL, LEFT$, LEN, LINE, LOAD, LOCATE, LOG, LPRINT, LPRINT USING, MID$, NAME, NEW, ON..GOTO, OPEN, OPTION BASE, PRINT, PRINT USING, PRINT#, RANDOMIZE, READ, REM, RESET, RESTORE, RIGHT$, RND, SAVE, SCREEN (modes 0 and 9), SIN, SPACE$, SQR, STOP, STR$, STRING$, SWAP, SYSTEM, TIMER, VIEW, WHILE..WEND, WIDTH, WINDOW, WRITE

5.4.3. Codes

The following special characters are used that appear in code tables 437 (e.g., DOS 3.30, U.S.A.), 860, 863, and 865:

ALT179–220 for frames in BASIC.ML
ALT224: α (Greek alpha)
ALT225: β (Greek beta)
ALT229: σ (Greek sigma)
ALT230: μ (Greek mu)
ALT241: $\pm$
ALT242: $\geq$
ALT243: $\leq$
ALT247: $\approx$ (approximately equal)
ALT253: ^{2} (square)

5.4.4. Troubleshooting Guide

If the menu page appears with a character string `0m a-[24; 1f` between the lower left corner and the following `C:\BASICM5` > prompt, the ANSI.SYS file must be activated in the CONFIG.SYS file by using the correct path, e.g., `device=C:\DOS\ANSI.SYS`.

If the full-screen graphics (menu item “Graph” in most programs), after pressing `PRT-SCR`, appears as pseudographics built up of, e.g., “z” or "ç" characters, check whether the GRAPHICS command was really executed before the CD\BASICM5 command, and whether the printer had been turned on at that time.

If the PC hangs up for some length of time and a "Device Fault in ####" message appears, the printer was not turned on when an LPRINT statement was to be executed. Turn on the printer and hit the `Alt-F5` keys to restart the program. If the error recurs, exit from QBASIC by selecting EXIT from the FILE menu. Restart your computer and try again.

A "File not found in ####" message appears if an improper file name was entered; hit `Alt-F5` to restart the program. If the problem recurs, check whether the default data derive has been correctly changed. Use program Y to change the path statement for your data.

Users who have loaded their PCs with many applications and were therefore obliged to strongly expand the CONFIG.SYS file with SET commands might run into conflict with the 256-byte default size forseen for configuration parameters. This occurs when, after loading a basic program from the menu page, an error message like the following is displayed:

```
List of existing DATA-Files in defined
DATA-Directory
C:\SMAC\BASICM5
File not found in 970
```
(e.g., program DATA)

This is rectified by accessing the CONFIG.SYS file (usually in the root C:\ or in C:\DOS) and adding the following instruction using an editor such as EDIT.COM from DOS.

```
shell=c:\dos\command.com c:\dcs/p/e:1000
```

Leave the editor, and restart your computer. The available space for SET-command parameters has now been expanded to 1000 bytes.

ATTENTION: If an error is committed during the modification of this very sensitive file, the PC will seize up on restarting. To recover, insert the DOS Start Diskette in the appropriate drive and reboot. It is a good idea to make a copy of the intact file before editing it; this is done, e.g., by typing

```
COPY C:\CONFIG.SYS C:\CONFIG.OLD ↵
```

5.5. LIST OF SYMBOLS AND ABBREVIATIONS

SYMBOL	PAGE	EXPLANATION
a	84	intercept
AL	74	action limits
assay		determination of content of, e.g., active principle
b	84	slope
Chi^2 (χ^2)	67	statistical indicator of similarity; χ^2 tables
CHN	36	Elemental analysis for C(arbon), H(ydrogen), and N(itrogen)
CI( )	26	confidence interval of quantity in parentheses
CL( )	28	confidence limits of quantity in parentheses
CL(X)	87	confidence limits of the estimate X
CL(Y)	87	confidence limits of the estimate Y
CL(y)	97	confidence limits of a measured value
c.o.v.	13	coefficient of variation; cf. RSD
CP	25	cumulative probability
d	42	difference between measured values, e.g., in paired t test
DVM	207	digital voltmeter
E( )	19	expected value
EMF	207	electro-motoric force, as found in pH electrodes
f or df	11	degrees of freedom (in some sections in Chapter 5 df is used instead of f to avoid confusion with the polynomial coefficient f
$f(\ldots)$	141	mathematical function
F	59	test statistic, F test
FWHM	143	full width at half maximum, width of a peak
GC		gas chromatography
GLP	17, 228	good laboratory practices
GMP	228	good manufacturing practices
GOF	137	goodness of fit, e.g., χ^2
H_0	77	"null" hypothesis
H_1	77	alternate hypothesis
HPLC	202	high pressure liquid chromatography
IHL	102	in-house limits
Inf (∞)		infinity, either $-\infty$ or $+\infty$
ICP	206	inductively coupled plasma spectrometry
ISE	207	ion-selective electrode
k, z	49	safety factors in detection of outliers (Huber's k, classical z)

k	113	normalization factor in weighed regression
LAR	249	largest absolute residual
LLS	194	laser-light scattering
LOD	105	limit of detection
LOQ	105	limit of quantitation
LR	84	linear regression
LRR	249	largest relative residual
LSD	207	least significant digit
m	46, 53	number of groups
m	87	number of repeat measurements
MAD	49, 216	median absolute deviation, Huber's outlier test
MC	145	Monte Carlo numerical simulation technique
mu (μ)	19	true value of mean
n	7	number of measurements, sample size
n	87	number of calibration measurements
n_1	39	number of samples, first series
n_2	39	number of samples, second series
ND(0, 1)	23	Normal Distribution with $\mu = 0$, $\sigma = 1$
PD	23	probability density
QC/QA		quality control/assurance
q	47	test statistic in multiple range test
q_c	47	critical q value
r	81	correlation coefficient
r^2	81	coefficient of determination
r_i	11	residual
R(n)	9	range of n values
RSD	13	relative standard deviation
s_d	39, 42	standard deviation of the mean difference
s_{res}	87	residual standard deviation
S_T	53	total sum of squares
s_x	11	standard deviation of distribution
$s_{\bar{x}}$	13	standard deviation of mean
S_{xx}	11, 87	sum of squares
	$S_{xx,w}$	113 weighed sum of squares
	S_{xy}	87 sum of squares
	$S_{xy,w}$	113 weighed sum of squares
	S_{yy}	87 sum of squares
	$S_{yy,w}$	113 weighed sum of squares
SI	102	specification interval
sigma (σ)	19	true value of standard deviation
SL	74, 102	specification limits

SOP	16	standard operating procedure
t	26	Student's t value, test statistic in t-test
t_c	30	critical t value
TAR	249	typical absolute residual
tau (τ)	15	time constant
TRR	249	typical relative residual
UV	90	Ultra violet part of spectrum
V_1	54	variance of first series, or within group
V_2	54	variance of second series, or between groups
V_a	87	variance of the intercept a
V_b	87	variance of the slope b
V_d	54	variance of the mean difference
V_p	39, 43	pooled variance
V_{res}	87	residual variance
V_x	11	variance of x
V_X	87	variance of the estimate $X = (y - a)/b$
V_Y	87	variance of the estimate $Y = a + b \cdot x$
w_i	112	weight assigned to individual measurement
$\bar{x}$	8	mean
$\bar{\bar{x}}$	52	grand average
x_i	8	ith x value
x_{ij}	53	element in array $x(,)$
x_m	7	median
x_{max}	9	largest value x_i
x_{min}	9	smallest value x_i
$\bar{x}_w$	113	weighed mean
$x(\ )$	11	array or vector of values
XRF	206	X-Ray fluorescence
y_i	87	measured value
$\bar{y}_w$	113	weighed mean
y^*	87	(average) value measured for unknown
z	23	standardized deviate in ND

REFERENCES

1. Hill, H. M., and Brown R. H., "Statistical Methods in Chemistry," *Anal. Chem.*, **40**, 376R–380R (1968).
2. Kratochvil, B., and Taylor, J. K., "Sampling for Chemical Analysis," *Anal. Chem.*, **53**, 924A–938A (1981).

3. Harris, W. E., "Sampling, Manipulative, Observational, and Evaluative Errors," *Int. Lab.*, 53–62 (Jan./Feb. 1978).
4. Boyer, K. W., Horwitz, W., and Albert, R., "Interlaboratory Variability in Trace Element Analysis," *Anal. Chem.*, **57**, 454–459 (1985).
5. Minkkinen, P., "Evaluation of the Fundamental Sampling Error in the Sampling of Particulate Solids," *Anal. Chim. Acta*, **196**, 237–245 (1987).
6. Krivan, V., and Haas, H. F., "Prevention of Loss of Mercury (II) During Storage of Dilute Solutions in Various Containers," *Fresenius Z. Anal. Chem.*, **332**, 1–6 (1988).
7. Hungerford, J. M., and Christian, G. D., "Statistical Sampling Errors as Intrinsic Limits on Detection in Dilute Solutions," *Anal. Chem.*, **58**, 2567–2568 (1986).
8. Yeung, E. S., and Synovec, R. E., "Detectors for Liquid Chromatography," *Anal. Chem.*, **58**, 1237A–1256A (1986).
9. Cooper, J. W., "Errors in Computer Data Handling," *Anal. Chem.*, **50**, 801A–812A (1978).
10. Horlick, G., "Reduction of Quantization Effects by Time Averaging with Added Random Noise," *Anal. Chem.*, **47**, 352–354 (1975).
11. Foley, J. P., "Systematic Errors in the Measurement of Peak Area and Peak Height for Overlapping Peaks," *J. Chromatog.*, **384**, 301–313 (1987).
12. Dose, E. V., and Guiochon, G., "Bias and Nonlinearity of Ultraviolet Calibration Curves Measured Using Diode-Array Detectors," *Anal. Chem.*, **61**, 2571–2579 (1989).
13. Horwitz, W., "Evaluation of Analytical Methods Used for Regulation of Foods and Drugs," *Anal. Chem.*, **54**, 67A–76A (1982).
14. Anscombe, F. J., "Rejection of Outliers," *Technometrics*, **2**(2), 123–147 (1960).
15. Davies P. L., "Statistical Evaluation of Interlaboratory Tests," *Fresenius Z. Anal. Chem.*, **331**, 513–519 (1988).
16. Worley, J. W., Morrell, J. A., Duewer, D. L., and Peterfreund, L. A., "Alternate Indexes of Variation for the Analysis of Experimental Data," *Anal. Chem.*, **56**, 462–466 (1984).
17. Shukla, S. S., and Rusling, J. F., "Analyzing Chemical Data with Computers: Errors & Pitfalls," *Anal. Chem.*, **56**, 1347A–1368A (1984).
18. Shatkay, A. and Flavian, S., "Unrecognized Systematic Errors in Quantitative Analysis by Gas-Liquid Chromatography," *Anal. Chem.*, **49**, 2222–2228 (1977).
19. Wanek, P. M. et al., "Inaccuracies in the Calculation of Standard Deviation with Electronic Calculators," *Anal. Chem.*, **54**, 1877–1878 (1982).
20. Thompson, M. R., and Dessy, R. E., "Use and Abuse of Digital Signal Processors," *Anal. Chem.*, **56**, 583–586 (1984).
21. Bauer, C. F., Grant, C. L., and Jenskins, T. F., "Interlaboratory Evaluation of High-Performance Liquid Chromatographic Determination of Nitroorganics in Munition Plant Wastewater," *Anal, Chem.*, **58**, 176–182 (1986).

22. Kateman, G., and Pijpers, *Quality Control in Analytical Chemistry*, Wiley, New York, 1981.
23. "Guide for Use of Terms in Reporting Data in *Analytical Chemistry*," *Anal. Chem.*, **58**, 269–270 (1986).
24. Wissenschaftliche Tabellen Geigy, 8th Ed., CIBA-GEIGY, Basel, 1980.
25. Taylor, J. K., "Quality Assurance of Chemical Measurements," *Anal. Chem.*, **53**, 1588A–1596A (1981).
26. Keith, H. K., Chairman, et al., "ACS Committee on Environmental Improvement, Principles of Environmental Analysis," *Anal. Chem.*, **55**, 2210–2218 (1983).
27. MacDougal, D. et al., "Guidelines for Data Acquisition and Data Quality Evaluation in Environmental Chemistry," *Anal. Chem.*, **52**, 2240–2249 (1980).
28. Solberg, H. E., Inaccuracies in Computer Calculation of Standard Deviation," *Anal. Chem.* **55**, 1611 (1983).
29. Bialkowski S. E., "Data Analysis in the Shot Noise Limit. 1. Single Parameter Estimation with Poisson and Normal Probability Density Functions," *Anal. Chem.*, **61**, 2479–2483 (1989).
30. Frazer J. W., "Computer Experimentation Techniques for the Study of Complex Systems," *Anal. Chem.*, **52**, 1205A–1220A (1980).
31. Kaye, W., and Barber, D., "Noise and Digital Resolution in a Microprocessor-Controlled Spectrophotometer," *Anal. Chem.*, **53**, 366–369 (1981).
32. Doerffel, K., *Statistik in der Analytischen Chemie*, 3rd Ed., Verlag Chemie, Weinheim, 1984.
33. Eckschlager, K., *Errors, Measurement & Results in Chemical Analysis*, Van Nostrand, New York, 1969.
34. Sachs, L., *Angewandte Statistik* (this book is also available in English), Springer-Verlag, Berlin, 1984.
35. Miller, J. C., and Miller, J. N., *Statistics for Analytical Chemistry*, Ellis Horwood Ltd., Chichester, 1986.
36. Renner, E., *Mathematisch-statistische Methoden in der Praktischen Anwendung*, 2nd Ed., Verlag Paul Parey, Berlin, 1981.
37. Hays, W. L., *Statistics*, Holt, Rinehart, and Winston, London, 1969.
38. Miller, R. G. Jr., *Simultaneous Statistical Inference*, Springer-Verlag, New York.
39. Harter, H. L., "Critical Values for Duncan's New Multiple Range Test," *Biometrics*, 671–685 (December 1960).
40. Kelly, P. C., "Outlier Detection in Collaborative Studies," *J. Assoc. Off. Anal. Chem.*, **73** (1) Vol. *73*, No. 1 (1990).
41. Rorabacher, D. B., "Statistical Treatment for Rejection of Deviant Values: Critical Values of Dixon's "*Q*" Parameter and Related Subrange Ratios at the 95% Confidence Level," *Anal. Chem.*, **63**, 139–146 (1991).

42. Doerffel, K., Herfurth, G., Liebich, V., and Wendlandt, E., "The Shape of CUSUM—An Indicator for Tendencies in a Time Series," *Fresenius J. Anal. Chem.*, **341**, 519–523 (1991).

43. Marshall, R. A. G., "Cumulative Sum Charts for Monitoring of Radioactivity Background Count Rates," *Anal. Chem.*, **49**, 2193–2196 (1977).

44. VanArendonk, M. D., and Skogerboe, R. K., "Correlation Coefficients for Evaluation of Analytical Calibration Curves," *Anal. Chem.*, **53**, 2349–2350 (1981).

45. Mitchell, D. G., Mills, W. N., Garden, J. S., and Zdeb, M., "Multiple-Curve Procedure for Improving Precision with Calibration-Curve-Based Analyses," *Anal. Chem.*, **49**, 1655–1660 (1977).

46. Ripley, B. D., and Tompson, M., "Regression Techniques for the Detection of Analytical Bias," *Analyst*, **112**, 377–383 (April 1987).

47. Bialkowski, S. E., "Data Analysis in the Shot Noise Limit. 2. Methods for Data Regression," *Anal. Chem.*, **61**, 2483–2489 (1989).

48. Ellerton, R. R. W., Strong, F. C. III, "Comments on Regression through the Origin," *Anal. Chem.*, **52**, 1151–1153.

49. Schwartz, L. M., "Effect of Constraints on Precision of Calibration Analyses," *Anal. Chem.*, **58**, 246–250 (1986).

50. Strong, F. C., "Regression Line that Starts at the Origin," *Anal. Chem.*, **51**, 298–299 (1979).

51. Strong, F. C., and Ellerton, R. R. W., "Comments on Regression through the Origin," *Anal. Chem.*, **52**, 1151–1152 (1980).

52. Wang, C. Y., Bunday, S. D., and Tartar, J. G., "Ion Chromatographic Determination of Fluorine, Chlorine, Bromine, and Iodine with Sequential Electrochemical and Conductometric Detection," *Anal. Chem.*, **55**, 1617–1619 (1983).

53. Shatkay, A., "Effect of Concentration on the Internal Standards Method in Gas–Liquid Chromatography," *Anal. Chem.*, **50**, 1423–1429 (1978).

54. Schwartz, L. M., "Rejection of a Deviant Point from a Straight-Line Regression," *Anal. Chim. Acta*, **178**, 355–359 (1985).

55. Bysouth, S. R., and Tyson, J. F., "A Comparison of Curve Fitting Algorithms for Flame Atomic Absorption Spectrometry," *J. Anal. At. Spectrom.*, **1**, 85–87 (1986).

56. Cardone, M. J., Palermo, P. J., and Sybrandt, L. B., "Potential Error in Single-Point-Ratio Calculations Based on Linear Calibration Curves with Significant Intercept," *Anal. Chem.*, **52**, 1187–1191 (1980).

57. Schwartz, L. M., "Statistical Uncertainties of Analyses by Calibration of Counting Measurements," *Anal. Chem.*, **50**, 980–985 (1978).

58. Porter, W. R., "Proper Statistical Evaluation of Calibration Data," *Anal. Chem.*, **55**, 1290A (letter) (1983).

59. St. John, P. A., McCarthy, W. J., and Winefordner, J. D., "A Statistical Method of Evaluation of Limiting Detectable Sample Concentrations," *Anal. Chem.*, **39**, 1495 (1967).

60. S. A. B., "Detection Limits, A Systematic Approach to Detection Limits is Needed," *Anal. Chem.*, **58**, 986A (1986).

61. Long, G. L., and Winefordner, J. D., "Limit of Detection," *Anal. Chem.*, **55**, 712A–724A (1983).

62. Taylor, J. K., "Limits of Detection," *Anal. Chem.*, **56**, 130A (letter) (1984).

63. Vogelgesang, J., "Limit of Detection and Limit of Determination: Application of Different Statistical Approaches to an Illustrative Example of Residue Analysis," *Fresenius Z. Anal. Chem.*, **328**, 213–220 (1987).

64. Currie, L. A., "Limits for Qualitative Detection and Quantitative Determination," *Anal. Chem.*, **40**, 586–593 (1968).

65. Williams, R. R., "Fundamental Limitations on the Use and Comparison of Signal-to-Noise Ratios," *Anal. Chem.*, **63**, 1638–1643 (1991).

66. Williams, T. W. and Salin, E. D., "Hazards of a Naive Approach to Detection Limits with Transient Signals," *Anal. Chem.*, **60**, 725–727 (1988).

67. Luthhardt, M., Than, E., and Heckendorff, H., "Nachweis-, Erfassungs- und Bestimmungsgrenze analytischer Verfahren," *Fresenius Z. Anal. Chem.*, **326**, 331–339 (1987).

68. Oppenheimer, L., Capizzi, T. P., Weppelman, R. M., and Mehta, H., "Determining the Lowest Limit of Reliable Assay Measurement," *Anal. Chem.*, **55**, 638–643 (1983).

69. Clayton, C. A., Hines, J. W., and Elkins, P. D., "Detection Limits with Specified Assurance Probabilities," *Anal. Chem.*, **59**, 2506–2514 (1987).

70. Schoonover, R. M., and Jones, F. E., "Air Buoyancy Correction in High-Accuracy Weighting on Analytical Balances," *Anal. Chem.*, **53**, 900–902 (1981).

71. Ratzlaff, K. L., and bin Darus, H., "Optimization of Precision in Dual Wavelength Spectrophotometric Measurement," *Anal. Chem.*, **51**, 256–261 (1979).

72. Lam, R. B., and Isenhour, T. L., "Minimizing Relative Error in the Preparation of Standard Solutions by Judicious Choice of Volumetric Glassware," *Anal. Chem.*, **52**, 1158–1161 (1980).

73. Schwartz, L. M., "Calibration of Pipets: A Statistical View," *Anal. Chem.*, **61**, 1080–1083 (1989).

74. Gernand, W., Steckenreuter, K., and Wieland, G., "Greater Analytical Accuracy through Gravimetric Determination of Quantity," *Fresenius Z. Anal. Chem.*, **334**, 534–539 (1989).

75. Snyder, L. R., and van der Wal, S. J., "Precision of Assays Based on Liquid Chromatography with Prior Solvent Extraction of the Sample," *Anal. Chem.*, **53**, 877–884 (1981).

76. Unadkat, J. D., Beal, S. L., and Sheiner, L. B., "Bayesian Calibration," *Anal. Chim. Acta*, **181**, 27–36 (1986).

77. Gardner, M. J., and Gunn, A. M. "Approaches to Calibration in GFAAS: Direct or Standard Additions," *Fresenius Z. Anal. Chem.*, **330**, 103–106 (1988).

78. Cardone, M. J., "New Technique in Chemical Assay Calculations. 1. A Survey of Calculational Practices on a Model Problem," *Anal. Chem.*, **58**, 433–438 (1986).

79. Cardone, M. J., "New Technique in Chemical Assay Calculations. 2. Correct Solution of the Model Problem and Related Concepts," *Anal. Chem.*, **58**, 438–445 (1986).

80. Franke, J. P., de Zeeuw, R. A., and Hakkert, R., "Evaluation and Optimization of the Standard Addition Method for Absorption Spectrometry and Anodic Stripping Voltametry," *Anal. Chem.*, **50**, 1374–1380 (1987).

81. Ratzlaff, K. L., "Optimizing Precision in Standard Addition Measurement," *Anal. Chem.*, **51**, 232–235 (1979).

82. Gardner, M. J., and Gunn, A. M., "Optimising Precision in Standard Additions Determinations," *Fresenius Z. Anal. Chem.*, **325**, 263–266 (1986).

83. Whang, C. W., Page, J. A., vanLoon, G., and Griffin, M. P., "Modified Standard Additions Calibration for Anodic Stripping Voltammetry," *Anal. Chem.*, **56**, 539–542 (1984).

84. Midgley, D., "Systematic and Random Errors in Known Addition Potiometry, A Review," *Analyst*, **112**, 557–572 (May 1987).

85. Horvai, G., and Pungor, E., "Precision of the Double Known Addition Method in Ion-Selective Electrode Potentiometry," *Anal. Chem.*, **55**, 1988–1990 (1983).

86. Schwartz, L. M., "Calibration Curves with Nonuniform Variance," *Anal. Chem.*, **51**, 723–727 (1979).

87. Garden, J. S., Mitchell, D. G., and Mills, W. N., "Nonconstant Variance Regression Techniques for Calibration-Curve-Based Analysis," *Anal. Chem.*, **52**, 2310–2315 (1980).

88. Watters, R. L. Jr., Carroll, R. J., and Spiegelman, C. H., "Error Modeling and Confidence Interval Estimation for Inductively Coupled Plasma Calibration Curves," *Anal. Chem.*, **59**, 1639–1643 (1987).

89. Thompson, Michael, "Variation of Precision with Concentration in an Analytical System," *Analyst*, **113**, 1579–1587 (Oct. 1988).

90. Phillips, G. R., and Eyring, E. M., "Comparison of Conventional and Robust Regression in Analysis of Chemical Data," *Anal. Chem.*, **55**, 1134–1138 (1983).

91. Thompson, Michael, "Robust Statistics and Functional Relationship Estimation for Comparing the Bias of Analytical Procedures over Extended Concentration Ranges," *Anal. Chem.*, **61**, 1942–1945 (1989).

92. Deming, S. N., and Morgan, S. L., "Simplex Optimization of Variables in Analytical Chemistry," *Anal. Chem.*, **45**, 278A–283A (1973).

93. Moody, J. R., Greenberg, R. R., Pratt, K. W., and Rains, T. C., "Recommended Inorganic Chemicals for Calibration," *Anal. Chem.*, **60**, 1203A–1218A (1988).

94. Proctor, A., and Sherwood, P. M. A., "Smoothing of Digital X-Ray Photoelectron Spectra by an Extended Sliding Least-Squares, (Analytical Approach)," *Anal. Chem.*, **52**, 2315–2321 (1980).

95. Subcommittee on Environmental Analytical Chemistry, ACS Committee on Environmental Improvement, Crummett, W. B., Chairman, "Guidelines for Data Acquisition and Data Quality Evaluation in Environmental Chemistry," *Anal. Chem.*, **52**, 2242–2249 (1980).

96. Pfeiffer, C. D., Larson, J. R., and Ryder, J. F., "Linearity Testing of Ultraviolet Detectors in Liquid Chromatography," *Anal. Chem.*, **55**, 1622–1624 (1983).

97. Dorschel, C. A., Ekmanis, J. L., Oberholtzer, J. E., Warren, F. V., and Bidlingmeyer, B. A., "LC Detectors: Evaluation and Practical Implications of Linearity," *Anal. Chem.*, **61**, 951A–968A (1989).

98. Lind, B., Elinder, C. G., Nilsson, B., Svartengren, M., and Vahter, M., "Quality Control in the Analysis of Lead and Cadmium in Blood," *Fresenius Z. Anal. Chem.*, **326**, 647–655 (1987).

99. Ayers, G., Burnett, D., Griffiths, A., and Richens, A., "Quality Control of Drug Assays," *Clin. Pharmacokinetics*, **6**, 106–117 (1981).

100. Carter, K. N., Scott, D. M., Salomon, J. K., and Zarcone, G. S., "Confidence Limits for the Abscissa of Intersection of Two Least-Squares Lines Such as Linear Segmented Titration Curves," *Anal. Chem.*, **63**, 1270–1278 (1991).

101. Leary, J. J., and Messick, E. B., "Constrained Calibration Curves: A Novel Application of Lagrange Multipliers in Analytical Chemistry," *Anal. Chem.*, **57**, 956–957 (1985).

102. Schwartz, L. M., "Nonlinear Calibration," *Anal. Chem.*, **49**, 2062–2068 (1977).

103. Kragten, J., "Least-Squares Polynomial Curve-Fitting for Calibration Purposes (STATCAL-CALIBRA)," *Anal. Chim. Acta*, **241**, 1–13 (1990).

104. Brubaker, T. A., Tracy, R., and Pomernacki, C. L., "Linear Parameter Estimation," *Anal. Chem.*, **50**, 1017A–1024A (1978).

105. Phillips, G. R., and Eyring, E. M., "Error Estimation Using the Sequential Simplex Method in Nonlinear Least Squares Data Analysis," *Anal. Chem.*, **60**, 738–741 (1988).

106. Christensen, M. K., "Determining the Parameters of First-Order Decay with a Nonzero End Point and Unequal Time Intervals," *Anal. Chem.*, **55**, 2324–2327 (1983).

107. Brubaker, T. A., and O'Keefe, K. R., "Nonlinear Parameter Estimation," *Anal. Chem.*, **51**, 1385A–1388A (1979).

108. Schwartz, L. M., "Lowest Limit of Reliable Assay Measurement with Nonlinear Calibration," *Anal. Chem.*, **55**, 1424–1426 (1983).

109. Beebe, K. R., and Kowalski, B. R., "An Introduction to Multivariate Calibration and Analysis," *Anal. Chem.*, **59**, 1007A–1017A (1987).

110. Frazer, J. W., Balaban, D. J., and Wang, J. L., "Simulation as an Aid to Experimental Design," *Anal. Chem.*, **55**, 904–910 (1983).

111. Rawlins, T. G. R., and Yrjonen, T., "Calculation of RIA Results Using the Spline Function," *Int. Lab.*, 55–66 (Nov./Dec. 1978).

112. Halang, W. A., Langlais, R., and Kugler, E., "Cubic Spline Interpolation for the Calculation of Retention Indices in Temperature-Programmed Gas–Liquid Chromatography," *Anal. Chem.*, **50**, 1829–1832 (1978).

113. Schwartz, L. M., and Gelb, R. I., "Statistical Analysis of Titration Data," *Anal. Chem.*, **50**, 1571–1576 (1978).

114. Ratzlaff, K. L., and Natusch, D. F. S., "Theoretical Assessment of Precision in Dual Wavelength Spectrophotometric Measurement," *Anal. Chem.*, **49**, 2170–2176 (1977).

115. Ratzlaff, K. L., and Natusch, D. F. S., "Theoretical Assessment of Accuracy in Dual Wavelength Spectrophotometric Measurement," *Anal. Chem.*, **51**, 1209–1217 (1979).

116. Stolzberg, R. J., "Uncertainty in Calculated Values of Uncomplexed Metal Ion Concentration," *Anal. Chem.*, **53**, 1286–1291 (1981).

117. Hampel, F., "Robuste Schaetzungen: Ein Anwendungsorientierter Ueberblick," *Biom. J.*, **22**, 3–21 (1980).

118. Danzer, K., "Robuste Statistik in der Analytischen Chemie," *Fresenius Z. Anal. Chem.*, **335**, 869–875 (1989) (in German, 65 references cited).

119. Bromba, M. U. A., and Ziegler, H., "Digital Smoothing of Noisy Spectra," *Anal. Chem.*, **55**, 648–653 (1983).

120. Bromba, M. U. A., and Ziegler, H., "Digital Filter for Computationally Efficient Smoothing of Noisy Spectra," *Anal. Chem.*, **55**, 1299–1302 (1983).

121. Bromba, M. U. A., and Ziegler, H., "Efficient Computation of Polynomial Smoothing Digital Filters," *Anal. Chem.*, **51**, 1760–1762 (1979).

122. Bush, I. E., "Fast Algorithms for Digital Smoothing Filters," *Anal. Chem.*, **55**, 2353–2361 (1983).

123. Bromba, M. U. A., and Ziegler, H., "Variable Filter for Digital Smoothing and Resolution Enhancement of Noisy Spectra," *Anal. Chem.*, **56**, 2052–2058 (1984).

124. Jones, R., "High-Pass and Band-Pass Digital Filtering with Peak to Trough Measurement Applied to Quantitative Ultraviolet Spectrometry," *Analyst*, **112**, 1495–1498 (Nov. 1987).

125. Lam, R. B., Wieboldt, R. C., and Isenhour, T. L., "Practical Computation with Fourier Transforms for Data Analysis," *Anal. Chem.*, **53**, 889A–901A (1981).

126. Horlick, G., "Digital Data Handling of Spectra Utilizing Fourier Transformations," *Anal. Chem.*, **44**, 943–947 (1972).

127. Doerffel, K., Wundrack, A., and Tarigopula, S., "Improving Signal-to-Noise by Evaluation of Correlation Functions," *Fresenius Z. Anal. Chem.*, **324**, 507–510 (1986).

128. Bialkowski, S. E., "Real-Time Digital Filters: Finite Impulse Response Filters," *Anal. Chem.*, **60**, 335A–361A (1988).

129. Bialkowski, S. E., "Real-Time Digital Filters: Infinite Impulse Response Filters," *Anal. Chem.*, **60**, 403A–413A (1988).

130. Enke, C. G., and Nieman, T. A., "Signal-to-Noise Ratio Enhancement by Least-Squares Polynomial Smoothing," *Anal. Chem.*, **48**, 705A–712A (1976).

131. Bromba, M. U. A., and Ziegler, H., "Application Hints for Savitzky-Golay Digital Smoothing Filters," *Anal. Chem.*, **53**, 1583–1586 (1981).

132. Ziegler, E., *Computer in der Instrumentellen Analytik*, Akademische Verlagsgesellschaft, Frankfurt am Main, 1973.

133. Madden, H. H., "Comments on the Savitzky-Golay Convolution Method for Least-Squares Fit Smoothing and Differentiation of Digital Data," *Anal. Chem.*, **50**, 1383–1386 (1978).

134. Gorry, P. A., "General Least-Squares Smoothing and Differentiation by the Convolution (Savitzky-Golay) Method," *Anal. Chem.*, **62**, 570–573 (1990).

135. Leach, R. A., Carter, C. A., and Harris, J. M., "Least-Square Polynomial Filters for Initial Point and Slope Estimation," *Anal. Chem.*, **56**, 2304–2307 (1984).

136. Kahn, A., "Procedure for Increasing the Accuracy of the Initial Data Point Slope Estimation by Least-Squares Polynomial Filters," *Anal. Chem.*, **60**, 369–371 (1988).

137. Edwards, T. R., "Two-Dimensional Convolute Integers for Analytical Instrumentation," *Anal. Chem.*, **54**, 1519–1524 (1982).

138. Ratzlaff, K. L., "Computation of Two-Dimensional Polynomial Least-Squares Convolution Smoothing Integers," *Anal. Chem.*, **61**, 1303–1305 (1989).

139. Kuo, J. E., Wang, H., and Pickup, S., "Multidimensional Least-Squares Smoothing Using Orthogonal Polynomials," *Anal. Chem.*, **63**, 630–635 (1991).

140. Güell, O. A., and Holcombe, J. A., "Analytical Applications of Monte Carlo Techniques," *Anal. Chem.*, **62**, 529A–542A (1990).

141. Poston, P. E., and Harris, J. M., "Maximum Likelihood Quantitative Estimates for Peaks: Application to Photoacoustic Spectroscopy," *Anal. Chem.*, **59**, 1620–1626 (1987).

142. Moler, G. F., Delongchamp, R. R., and Mitchum, R. K., "Estimation of the Variance of the Area of a Single Chromatographic Peak," *Anal. Chem.,*, **55**, 842–847 (1983).

143. Johansson, E., and Wold, S., "Minimizing Effects of Closure on Analytical Data," *Anal. Chem.*, **56**, 1685–1688 (1984).

144. Hall, P., and Selinger, B., "A Statistical Justification Relating Interlaboratory Coefficients of Variation with Concentration Levels," *Anal. Chem.*, **61**, 1465–1466 (1989).

145. Helsen, J. A., and Vrebos, B. A. R., "Quantitative X-Ray Fluorescence Analysis: Limits of Precision and Accuracy," *Int. Lab.*, 66–71 (Dec. 1986).

146. Helm, Volker, Ph.D. thesis submitted by, "Stabilitätsuntersuchungen an wässrigen Peptidlösungen," Hochschulschriften Bd.13, Lit Verlag, Münster + Hamburg, B.R.D., Referent Prof. Dr. B. W. Müller, ISBN 3-88660-723-2, DBN 90.135370.1, 90.09.26.

147. Abramowitz, M., and Stegun, I. A., (Eds.), *Handbook of Mathematical Functions*, Dover Publications, New York, 1970, Chap. 26.

148. Gardiner, M. J., and Bombay, B. F., "An Approximation to Student's *t*," *Technometrics*, **7**, 71 (1965).

149. Chambers, W. F., "Comment on Calculator Program Yielding Confidence Limits for Least Squares Straight Line," *Anal. Chem.*, **49**, 884 (correspondence) (1977).

150. Dudewicz, E. J., and Dalal, S. R., "On Approximations to the *t*-Distribution," *J. Qual. Technol.*, **4**, 196–198 (1972).

151. Johnson, E. E., "Empirical Equations for Approximating Tabular *F* Values," *Technometrics*, **15**(2) 379–384 (1973).

152. Kisner, H. J., Brown, C. W., and Kavarnos, G. J., "Multiple Analytical Frequencies and Standards for the Least-Squares Spectrometric Analysis of Serum Lipids," *Anal. Chem.*, **55**, 1703–1707 (1983).

153. Borman, S. A., "Math is Cheaper than Physics," *Anal. Chem.*, **54**, 1379A–1380A (1982).

154. Ferrus, R., and Torrades, F., "Bias-Free Adjustment of Analytical Methods to Laboratory Samples in Routine Analytical Procedures," *Anal. Chem.*, **60**, 1281–1285 (1988).

155. Royston, G. C., "Comments on Unrecognized Systematic Errors in Quantitative Analysis in Gas Chromatography," *Anal. Chem.*, **50**, 1005 (1978).

156. Schwartz, L. M., and Gelb, R. I., "Statistical Uncertainties of End Points at Intersecting Straight Lines," *Anal. Chem.*, **56**, 1487–1492 (1984).

157. Betti, M., Papoff, P., and Meites, L., "Factors Affecting the Precisions of Analyses, by Potentiometric Titrimetry, of Solutions Containing Two Weak Acids," *Anal. Chim. Acta*, **182** 133–145 (1986).

158. Steinier, J., Termonia, Y., and Deltour, J., "Comments on Smoothing and Differentiation of Data by Simplified Least Square Procedure," *Anal. Chem.*, **44**, 1906–1909 (1972).

159. Green, J. R., and Margerison, D., *Statistical Treatment of Experimental Data*, Elsevier Scientific, Amsterdam, 1978.

INDEX

A

Absorbance, 42, 70, 81, 201
Accuracy, 116, 140, 210, 232
Adcock's solution, 85
Algorithm, 2, 119, 145f
Analyte, 118
Analytical protocol, 2, 36,116, 141, 159, 202
Arrhenius diagram, 223
Artifacts, 2f, 12, 207
Assumption, 4, 36, 85f, 105, 123, 125, 240
Asymptotic behavior, 162

B

Back-calculation, 230ff
Background, 117
Bar chart, *see* histogram
Baseline, 117
BASIC statements, 277, 315
Benchmark, 51, 69, 138, 200
Bias, *see* (systematic) error, accuracy
Bin, 24, 64, 144, 194, *see* class

C

Calibration:
  3, 17, 82, 89, 107ff, 116, 118, 164ff, 229
  Bayesian, 109
  scheme, 100, 166
  standards, 229
Catastrophe theory, 41
Center of mass, 88, 115
Chemical + physical test methods:
  aliquot, 155, 164
  blank matrix, 118
  color, 225
  control samples, 116
  dilution, 15, 155
  extraction, 118, 159, 164, 231
  Internal Standard, 159, 204, 229
  isotope labelling, 112
  Karl Fischer, 169, 212, 312
  matrix effect, 119
  pH, 81, 225
  precipitation, 206
  purity, 173
  QC samples, 229
  round-robin test, 81
  sample work-up, 2, 100, 118
  specificity, 189
  spiked sample, 109, 118
  viscosity, 180
  weighing, 15, 155, 181
  weight-%, 173
Chi^2, 26f, 36, 61ff, 137f, 196, 243, 248ff, 142
Class, 24, 64, 66, 194, *see* bin
Closed algebraic solution, 86, 135f, 146
Closure, 194
Cluster analysis, 197
Coefficient of determination, 13, 43, 81ff, 96, 99, 114, 139, 144, 230, 235
Coefficient of variation, 13, 91
Confidence Interval, 28–32, 92, 100, 102, 165, 180
Confidence level, 27, 55
Constraint, 120
Control chart, 72ff, 156, 180ff
Convergence, 162f
Correlation, 183, 187
Correlation coefficient, *see* coefficient of determination
Cost components 101, 107ff, 164ff
Critical q-value, 47
Cumsum chart, 72ff, 145, 180
Cumulative frequency, 25
Cumulative Probability, 25, 28, 68, 70, 146–148, 158, 239ff
Curvature, 120
Curve fitting:
  160, 185
  algebraic solution 86, 130, 135f, 146
  brute force, 130, 135, 304
  graphical, 135f

D

Data
- acquisition, 2, 119
- exploratory analysis, 82, 138f, 226
- multi-dimensional, 125ff, 129ff
- reduction, 2, 14, 119, 173
- smoothing, *see* filters
- visualization, 125ff

Data files:
- 53, 68,
- format, 260–261
- using, 259–260
- Lotus 1-2-3, 268–269
- Symphony, 268–269
- Microsoft Excel, 270
- SuperCalc, 270
- dBASE III + , 270
- dBASE IV, 270
- ASCII, 270
- ARRHEN 294
- CALIB, 295
- CREAM, 295
- CYANIDE, 295
- FILLTUBE, 312
- GLASS, 295
- HISTO, 50, 295
- HPLC, 296
- JUNGLE, 296
- LRTEST, 296
- MOISTURE, 64, 296
- MSD, 296
- PARABOLA, 297
- QRED.TBL, 48, 299
- SHELFLIF, 297
- SIEVE, 198, 297
- SIM, 297–298
- SMOOTH, 298
- TABLET, 215ff, 226, 298
- UV, 88, 298
- VALID, 84, 299
- VOLUME, 299
- WLR, 299

Decision, 2f

Decomposition, 221

Detector, *see* sensor

Differentiation, 146

Diffusion, 124

Digitization, 22, 70, 207

Dispersion, 9

Distribution:
- 20
- Binomial-, 22
- Chi-square-, 26
- function, 3
- Normal- (Gaussian-) 22ff, 58, 66, 71, 86, 146, 156ff, 226, 239ff
- Poisson, 22, 86
- skewed, 21
- t-, 22, 26
- width of, 58

Documentation, 228

Double-precision math, 13, 120

Dynamic range, 103, 117, 143

E

Equilibrium, 184, frozen-, 119

Error:
- experimental, 138
- of sampling, 218
- probabilities of α, β, 76, 104
- propagation, 11, 139ff, 145, 212
- random or stochastic, 19, 141, 146ff
- recovery procedure, 117, 159
- systematic, 19, 116, 139, 140f, 210
- type I, II, 76, 104

Estimate, 8, 11, 31, 35, 61, 94ff

Euclidian distance, 197ff

Examples:
- alternative hypotheses, 158
- ANOVA test, 54
- artifacts in calculation of r^2, 84
- asymmetrical CL, 121
- bins and histograms, 64f
- calculate CL(std.dev), 62
- calculation of CP, 25
- calculation of SD, truncation, 11
- Chi^2-test, 67, 248
- CL of estimated SD, 254
- CL($\bar{x}$), 27, 31
- components of variance, 91
- CP = $f(z)$, 243
- detection limit, 232
- error propagation, 213
- estimating yield, 132
- F-test, 60, 248
- interpolation of tables, 178
- interpreting a situation, 77
- mean and median, 7f

number of determinations, 57
outlier detection, 50, 215
$p = f(f, t)$, 246
parameter optimization, 69
projection into a plane, 198
reliability of a standard deviation, 12
result, reporting and rounding of, 13, 18
significance of slopes, 93
significant digits, 140
specification limits, 214, 219
t-test, 32, 41–45, 246
target concentration, 175
variance decomposition, 202, 205, 218
weighed regression, 113
$z = f(\text{CP})$, 241
Experimental protocol, 36, 49
Expert system, 72
Exponential equation, 161ff
Extrapolation, 110, 196
False positive, negative, 48, 77, 104, 158

F

Filter:
box-car, 144ff
correlation analysis, 144
Fourier transformation, 144
moving average, 144ff
Savitzky-Golay, 144ff
techniques, 3, 143
width, 144
First derivative, 145
Flow sheet, 2, 33, 139, 149

G

Good Laboratory Practices, 17, 33ff, 51, 90, 101, 109, 116, 153, 210, 228, 238
Good Manufacturing Practices, 33f, 41, 90, 116, 153, 159, 164, 176, 179, 181
Goodness of fit, 69, 136ff, 206f
Grand average, 53

H

Hardware, 2, 3, 5, 101, 150ff, 276ff, 316
Heteroscedacity, 86, 90, 111
Histogram, 22, 24, 64, 66
Homogeneity, 17, 111, 164, 181, 218
Homoscedacity, 86
Huber's *k*, 215ff
Hypothesis, 30, 36, 37ff, 40f, 44f, 57, 63, 68, 76

I

Impurity profile, 173, 187, 225
Initial estimates, 162f
Intercept, 87f, 229
Interference, 116, 139, 204
Interpolation, 94ff, 105, 110, 117, 178
Interpretation, 2, 128f, 185, 226
see hypothesis
Intersection, 120
Inverse function, 87, 98, 121

L

Laboratory equipment and techniques:
AAS, 117
aspirating cuvette, 168
automation, 202
autosampler, 168
balance, 181
CHN analysis, 36
chromatography, 119, 128, 151
densitometry, 142, 159ff
diffuse reflection, 159
digital volt meter, 22, 33f, 70, 119, 207ff
diode-array detector, 159
dispenser, 157, 209
firmware, 124
flame photometry, 16
Flame ionization detector, 123
FT-NMR, 143
GC, 111, 173, 187, 213, 222, 228
GC-MS, 222
HPLC, 15, 17, 42, 116, 151, 159, 164, 173, 187, 192, 202, 214, 218, 226 228, 230, 296
ICP, 206f
instruments, 2f, 228
integrator, 120
ion chromatography, 128, 212
Least Significant Digit, 209, 214
Liquid Crystal Display, 181
LLS, 194
maintenance, 3
MS, 26, 222
optical spectroscopy, 26
pH/ion meter, 118, 207ff, 211
photometry, 42, 70, 112, 155, 159, 164, 199
polarography, 128, 206f
sensor, 2f, 14, 119f

Laboratory equipment and techniques: (*Continued*)
  stray light, 201
  system suitability test, 18, 60, 116
  time constant, 15
  titration, 100, 120, 128, 187ff, 212, 226
  TLC, 142, 159
  transducer, 2
  UV, 15, 17, 90f, 159, 187
  XRF, 206f
Lambert-Beer plot, 81, 84
Limits:
  action-, 73f, 102, 159
  confidence-, 28–32, 36f, 92, 100, 102, 114, 122, 221f, 237f
  detection-, Quantitation-, 103ff, 235
  in-house-, 102, 157
  in-process control-, 184, 187, 206, 220
  specification-, 73f, 100, 102, 157f, 164, 205, 213f, 218ff
Linear algebra, 146
Linear range, 117
Linear regression:
  84ff, 129, 177
  intersection of two lines, 120
  through origin, 88, 177
  weighed, 111ff
Linearity, 117, 162, 230, 92ff
Linearization, 121
Logarithmic scale, 230
Lotus 1-2-3, 268–269

M

Matrix, 118, 229
Matrix inversion, 146
Maximum likelihood method, 85f
Measurements, independency of, 14
Median, 7, 48, 70
Median Average Deviation, 48, 216ff
Method justification, 229
Method transfer, 230
Model:
  36f, 120, 123, 125, 135, 149f, 184f, 196
  additive effect, 52
  deterministic, 33, 149
  least-squares, 68, 85, 123, 137
  nonlinear, 120ff
  statistical, 3
  stochastic, 33
  weighing, 68, 86, 111ff, 123, 144
Monte Carlo, 22f, 51, 67, 124, 139, 142ff, 214, 240, 243
Moving average smoothing, 144ff
Moving Average, 25
Multilinear regression, 122f, 133
Multiple Range Test, 46, 54, 157, 170

N

Newton's interpolation, 136, 146
Noise:
  3, 19, 34, 82, 102f, 117
  baseline, 104
  filter, 3
  Poisson-distributed, 26
  signal jitter, 209
Nonlinearity, 116
Normal probability chart, 64, 156
Normal Distribution, 22ff, 58, 66, 71, 86, 146, 156ff, 226, 239ff
Normalization, 160ff
Normalized deviate, 68, 74, 147, 240ff
Number of determinations, 55ff, 101
Numerical:
  accuracy, 281
  artifacts, 11, 12, 87, 140
  iterative technique, 68, 121, 135f, 146
  precision, 140
  technique, 85, 136

O

Observed frequency, 21, 22
Optimization, 68, 85, 107, 130, 149, 164ff, 186, 207ff
Optimization, by simplex, 70, 124, 130, 134, 136, 162ff
Organization of work, 101, 107
Outlier, 49, 94, 97, 157
Overdetermination, 85

P

Parabola, 142f
Parsimony, 128
Partial differential, 88, 141
Peak:
  area, 42
  height, 160ff
  width, 144
Percentiles, 10

Pharmaceutical dosage form:
active principle, 118
cream, 221
placebo, 118
tablet, 164, 169, 181, 213ff
injectables, 220
ovules, 220
Pharmaceutical practices:
accelerated test, 177
blinded samples, 229
Content Uniformity, 165, 181, 213ff
dissolution, 181
dry residue, 180
friability, 181
GLP, 17, 33, 36, 51, 90, 101, 109, 116, 153, 210, 228, 238
GMP, 41, 90, 101, 116, 153, 159, 164, 176, 179, 181, 213
IPC, 187, 206, 220
moisture, 221
stability test, 81, 176, 178
stress test, 176, 220, 223
Poisson-distribution, 22, 86
Polynomial, 121f, 145, 159, 161ff, 240–243, 247
Power curve, 159
Precision, 107, 164, 208
Precision, numerical, 13, 84
Primary container, 220
Printers, 293
Probability:
of error α, β, 76, 104
chart, 70
Density, 21, 26, 147, 239ff
Process parameters, 20
Program:
ARRHENIUS, 223, 290
CHI, 61–62, 66ff, 169, 248ff, 292–293
CONV, 27–35, 293
CORREL, 82, 139, 187, 193, 226ff, 282–283
CPVAL, 25, 70, 102, 240, 291
DATA, 51, 199, 226, 263–267
EUCLID, 198, 285–286
FACTOR, 8, 130ff, 266, 283–284
FVAL, 61–62, 170, 246ff, 292
HISTO, 64ff, 72, 273–274
HUBER, 50ff, 215, 226, 274–275
HYPOTHES 279–280
IMPORT, 267–271
LR, 286–287
MSD, 12f, 60, 62, 215, 218, 273
MULTI, 48, 64, 157, 169, 280–282
NVAL, 292
PVAL, 291–292
PDVAL, 291
PRI, 293
SHELFLIF, 222, 289
SIMGAUSS, 124, 148, 215, 272
SIMILAR, 124, 142, 271–272
SMOOTH, 75, 145, 277–279
TESTFIT, 162, 185, 177, 197, 276–277
TTEST, 40, 59, 215, 275–276
TVAL, 30ff, 171, 222, 243, 291
VALID, 84, 232, 235, 287–288
VALIDLL, 232, 235, 288
VIEW_XY, 139, 226, 284
WLR, 288–289
XYZ, 139, 284–285
Y, 293
Z, 294
ZVAL, 25, 146f, 241, 290–291
Programming, 152ff

Q

QBASIC Commands, 257
q-value, 170
Quadratic regression, 200f
Qualifications, 101
Quality Control, 17, 90
Quantiles, 10, 58
Quantitation, 159ff
Quantization noise, 103ff, 207

R

Range, 9, 48, 59
Recovery, 118
Recursive algorithm, 146
Reduced critical q-value, 48
Regula falsi, 136
Regulations, 173, 213f, 218f, 222, 228, 238
Relative Standard Deviation, 3, 13, *see* c.o.v.
Repeatability, 16–18, 138, 157, 203f, 219
Reproducibility, 16–18, 118, 138, 157, 203f, 210
Residual, 11, 90, 93, 118, 125, 137
Residual Standard Deviation, 28, 43, 83, 96, 111, 114, 118, 138, 179, 183,
Resolution, 22, 90, 116, 209ff

Robust, *see* rugged
Rounding of results, 18, 19
RSD, 200, 203f, 212
Ruggedness, 48, 58, 116, 137, 141ff, 231

S

Sample, size, 3
Sampling:
  2f, 14
  error, 218
  variability, 19
  volume, 15
Saturation effect, 103, 162, 184
Saturation level, 117
Savitzky-Golay, 144ff
Selectivity, 116, 128
Sensitivity, 189
Signal:
  118f
  aquisition, 34
  area, 3, 173
  clipping of, 33
  distortion, 122, 143ff
  drift, 19
  electrical, 3
  FWHM, 143f
  integration, 15
  path, 120
  shape, 143
  static, 143
Signal/Noise ratio, 103ff, 143, 145
Significant digits, 12, 140
Simulation 33–35, 139, 145, 148ff, 207ff, 215, 225
Single-precision math, 13, 120
Skills, 101
Slope, 87, 92, 145, 229
Smoothing of signals, 136, 143ff, *see* filters
Software:
  see Program
  120, 123, 150ff
  ANSI.SYS, 251
  AUTOEXEC.BAT, 251
  DOS 3.3, 251
  DOS 5.0, 251
  GRAPHICS.COM, 251
  LOTUS 1-2-3, 268–269
  Mircosoft Excel, 270
  dBASE III + , 270
  dBASE IV, 270
  QBASIC, 257–290
  Instructions, 250–263
  Reference, 263–299
  SuperCalc, 270
  Symphony, 125, 268–269
Solubility, 184
Solvent residues, 187
Specifications, 55, 72
Specifications, of design, 3
Spline function, 136
Spread, 9
Stability, 116, 220
Standard (solutions), 116
Standard addition, 109ff, 207
Standard error, *see* standard deviation of the mean
Standard Deviation, 9, 11, 118, 141
Standard Operating Procedures, 16, 18, 37, 92, 210
Statistical Ccontrol, 230
Statistical method 141 and theory 2
Statistical test:
  ANOVA, 46ff, 51ff, 81, 128f, 157, 170
  Bartlett, 46ff, 63f, 157, 169
  Chi^2, 66
  counting (e.g. impulses), 112
  Dixon, 48
  F-, 58
  factorial, 129f
  Huber, 48ff, 215ff
  multiple range-, 46ff
  one- and two-sided analysis, 30, 41, 68
  paired samples, 43
  power of a test, 77
  t-, 40, 128, 143, 178
Step change, 15
Student's *t*, 243
Subrange ratio, *see* statistical tests, Dixon's
Sum of squares, 53, 87, 101, 113

T

Target level, 175
Taylor series, 162
Terminology, 6
Time average, 3
Trace analysis, 3, 173, 199
Transformation, 121ff, 137f
Truncation, 11, 140, 207

V

Validation, 17, 81, 94, 141, 152, 164, 169, 214, 228, 230ff
Variability, of population + sampling, 19
Variance:
11, 38
additivity of, 18
analysis of components, 51f, 170
decomposition, 165, *see* ANOVA
pooled, 177
residual, 90ff, 114

Y

Youden plot, 43, 52, 81